Monographs of the Physiological Society No. 31

Physiological Aspects of Deep Sea Biology

Sounding and trawling on board the *Challenger* *c.* 1872 (Murray & Hjort, 1912)

PHYSIOLOGICAL ASPECTS OF DEEP SEA BIOLOGY

A.G. MACDONALD
Lecturer in Physiology, University of Aberdeen

CAMBRIDGE UNIVERSITY PRESS
CAMBRIDGE
LONDON · NEW YORK · MELBOURNE

Published by the Syndics of the Cambridge University Press
The Pitt Building, Trumpington Street, Cambridge CB2 1RP
Bentley House, 200 Euston Road, London NW1 2DB
32 East 57th Street, New York, NY 10022, USA
296 Beaconsfield Parade, Middle Park, Melbourne 3206, Australia

Library of Congress catalogue card number: 73-90652

ISBN: 0 521 20397 X

First published 1975

Printed in Great Britain
at the University Printing House, Cambridge
(Euan Phillips, University Printer)

To

J, K, J & H

CONTENTS

PREFACE

Monographs of the Physiological Society are normally advanced texts on intensively studied and discrete areas of physiology. This monograph could hardly be more different. In it I have attempted to discuss many physical and physiological features of all grades of organisms which inhabit the largest of biological environments – the deep sea.

That one biologist has undertaken such a task single handed is a good indication of the paucity of knowledge in this subject. However, despite our massive ignorance of deep sea physiology my attempt to write a reasonably comprehensive account has involved me in subjects of which I have no research experience. I hope experts in these numerous fields will accept the spirit of the enterprise.

This book is partly the outcome of my own participation in experimental work at sea and I should like to thank the Natural Environment Research Council for its support.

I am grateful to friends and colleagues who helped me with their comments on selected chapters at an early stage in the preparation of this book. They are; Drs J. H. S. Blaxter, R. J. Conover, P. C. Croghan, A. D. Hawkins, J. W. Porteous, J. R. Sargent, F. B. Williamson, A. M. Zimmerman, and Professors G. R. Kelman and J. A. Kitching FRS. I am particularly indebted to Professor E. J. Denton FRS, who commented most helpfully on the entire manuscript at a later stage in its preparation. I also express my gratitude to Jennifer Macdonald for typing the manuscript, to those authors who made photographs available to me, and to the staff of Cambridge University Press.

Aberdeen, 1974 A. G. MACDONALD

1 INTRODUCTION TO THE DEEP SEA AND ITS INHABITANTS

The deep sea is the largest of biological environments. Fig. 1.1 shows the hypsographic curve for the earth, in which the area of the surface is plotted against height relative to sea level. It makes the comparison between the area of the terrestrial and marine environments, but further, it contrasts the enormous volume of the deep sea with the relatively shallow surface layer where familiar physical conditions prevail.

The average depth of the oceans is 3800 metres. The greatest depth is the Challenger Deep in the Marianas Trench, where in 1952 a depth slightly in excess of 10790 m was measured by sounding wire and acoustic means (Gaskell, Swallow & Ritchie, 1953). According to Table 1.1 taken from Sverdrup, Johnson and Fleming (1942), the combined volume of the seas and oceans is 1370×10^6 km^3, of which the vast bulk is the deep sea. There is no other biological environment on a scale remotely comparable to this.

Fig. 1.1 shows that the great depths, the deep sea trenches, occur in limited areas. The restricted nature of the trenches is of considerable geological interest. They are also of interest biologically because they are populated with a special group of organisms, the so-called hadal fauna and bacterial flora. The deep sea clearly divides itself into two, the mass of it stretching around the world with an average depth of nearly 4000 m, in contrast to the highly localised trenches, many of which have a depth of two or three times that value.

The depth at which the deep sea begins is still open to discussion. The author has followed Hedgpeth's (1957) terminology but has preferred to use the term bathypelagic to describe animals which live at depths greater than 1000–2000 m and mesopelagic for those at depths of between 200 and 1000 m. These terms are based on ecological groupings of animals and it will be of some interest to see if they correspond to physiological criteria. It will be shown

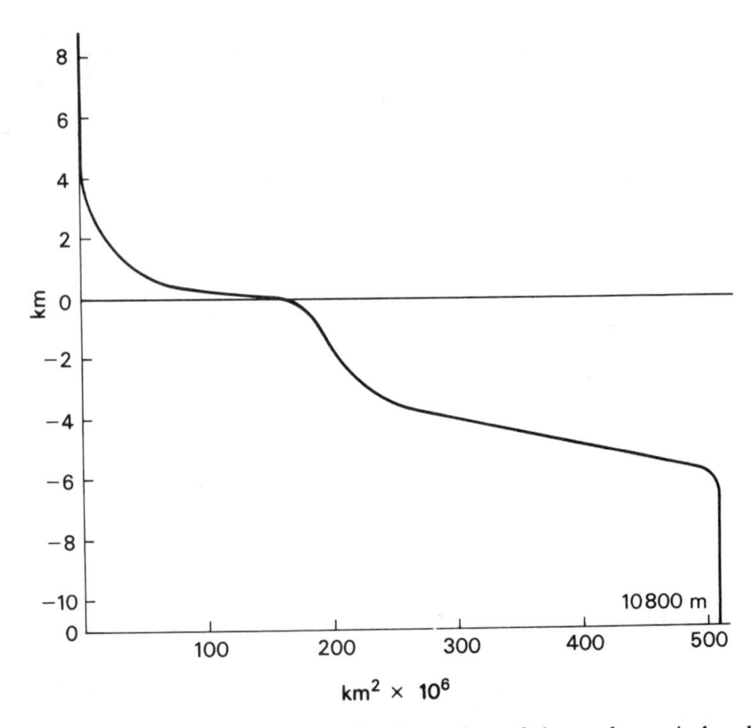

Fig. 1.1. The hypsographic curve for the surface of the earth, vertical scale, height relative to sea level, horizontal scale, area of the earth's surface. The area of the deep sea trenches is not to scale. (McLellan, 1965)

that the deep sea has a number of distinctive physical features, notably low temperature, high pressure, a lack of sunlight and an attenuation of other stimuli, distance from primary production and constancy in its physical conditions. Each factor gradually attains its characteristic level independently as the water column is descended and we should accept that there is no sharp upper physical boundary to the deep sea. Furthermore, there are anomalous regions in deep water where low temperatures do not prevail, or which lie beneath waters of widely differing fertility. The most predictable feature of the deep sea is its high pressure, which is closely related to the height of the water column.

As we start to consider the physical features of this unique environment in more detail, the reader may be assured that animals and micro-organisms are found down to the greatest

TABLE 1.1. *Area, volume and mean depth of the sea.* (Sverdrup, Johnson & Fleming, 1942)

Sea	Area (10^6 km^2)	Volume (10^6 km^3)	Mean depth (m)
Atlantic Ocean (excluding adjacent seas)	82.441	323.613	3926
Pacific Ocean (excluding adjacent seas)	165.246	707.555	4282
Indian Ocean (excluding adjacent seas)	73.443	291.030	3963
All oceans (excluding adjacent seas)	321.130	1322.198	4117
Arctic mediterranean	14.090	16.980	1205
American mediterranean	4.319	9.573	2216
Mediterranean Sea and Black Sea	2.966	4.238	1429
Asiatic mediterranean	8.143	9.873	1212
Large mediterranean seas	29.518	40.664	1378
Baltic Sea	0.422	0.023	55
Hudson Bay	1.232	0.158	128
Red Sea	0.438	0.215	491
Persian Gulf	0.239	0.006	25
Small mediterranean seas	2.331	0.402	172
All mediterranean seas	31.849	41.066	1289
North Sea	0.575	0.054	94
English Channel	0.075	0.004	54
Irish Sea	0.103	0.006	60
Gulf of St Lawrence	0.238	0.030	127
Andaman Sea	0.798	0.694	870
Bering Sea	2.268	3.259	1437
Okhotsk Sea	1.528	1.279	838
Japan Sea	1.008	1.361	1350
East China Sea	1.249	0.235	188
Gulf of California	0.162	0.132	813
Bass Strait	0.075	0.005	70
Marginal seas	9.079	7.059	874
All adjacent seas	39.928	48.125	1205
Atlantic Ocean (including adjacent seas)	106.463	354.679	3332
Pacific Ocean (including adjacent seas)	179.679	723.699	4028
Indian Ocean (including adjacent seas)	74.917	291.945	3897
All oceans (including adjacent seas)	361.059	1370.323	3795

depths. In the early part of the nineteenth century Ross retrieved molluscs, echinoderms and worms from a depth of 1800 m, in 1860, corals encrusting a telegraph cable were hauled from a sea floor depth of 2160 m; and by 1869 benthic molluscs had been trawled from 4430 m in the Bay of Biscay (Scheltema & Scheltema, 1972). The celebrated cruise of the British research vessel *Challenger* from 1872 to 1876 consolidated these discoveries and stimulated the growth of oceanographic science generally (Frontispiece). For an excellent almost contemporaneous account of the early development of deep sea biology see Murray and Hjort (1912).

According to Ekman (1953) there was a period in the middle of the nineteenth century when the existence of life in the deep sea was seriously doubted by some biologists, chiefly because Forbes's work in the Aegean Sea had indicated a paucity of life on the sea floor. Forbes later found abundant animals at depth. The azoic region of Forbes may conceivably exist in certain very deep pelagic regions of the oceans through nutritional reasons. Certainly the existence of animals and bacteria throughout the oceans, including the floors of the deepest trenches, should be seen as something quite remarkable, raising a host of questions basic to the life sciences.

The physical environment

The deep sea is not stagnant. Currents in deep water allow dissolved atmospheric gases and the other chemical constituents of seawater to be fairly well mixed. Fig. 1.2 shows a block diagram of the north–south flow of water in the Atlantic Ocean. Cold water from the polar regions sinks and flows towards the equator. Typical horizontal velocities are 10–15 cm sec^{-1}, but individual currents may flow at 60 cm sec^{-1}. In favourable cases photographs of the deep sea floor show how the sediment is scoured or rippled (Fig. 1.3). On rare occasions turbidity currents descend the continental slope at great speed.

The vertical velocity of deep water is usually much slower than that of water flowing horizontally. The age of deep seawater, measured from the time when it left the surface, has been variously estimated as many hundreds of years. The deep sea is composed of large, slowly moving water masses, in which there may be an additional small structure. Water masses differ very slightly in physical and chemical properties and may conceivably provide deep sea mid-water animals with useful sensory information in what would otherwise be a vast homogeneous environment.

Hydrostatic pressure increases by 1 atm (atmosphere) for each 10 metres of depth. Thus at 100 m, the total pressure is 11 ATA (atmospheres absolute) or 10 atm above atmospheric pressure.

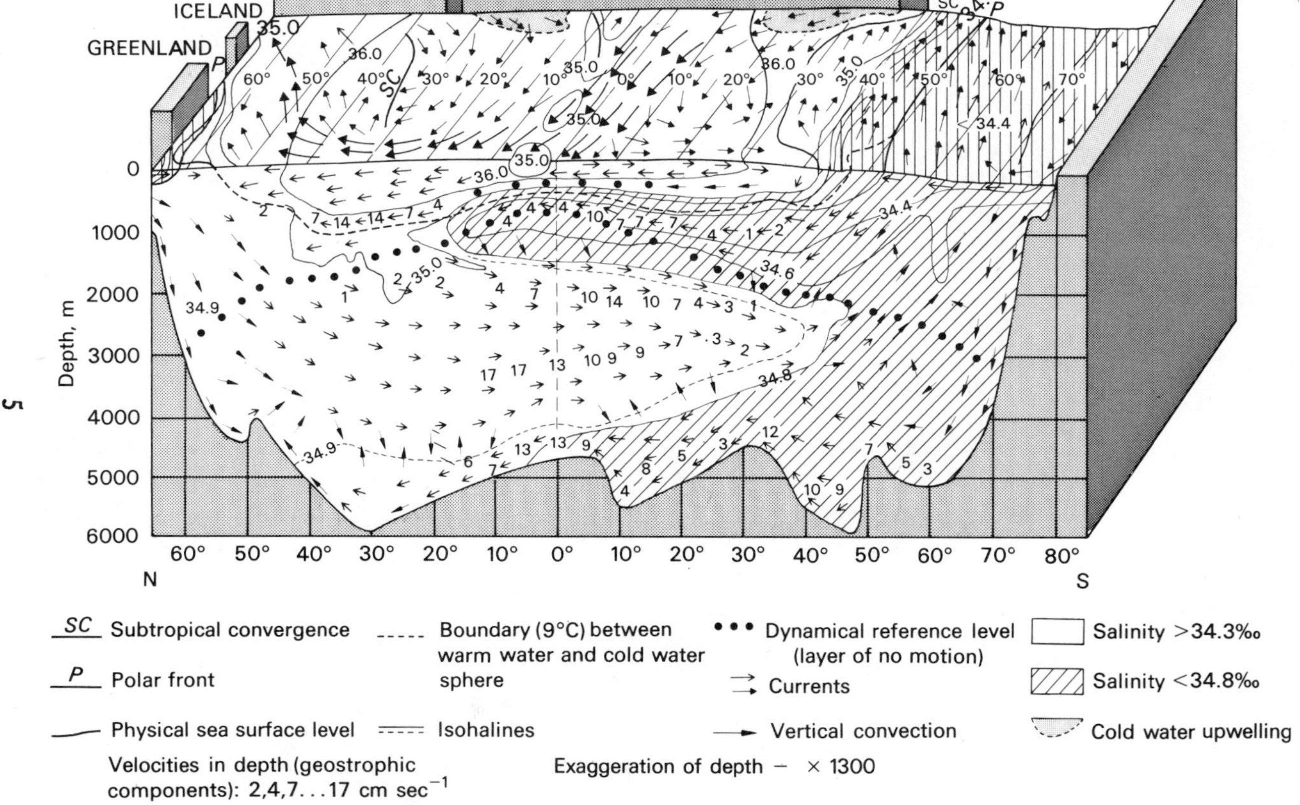

Fig. 1.2. Abyssal circulation. Schematic block diagram of deep-sea circulation in the western Atlantic. (Heezen & Hollister, 1971)

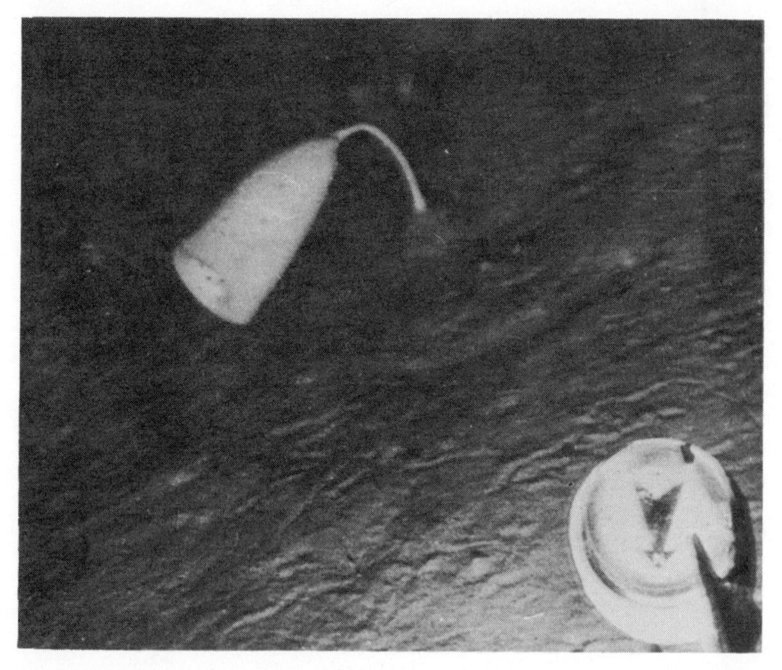

Fig. 1.3. Strong northerly flowing currents have forced the glass sponge to bend over on the sea bed. Photographed on the western part of the Bermuda Rise at a depth of 5378 m. (Heezen & Hollister, 1971; photo: E. Schneider)

Pressure in 'SI' units is expressed as newtons per square metre, or pascals ($1\ \mathrm{N m^{-2}} = 1\ \mathrm{Pa}$). A newton is the force which, when applied to a mass of a kilogram, produces an acceleration of a metre per second per second. The SI system is novel in biological high pressure work and in this book, which reviews past work, the atmosphere is used as the principal unit of pressure. A standard atmosphere is equal to $14.696\ \mathrm{lb\,in^{-2}}$, $1.033\ \mathrm{kg\,cm^{-2}}$ or $101.325\ \mathrm{kN\,m^{-2}}$.

Pressure increases the density of seawater which, in turn, slightly affects the pressure generated by a given column of water. Thus $P = \mathbf{g}D_{\mathrm{m}}d$ where P is pressure, $\mathbf{g}$ the acceleration due to gravity, D_{m}, mean water density and d its depth (McLellan, 1965). For biological purposes we ignore all the terms other than d. The maximum pressure in the deep sea is approximately 1100 atm.

Seawater of normal salinity and at low temperature is compressed by some 4 per cent at this pressure so the concentration of salts in the water at the greatest ocean depths is that much more concentrated than in surface water. The molecular structure and electrical conductivity of deep sea water is outlined in Chapter 2.

The circulation of the oceans is too slow to achieve a uniform distribution of temperature. A series of profiles (Fig. 1.4) shows that cold water prevails at depth. Any definition of the deep sea should certainly make reference to its uniformly low temperature. Bruun (1957) suggests the 10 and 4 °C isotherms have particular significance. 10 °C commonly marks the lower limit of the epipelagic region and 4 °C the top of the bathypelagic zone. These are useful temperatures to have in mind.

By familiar standards the deep sea is not only a low-temperature environment but it is also a highly constant one. Fig. 1.5 shows how the temperature of the Kuroshio current in the North Pacific varies throughout the year. At 100 m depth the annual variation in temperature is approximately 5 degC whereas at 1000 m it is 0.2 degC. One would expect the temperature of deep waters to vary little from year to year and in general this is confirmed by a number of long term studies (see Muromtsev, 1963, for a detailed confirmation of this point).

Physical oceanographers distinguish between observed temperatures and potential temperatures. The latter is the temperature which a volume of seawater would reach if it were raised to the surface and allowed to decompress without heat flowing in from the environment. A convenient form of the equation relating depth change (pressure) and adiabatic warming or cooling is:

$$\Delta T = 10^5 \frac{T\alpha}{J C_p} \mathbf{g}\rho \text{ deg cm}^{-1}$$

in which T is the absolute temperature, $\mathbf{g}$, acceleration due to gravity, ρ, density, α, the coefficient of thermal expansion, C_p the specific heat and J the mechanical equivalent of heat (Horne, 1969). The effect is a small one by physiological standards. Seawater raised adiabatically from a depth of 10000 m and at an initial temperature of 0 °C will cool slightly less than 2 degC.

Normal seawater freezes at about −1.8 °C and temperatures of less than −1 °C are rare in the deep sea. Pressure depresses the freezing point of pure water; at 1000 atm it freezes at −9 °C (Hamann, 1957).

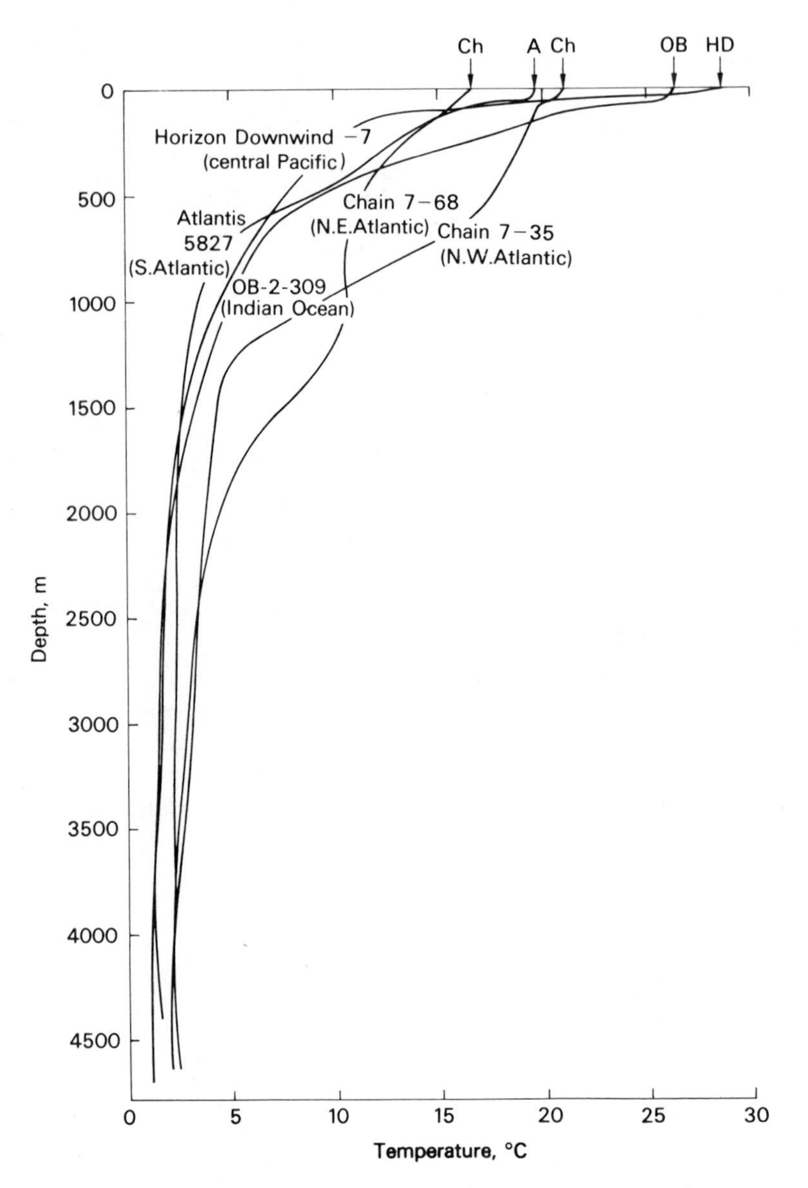

Fig. 1.4. Vertical distribution of temperature in the deep sea. (McLellan, 1965)

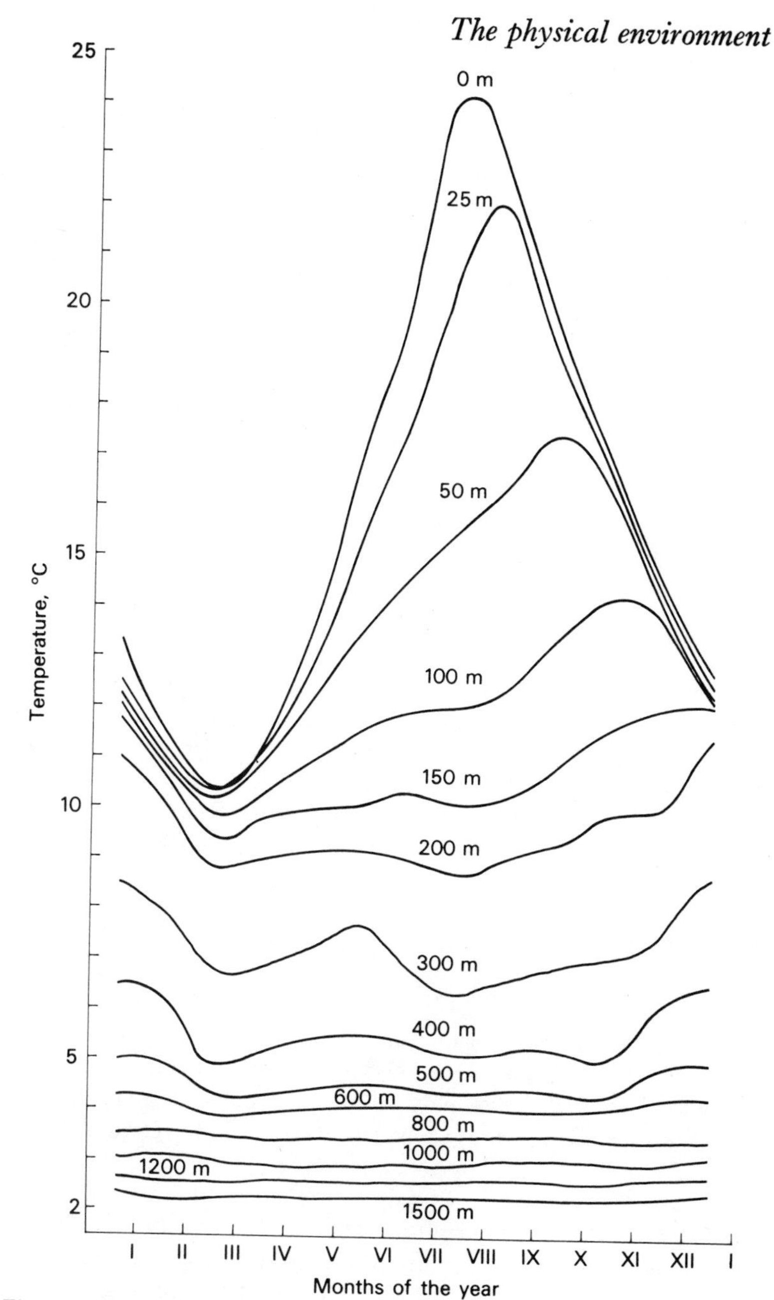

Fig. 1.5. Monthly temperature changes at various depths in the Kuroshio region (39 °N, 153 °E). (Muromtsev, 1963)

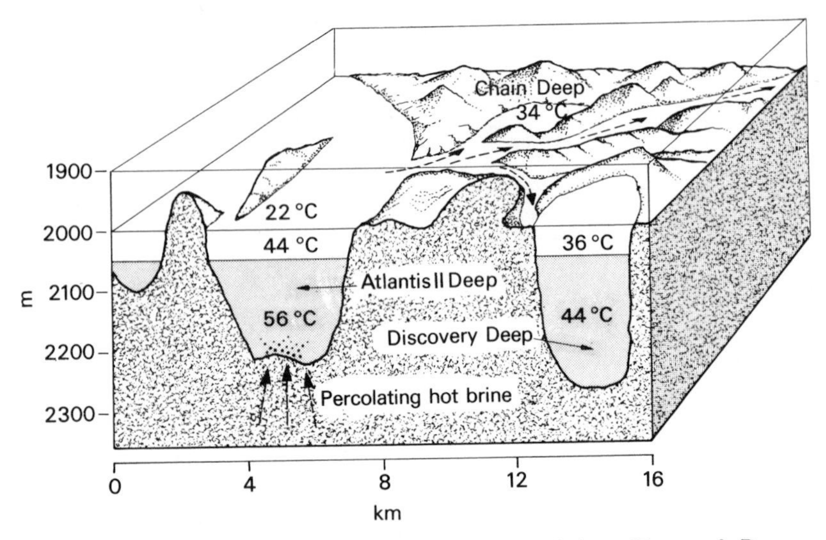

Fig. 1.6. Regions of the Red Sea which contain hot brines. (Degens & Ross, 1970)

An exception to the rule that the deep sea is cold has been discovered in the Red Sea (Fig. 1.6). Several depressions in the sea floor extending to depths in excess of 2000 m contain water in the temperature range of from 44 to 56 °C and concentrations of salts tenfold normal (Degens & Ross, 1969). Bacteria are absent from the hottest parts of these depths and no animals are reported in the locality.

Light is rapidly absorbed by seawater. At depths below 1000 m it is typically attenuated to 10^{-10} of the surface intensity and is insufficient for vision (Chapter 5). Photosynthesis rarely achieves a net production of organic material at depths below 100 m (Chapter 6). The bulk of the deep sea is therefore dark with only a dimly illuminated upper layer. It sustains no photosynthesis. Radiation levels are presumably low in the mid-water but on the deep sea floor higher levels may prevail where radioactive material accumulates.

The solutes in seawater may be classified as organic materials, salts and dissolved atmospheric gases. The variety of organic solutes present and their significance are described in Chapter 6 in the context of deep sea nutrition. The atmospheric gases which have particular biological significance, namely oxygen, nitrogen

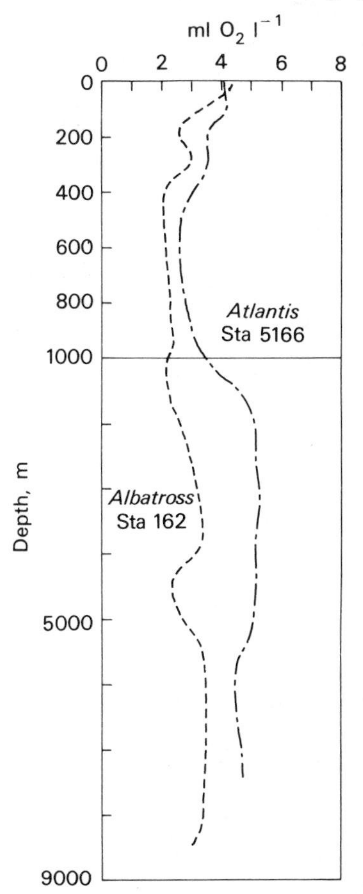

Fig. 1.7. Representative vertical distribution of dissolved oxygen in the deep sea. *Albatross* station, Mindanao Deep: *Atlantis* station, Brownson Deep. (Richards, 1957)

and carbon dioxide, are present in moderate amounts throughout deep water.

Oxygen enters the sea at the air–water interface and is released from photosynthesising plants. It is consumed by the respiration of animals and micro-organisms. Thus the steady state concentration of dissolved oxygen at depth is determined by its rate of supply and its rate of consumption. In general the oxygen demand of the deep sea is very small and oxygen is available in concentrations sufficient for the needs of most organisms (Fig. 1.7).

It is often the case that the concentration of dissolved oxygen passes through a minimum at a depth of several hundred metres. In the Pacific Ocean in particular, the low levels of dissolved oxygen may reach 1 per cent of the concentration of oxygen in seawater in equilibrium with the atmosphere. Where oxygen minima occur at considerable depth however, as at 1000–2000 m in the South Pacific (Muromtsev, 1963), the concentration of oxygen is still quite high, and usually in excess of 20 per cent of the surface saturation value. In locally stagnant regions near the sea floor much lower concentrations of oxygen may be found (Richards, 1957).

Oxygen concentration has to be distinguished from oxygen partial pressure. We find two main factors in the deep oceans which affect the partial pressure of oxygen. Henry's Law states that M, the mass of gas dissolved, is equal to the product of P, the partial pressure of the gas, and S its solubility. Thus $M = PS$. The low temperature of the deep sea lowers the partial pressure of oxygen dissolved in a given volume of seawater, by raising its solubility. Conversely, the high hydrostatic pressure, which diminishes the solubility of gas in water, raises the partial pressure of dissolved gas in a given volume of water. Note that we are not here concerned with the effect of high partial pressures of gases in the gaseous phase. Thus the data of the chemical oceanographers, who are mainly concerned with measurements of dissolved gas concentration, have also to be interpreted as partial pressure or activity for some biological purposes (see p. 145, Chapter 4).

The concentration of dissolved nitrogen throughout the deep sea is close to the level attained in equilibrium with the atmosphere at the surface. Hydrostatic pressure appears to raise its equilibrium pressure as in the case of oxygen. Nitrogen is a constituent of the gaseous contents of the buoyancy organs of some deep sea animals (Chapter 5).

Carbon dioxide enters the sea from the atmosphere and as the end product of the respiratory metabolism of marine organisms. The behaviour of dissolved carbon dioxide and the bicarbonate and carbonate ions with which it is in equilibrium is affected by changes in temperature and pressure

$$CO_2(\text{solution}) + H_2O \rightleftharpoons H^+ + HCO_3^- \qquad (1)$$

$$HCO_3^- \rightleftharpoons H^+ + CO_3^{2-} \qquad (2)$$

TABLE 1.2. *Ratio of apparent solubility products at 1 atm to apparent solubility product at stated pressure and 22 °C.* (Pytkowicz, 1969)

P (atm)	Calcite	Aragonite
1	1.0	1.00
100	1.10	—
200	1.25	—
500	1.88	1.69
750	2.62	—
1000	3.56	2.71

A reduction of the temperature of seawater shifts both equilibria to the left, but as a lower temperature increases the solubility of carbon dioxide in water a greater amount of dissolved inorganic carbon can be simultaneously achieved (Horne, 1969). At elevated pressures both equilibria are shifted to the right. In the cold waters of the deep sea the pressure effect predominates. Thus seawater is slightly more acid at depth, the ΔpH per 100 atm being 0.02 and 0.035 units at pH 8.5 and 7.5 respectively. However, the biological implications of this are not yet clear.

Other equilibria are affected by high pressure, notably

$$CaCO_3 \rightleftharpoons Ca^{2+} + CO_3^{2-} \quad (3)$$

Surface seawater is normally supersaturated with calcium carbonate but deep seawater is undersaturated. Table 1.2 shows the effect of pressure on the solubility products of the two crystalline forms of calcium carbonate. The extent of the pressure effect in nature is evident in Fig. 1.8 which shows saturation is attained at depths of from 300 to 3300 m in the Pacific Ocean.

The ocean floor provides a living space far greater than the terrestrial environment. Fig. 1.1 shows the area of dry land to be about 150×10^6 km^2, whereas the area of the abyssal plains and continental slope, below a depth of 1000 m is about 300×10^6 km^2. As might be predicted from the effects of pressure on solubilities, the sediments in the deeper waters lack calcareous (or siliceous) material and they are described as clays. At lesser depths the sediments are predominantly calcareous and siliceous oozes derived from animal remains. However, plenty of calcareous material is to be found at depths below the level at which the waters are undersaturated with calcium carbonate because solution processes are

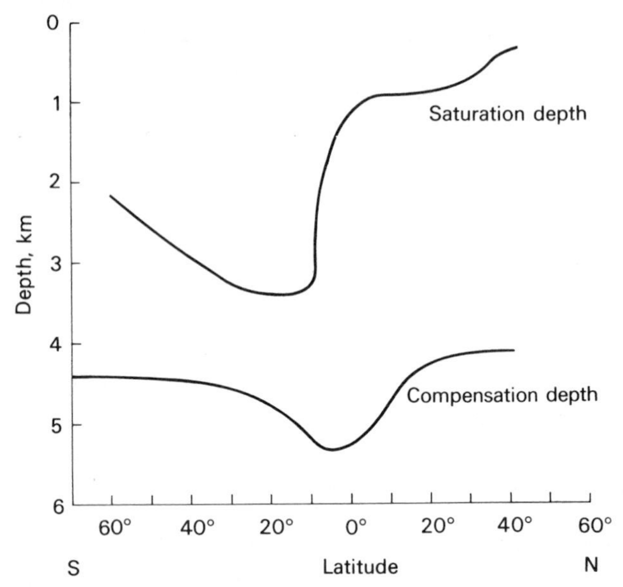

Fig. 1.8. The saturation depth and the compensation depth for carbonate in the Pacific Ocean. (Pytkowicz, 1970)

extremely slow and equilibrium is never achieved (Fig. 1.8). The term compensation depth is used in this context, referring to the depth at which the rate of supply of calcareous material equals the rate at which it dissolves.

The inhabitants

Representatives of many of the groups of animals and micro-organisms which are commonly found in shallow seas are also to be found in the deep sea. In comparison with those living in shallow seas, deep water organisms are few in number and in general their total biomass declines with further increase in depth. Several life-styles are normally distinguished; animals or micro-organisms living on the sea floor are referred to as the benthos or as benthic while mid-water organisms are pelagic and variously labelled according to living depth. The distinction between benthic and pelagic organisms is not physiologically trivial and it will be argued in Chapter 6 that the former live in a two dimensional world relatively well stocked with nutrients whereas bathypelagic animals inhabit a three dimensional environment singularly devoid of food.

PELAGIC ANIMALS

A mid-water trawl from a depth of a thousand metres or so will normally recover a wide range of small animals, many of striking form and colour. Small fish will be seen, the silver ones characteristic of mid-water depths of 300–500 m with black fish more commonly obtained from greater depths. Many red pelagic animals will also be seen, most of them are likely to be crustacea from depths around 500 m. The crustacea are frequently as big as the fish. Other strange objects will be found in the trawl which turn out to be worms, jellyfish, molluscs and all manner of animals in unfamiliar guises. A fairly common feature is an absence of pigment and many soft-bodied animals appear no more than a translucent jelly.

Rarely a trawl will bring up squids measuring a metre in length or fishes longer than 20 cm. The catch is much reduced if the trawl is deployed at depths of several thousand metres. If a fine mesh net is used, small animals such as copepods and amphipods will be caught, and many of these will be red in colour if they come from the mid-depths.

The writer is not competent to discuss this assemblage of deep sea animals in any detail. The reader should consult books and papers by Hardy (1956) and Marshall (1954) for a vivid and expert introduction. Here a few notes on some animals, many of which the author has observed either at sea or in museum collections, are provided to illustrate features which are distinctive of, if not unique to, the deep sea.

Crustacea occur in great numbers in the sea both in shallow and deep water. Fig. 1.9 shows two examples of mesopelagic scarlet prawns whose colour renders them inconspicuous in the dim blue–green light of their environment. Many are robustly constructed and would appear to be powerful swimmers, in contrast to many of the fish. *Sergestes*, *Systellaspis* and *Acanthephyra* survive trawling and remain active in seawater at low temperature for days. They beat their ventilatory pleopods powerfully and nearly continuously. Some of these large Crustacea are very elegantly constructed as can be seen from Fig. 1.10.

Smaller crustacea are as numerous as the large prawn-like animals. Copepods, amphipods and ostracods are common (Fig. 1.11). A particularly interesting ostracod is *Gigantocypris* which shows several features distinctive of the deep sea fauna.

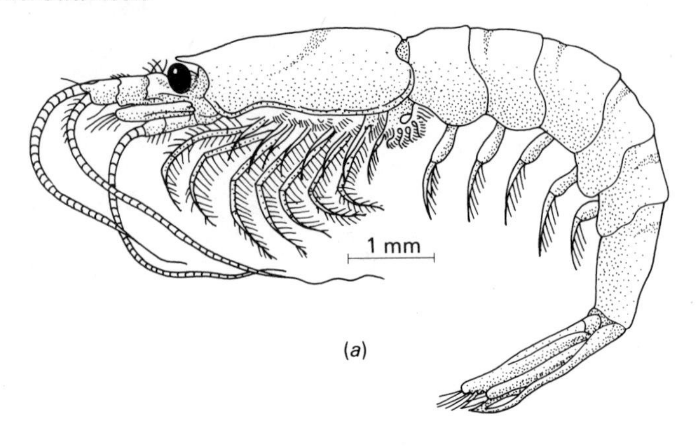

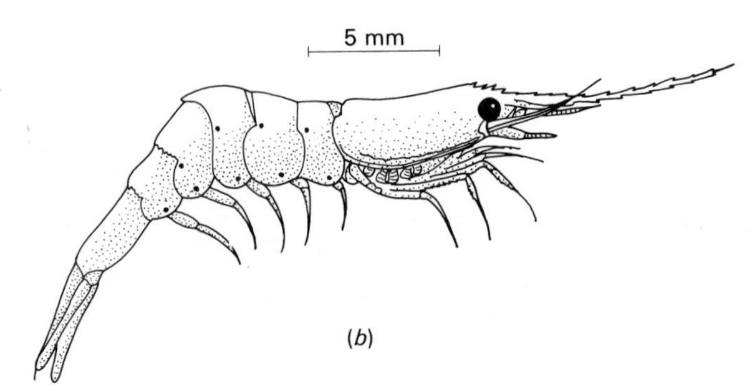

Fig. 1.9. Large mesopelagic Crustacea. (*a*) *Thysanopoda acutifrons*, a young specimen 16 mm long seen from the side. Adults attain a length of 45 mm. (Einarsson, 1945). (*b*) *Systellaspis debilis*. (Coutière, 1938)

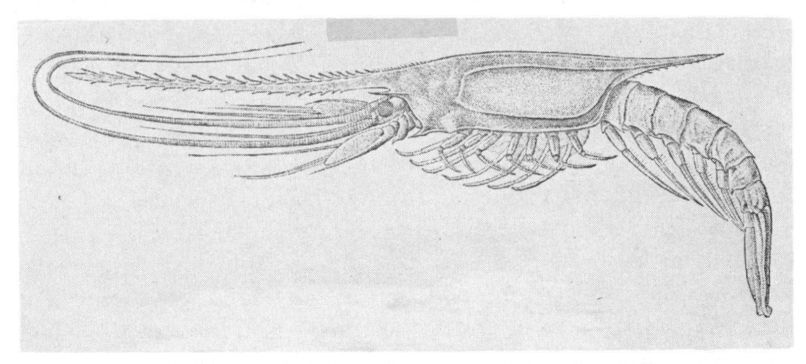

Fig. 1.10. *Gnathophausia zoea* ×2. (Agassiz, 1888)

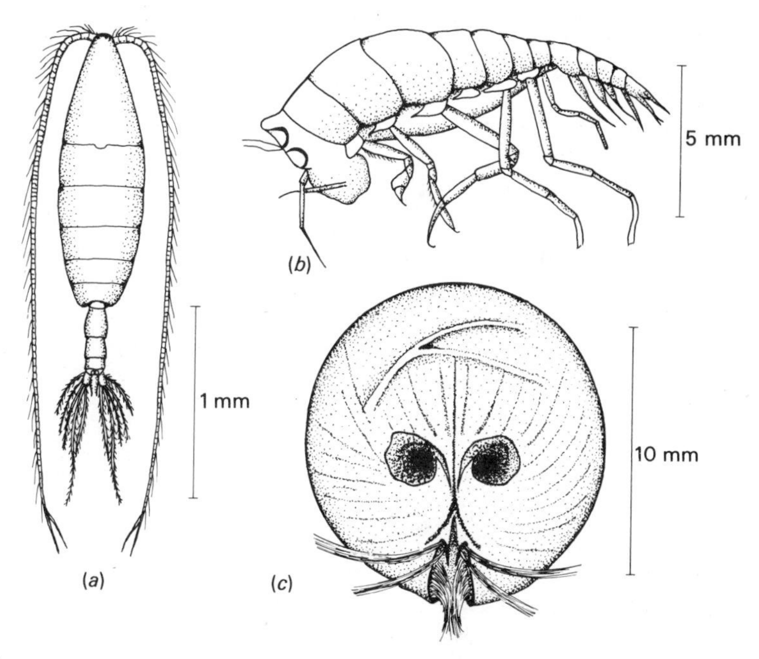

Fig. 1.11. Smaller deep water Crustacea. (*a*) *Bathycalanus richardi*: body 12 mm long. This is a very large deep water copepod. (Rose, 1933) (*b*) *Scypholanceola agassizii*, a deep sea amphipod. (Murray & Hjort, 1912) (*c*) *Gigantocypris* sp., a giant deep sea ostracod, coloured red–orange, with a body diameter of up to 2 cm. Several species inhabit the oceans; see text. (Murray & Hjort, 1912)

Gigantocypris, as its name suggests, is a giant among ostracods. It may grow to a size of two centimetres in diameter and is coloured varying shades of red–orange. *Gigantocypris* possesses the normal complement of ostracod limbs, two anterior trunk limbs and head limbs, which are relatively small. The posterior portion of the body is disproportionately large. The carapace is fused dorsally for over half the circumference and is unable to hinge open in the usual way of ostracods. Ventrally there is a slit between the two halves of the carapace and at both the anterior and posterior ends the slit widens to an oval hole. Swimming limbs, the second antennae, protrude through the anterior opening and appear to use the carapace as a fulcrum, or as Cannon (1940) puts it, as rowlocks.

Freshly caught *Gigantocypris* are often nearly buoyant and swim actively although they seem to lack a sudden escape response. No giant nerve fibres are reported to innervate their limbs as in other smaller ostracods.

Gigantocypris' body is suspended within the protective carapace and is moved by retractor muscles, including the retractor muscle for the large antennae, which work in opposition to protractor, body wall muscles. The former pull the body up into the carapace and away from the ventral opening; the latter act by compressing the fluid contents of the body, forcing it down and forward to the anterior opening. Cannon (1940) suggests that the disproportionately large heart, presumably a powerful organ, may be required to sustain an adequate blood circulation during this sort of manoeuvre. The cardiac neurone is innervated by a nerve which connects to nerve bundles which also supply both retractor and protractor muscles.

Another enlarged organ in this animal is the nauplius eye which is divided into two lobes. The upper part of the animal's vertical range lies in the dim twilight zone of the sea.

Feeding has not been described but gut contents include arrow-worms, the large copepod *Pleuromamma robusta*, and small fish. These prey are active animals and it would be of interest to know how they are captured. Young *Gigantocypris* do not grow up through larval stages but hatch direct from the brood pouch in the carapace as miniature adults. According to Poulsen (1962) embryonic development proceeds in the normal crustacean manner with segments developing fastest at the anterior. The organism hatches when it has developed its first pair of trunk limbs and the second trunk limbs develop in the free living young. Among the adult population females predominate being 3–6 times more abundant than males.

There are five species of *Gigantocypris* (Table 1.3).

Knowledge of this animal's geographical distribution is incomplete but the evidence clearly indicates that morphologically distinct types predominate in certain regions. The information available on the vertical distribution of *Gigantocypris* is even less satisfactory. Poulsen (1962) discusses data obtained from successive trawls carried out with open nets and without a depth recorder. The evidence indicates that *Gigantocypris* normally lives well below the densely populated sunlit layers of the ocean. The limited data show that it is distributed at depths between a few

TABLE 1.3. *Distribution of* Gigantocypris *in the oceans.* (Poulsen, 1962)

Species	Geographical distribution
G. agassizii	E. Pacific, some records from W. Pacific
G. mülleri	Throughout the N. Atlantic Ocean and Caribbean Sea
G. dracontovalis	Tropical Atlantic, Indian Ocean and W. Pacific
G. danae	Indian Ocean, E. Indies
G. australis	S.W. Pacific

TABLE 1.4. *The vertical distribution of different* Gigantocypris *species.* (Poulsen, 1962)

Species	Depth range (m) (assuming depth fished = $\frac{1}{3}$ wire paid out)	Temperature range (deg C)
G. agassizii	900–1600	2.1–3.5
G. mülleri	260–2300	2.6–4.9
G. dracontovalis	—	—
G. danae	600–2000	2.4–4.4
G. australis	—	—

hundred metres and 2300 metres (Table 1.4), which is broadly consistent with results currently being obtained by workers at the National Institute of Oceanography, UK, who are using modern methods of sampling (see also p. 194).

Generally the information we have about *Gigantocypris* adds up to an interesting example of a deep sea animal. Its large size, red–orange coloration, preponderance of females in the adult stages and viviparous development are not uncommon in animals which live well below the surface of the sea. As a genus it is distributed around the world's oceans in water cooler than 5 °C.

Preliminary studies of this animal's buoyancy, respiration, circulation and locomotor performance in conditions simulating the deep sea are described in Chapters 4 and 5.

Both cartilaginous fish (Elasmobranchii) and true bony fish (Actinopterrygii) live in the deep sea. Many bony fish are very small and some show extraordinary features. Fig. 1.12 shows a typical mesopelagic fish. It is coloured silver whereas bathy-

Fig. 1.12. Deep sea fishes. (*a*) *Sternoptyx diaphana*, mesopelagic, typical length 5 cm. (*b*) *Gonostoma elongatum*, 500–1500 m, typical length 12 cm. ((*a*) (*b*) Gunther, 1887)

pelagic fish are more often black, and many of both categories are small and feebly built (Fig. 1.13). The hatchet fish *Sternoptyx* is unlike any fish from the surface layers. The example shown in Fig. 1.12 is 4 cm long and approximately 1 cm wide (i.e. thick). Much of the body is taken up with large eyes, jaws, a swimbladder and photophores which shine a camouflaging light downward (Chapter 5). These animals swim slowly along in schools at depths where the sunlight is very dim; they live in the very topmost part of the deep sea.

Some bathypelagic fish exhibit enormous jaws and an ability to take in relatively large prey. Feeding is but one of the problems of the dispersion of life in the deep sea. Reproduction is another and in the angler fishes we see some spectacular morphological adaptations related to both feeding and reproduction. *Ceratias holböelli* is a rare deep sea angler fish in which females attain a length of 1 m whereas adult males grow to approximately 10 cm in length and live parasitically on the female. Fig. 1.14 from Bertelsen (1951) shows the extraordinary metamorphosis of the male.

The adult female consists of three approximately equal parts: mouth parts, body and tail fin. The jaws may open 130° and are directed upwards towards a lure, which is probably luminescent,

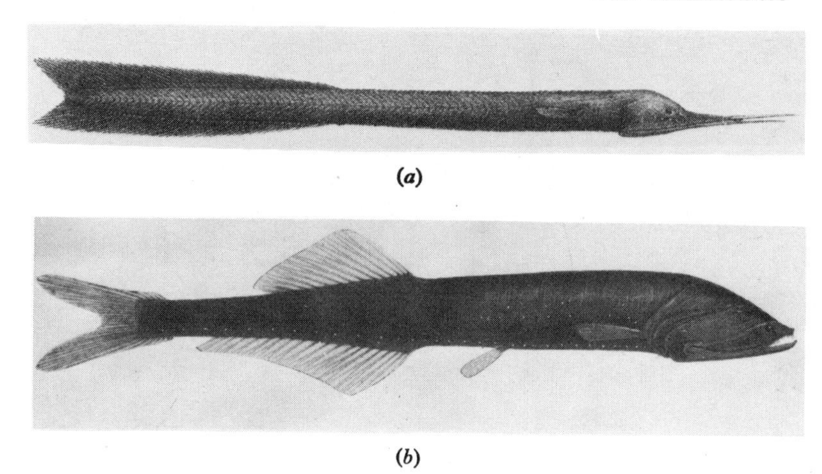

Fig. 1.13. Bathypelagic fishes. (*a*) *Cyema atrum*, 1000–3000 m, typical length 10 cm. (Gunther, 1887) (*b*) *Cyclothone microdon*, 500–2000 m, typical length 10 cm. (Murray & Hjort, 1912)

as is the case in shallow water angler fish. The eyes are small and lie beneath the skin in the adult stage but are relatively large in the juveniles and show some of the features typical of deep sea fish discussed in Chapter 5. The animal has no pelvic fins, its body musculature is much reduced, and a swimbladder is absent. The lure is suspended on the illicium which is the cephalic spine of the dorsal fin and according to Bertelsen, works in a mechanically satisfactory way.

Feeding has not been observed in *Ceratias* but Bertelsen suggests the following. Despite the absence of a swimbladder the animal is probably close to neutral buoyancy and can maintain its level quietly in the water (see Chapter 5). The lure may well be 'fished' to mimic luminescent prey organisms and animals which are attracted to it are sensed by the lateral line organs and perhaps also visually. A rapid gulping action of the jaws causes water to sweep the prey into the mouth while inwardly directed teeth ensure it does not escape.

The adult male feeds parasitically on the female. The initial attachment is made by the male's jaws; subsequently the mouth fuses with the female tissue and the male gut atrophies. It appears that males grow in size after becoming attached to the female so assimilation of female nutrients takes place.

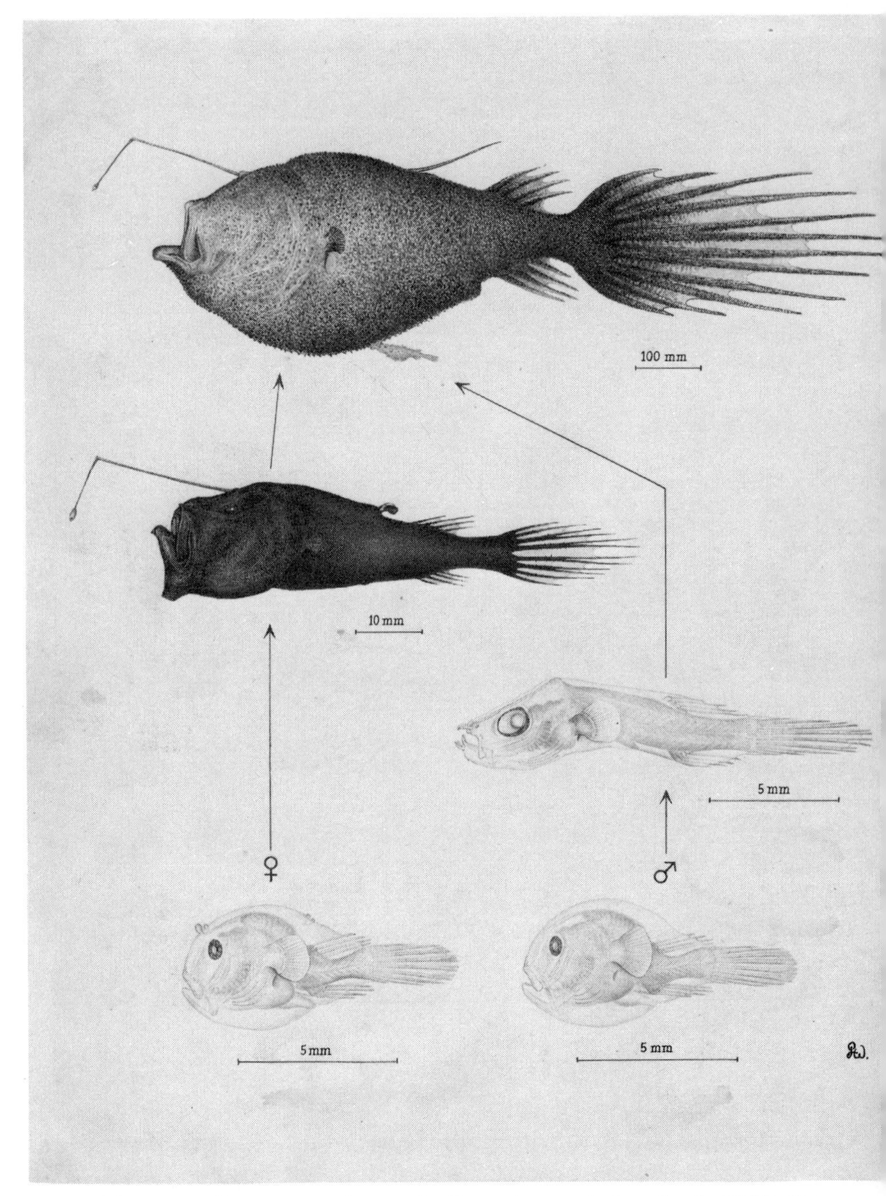

Fig. 1.14. Ontogeny and sexual dimorphism of *Ceratias holböelli* Kröyer. Adult female, *c.* 100 cm with adult parasitic male, *c.* 100 mm. Adolescent female, 70 mm, larval female, 8.5 mm. Adolescent male, oldest freeliving stage, 16 mm, larval male, 8.5 mm (Bertelsen, 1951)

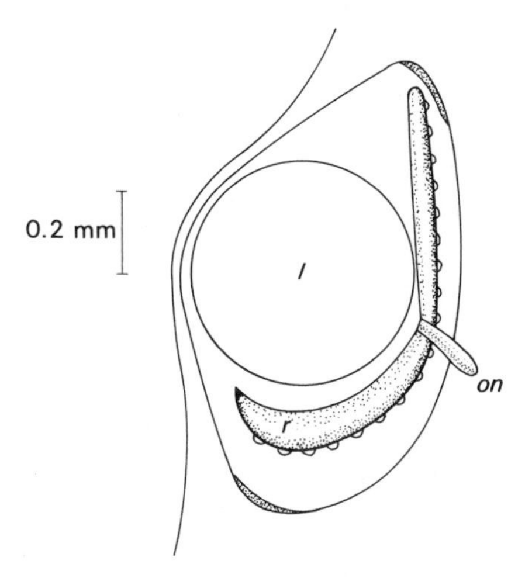

Fig. 1.15. Eye of juvenile (free living) male *Ceratias holböelli* (×45). *l*, lens; *on*, optic nerve; *r*, retina. See Chapter 5. (Munk, 1966)

Behind the mouth of the parasitic male there is a single gill slit through which a respiratory current of water may flow. However, according to Bertelsen the blood circulation of the male is continuous with that of the female.

Ceratias holböelli has been found in all the world's oceans, over a wide range of latitude. Larvae are restricted to 40 °N–35 °S, and are thought to frequent shallower waters than adults. The vertical distribution of *Ceratias* is not known, but it is presumed to be bathypelagic because it is not caught in mesopelagic depths.

The life cycle in Fig. 1.14 makes sense if the young feed and grow in the well stocked shallow waters when their vision is good (Fig. 1.15) and subsequently descend to the impoverished depths. Descent is associated with a reduction in the eyes of both sexes, the growth of the elaborate fishing mechanism in the female and male parasitism. *Ceratias* exemplifies in extreme form the passive and economical way of life common among deep sea fish.

Cephalopods from the mesopelagic depths may be small and delicate (e.g. *Calliteuthis reversa*), of intermediate size, or in the case of *Architeuthis princeps*, very large indeed (Fig. 1.16). All are carnivores and, like fish, hunt either actively or passively suspended in the water column.

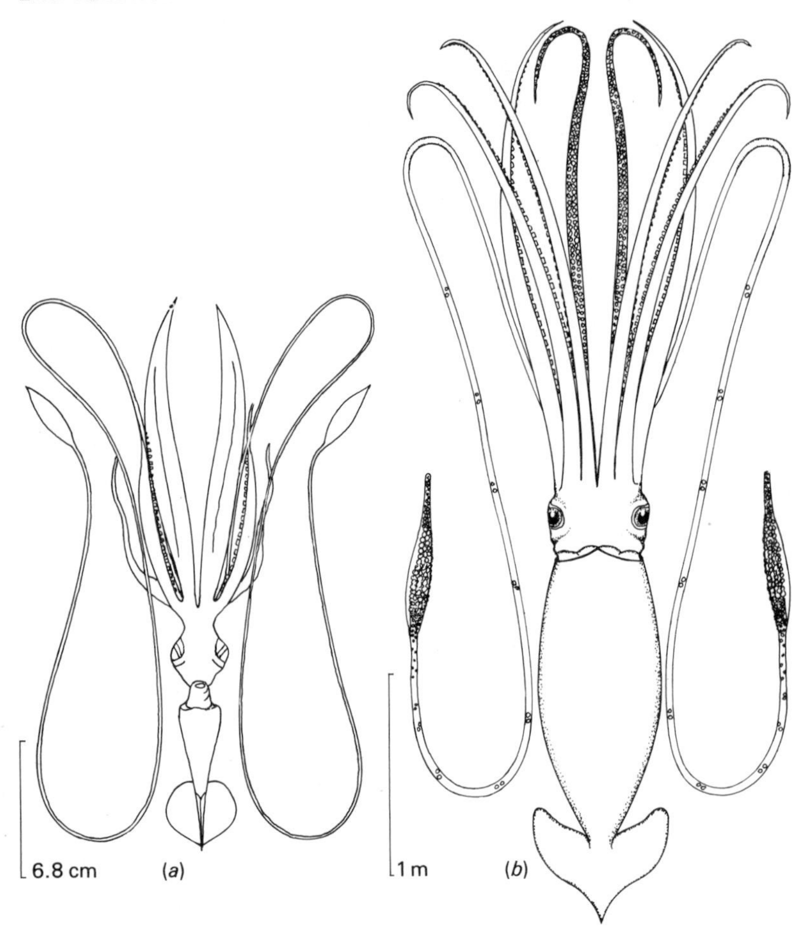

Fig. 1.16. Deep-sea cephalopods. (*a*) *Chiroteuthis veranyi*, ventral side. This animal is pelagic and is most often caught between 300 and 1000 m depth. (Clarke, 1966). (*b*) *Architeuthis princeps*. (Agassiz, 1888)

BENTHIC ANIMALS

Animals which live on the sea floor are relatively easy to photograph *in situ* (Chapter 7) and their depth distribution is readily determined.

The style of life of deep sea benthic animals is probably little different from that of the shallow water benthos. We may usefully distinguish the epifauna which lives on the surface of the sea bed, such as echinoderms and sponges, from the infauna which lives partly or totally submerged in the bottom sediment, such as worms or bivalve molluscs. There is also a bentho-pelagic fauna feeding off the benthos. Fig. 1.17 shows octopods swimming just above the sea bed at a depth of 3800 m and Fig. 1.18 shows sable and hag fishes in the Coronado submarine canyon, at 1200 m depth. Fig. 1.19 shows the tripod fish *Bathypterois* also at a depth of 1200 m, photographed from a deep research submarine (Chapter 7). A very large shark is seen at 1949 m depth in Fig. 1.20. The rat tail, common on the sea floor at moderate depths has been the subject of biochemical studies which are discussed in Chapter 3; Fig. 1.21 shows a typical specimen. The bottom-living fish shown in Fig. 1.22 is of special interest: *Bassogigas profundissimus* has been credited with a gas-filled swimbladder at a depth of 7160 m where the ambient pressure is in excess of 700 atm (Anon, 1970). It is the deepest living identified bony fish.

Some examples of epibenthic invertebrates are shown in Fig. 1.23. Epibenthic Crustacea show very long slender appendages in a number of cases (Fig. 1.24) while others are morphologically indistinguishable from shallow water forms (Fig. 1.25).

The infauna is rarely caught in photographs but Fig. 1.26 shows an exceptional picture of an acorn worm (not a true worm, an enteropneust) in the process of extruding its faecal trail. Many photographs have been taken which show the tracks of animals on the deep sea floor (Fig. 1.27), and also debris from land.

One of the most remarkable groups of animals of the deep sea infauna is the Pogonophora, which also occur in fairly shallow water. They were established as a distinct taxonomic group of animals only some twenty years ago, chiefly through the work of Ivanov (1963). It now seems likely that earlier marine biologists failed even to recognise specimens dredged from the sea floor as animals. It is possible that the deep sea still contains undiscovered animals of major taxonomic or phylogenetic interest. It certainly contains organisms of outstanding physiological and biophysical interest. The list of so-called living fossils found in the deep sea includes stalked crinoids, hexactinellid sponges, the crustacean family Eryonidae, the cephalopod *Vampiroteuthis*, the segmented mollusc *Neopilina* and the coelocanth *Latimeria* (the last two do not live very deep). We should not, however, conclude that ancient species seek refuge in the deeps; the deep sea is an

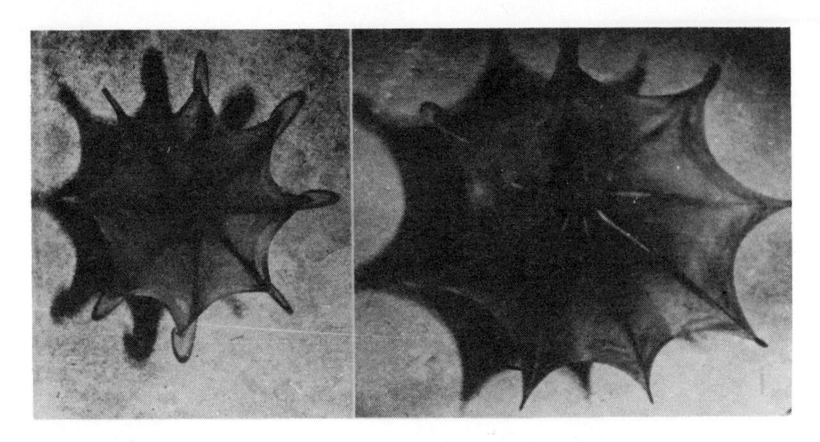

Fig. 1.17. Octopods hovering above the sea floor at a depth of 3800 m off the Saint Croix Virgin Islands in the Caribbean. (Heezen & Hollister, 1971; photo: W. Brundage)

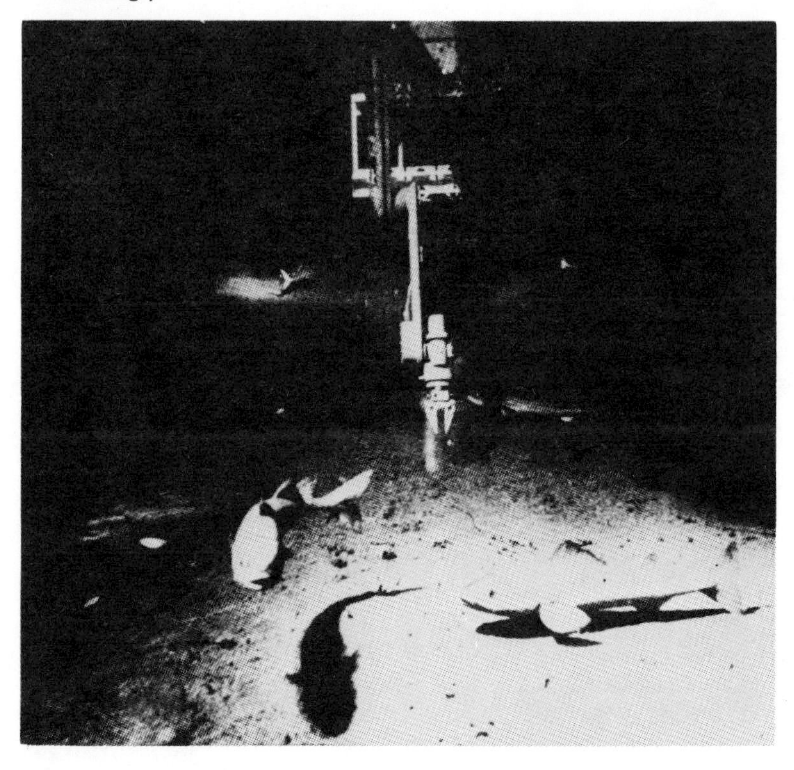

Fig. 1.18. Sable fish and hag fish (arrowed) at a depth of 1200 m in the Coronado submarine canyon off California, with part of a deep submersible vehicle in the background. (Heezen & Hollister, 1971; photo: R. Dill)

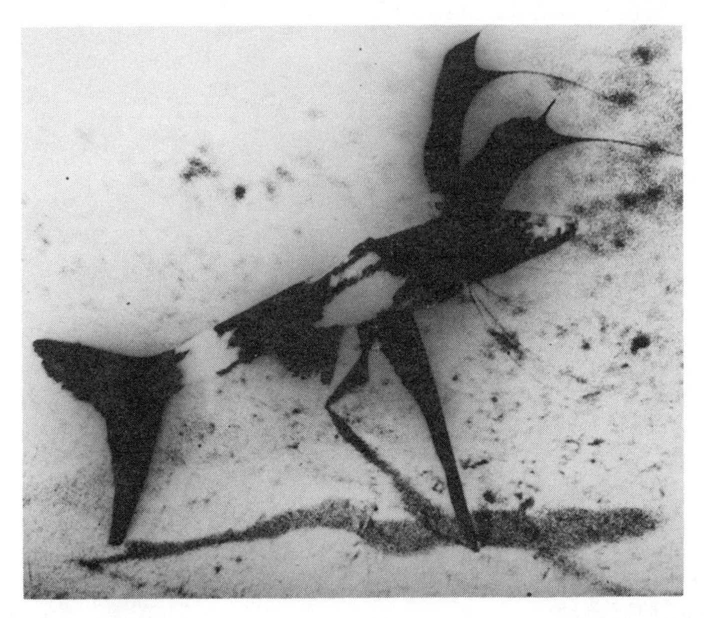

Fig. 1.19. *Bathypterois bigelowi*, the tripod fish, photographed from a deep submersible vehicle at a depth of 1200 m, in the De Soto submarine canyon (Gulf of Mexico). (Heezen & Hollister, 1971; photo: R. Church)

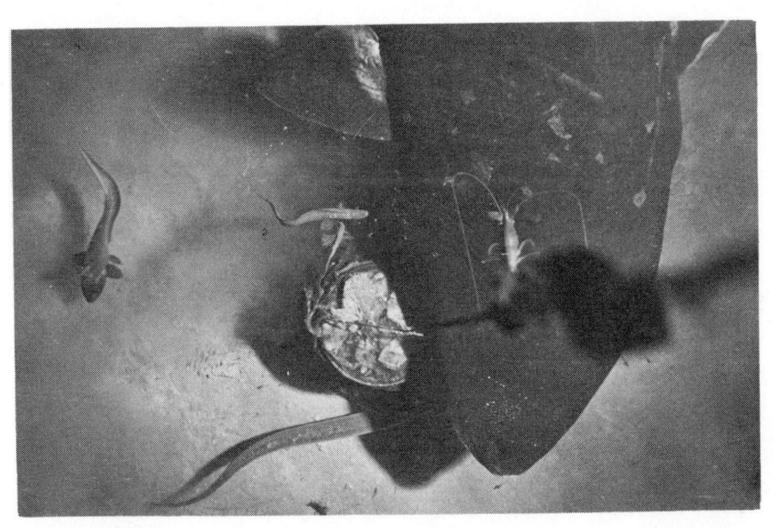

Fig. 1.20. Sleeper shark attracted to a can of bait at 1949 m depth off Hawaii The head of the animal is approximately 1.5 m wide. (Heezen & Hollister, 1971; photo: J. Isaacs)

Fig. 1.21. Rat tail fish photographed at 2000 m depth in the N.W. Atlantic. The camera was suspended 6 m above the sea floor and the fish is approximately 1 m long. (Photograph by courtesy of Dr R. Y. George)

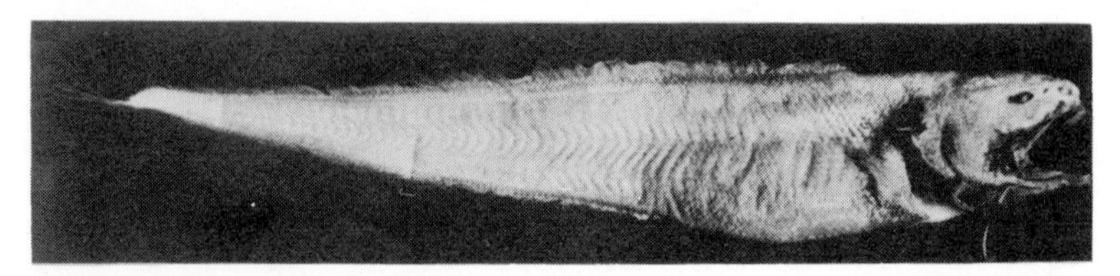

Fig. 1.22. *Bassogigas profundissimus*, a fish which lives close to the sea floor down to depths of 7000 m. (Nybelin, 1957)

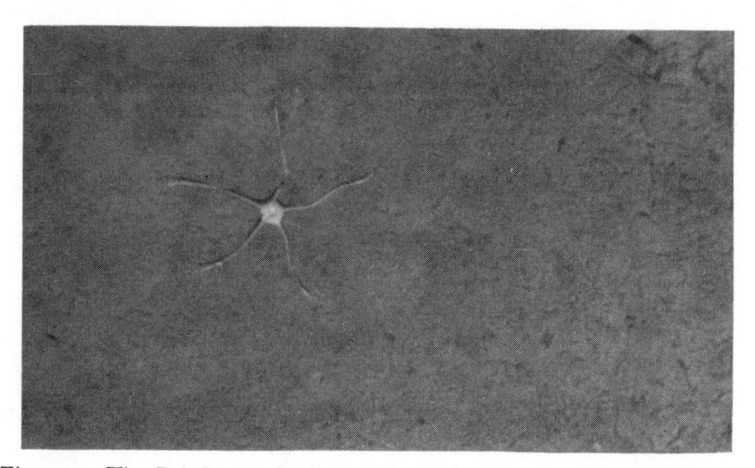

Fig. 1.23. The Brittle star *Bathypectinura heros* and the sea urchin *Echinus affinus* photographed at a depth of 2000 m in the NW Atlantic. (Photograph by courtesy of Dr R. Y. George)

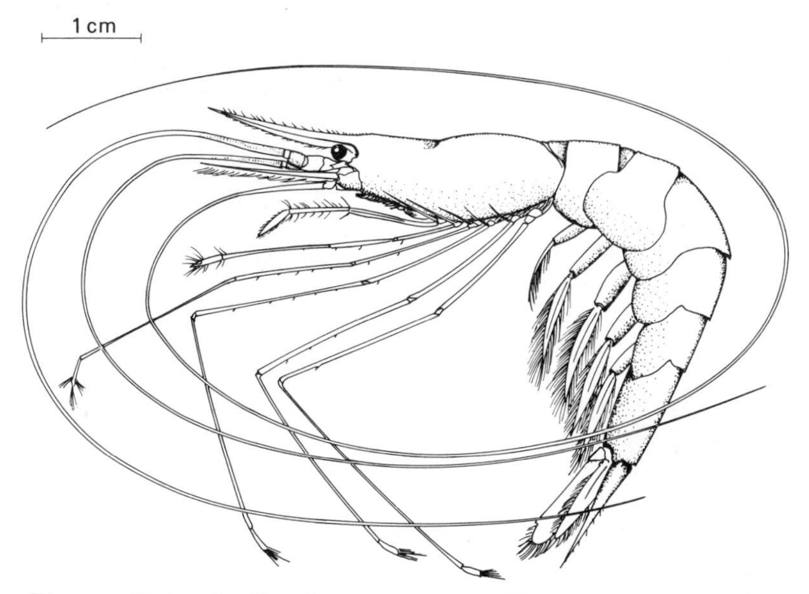

Fig. 1.24. Bottom dwelling deep sea crustacean, *Nematocarcinus ensiferus* × 1. (Agassiz, 1888)

enormous and little explored environment in which both ancient and recently evolved forms live.

It is easy to see why the Pogonophora were not recognised as animals until only recently. Their thin body, usually only a few centimetres long, lies within an insignificant string-like tube. Their internal structure suggests they have affinities with the Deuterostomia which is a major phylogenetic division of animals. We need not be concerned with the criteria used in classifying the Pogonophora but if they are put with the deuterostomes then they are bracketed with the echinoderms, enteropneusts and primitive chordates and set apart from such major invertebrate classes as annelids, arthropods and molluscs. Their worm-like appearance may therefore be superficial and is doubtless related to their life in mud.

The body is divided into three parts as shown in Fig. 1.28, which illustrates *Siboglinum caulleryi*. A tentacle (or in many species a group of tentacles) grows on the first segment, the prosoma. The second part of the body, the mesoma, is a compact cylindrical unit, often grooved and sharply marked off from the other body sections. The trunk or metasoma constitutes the major

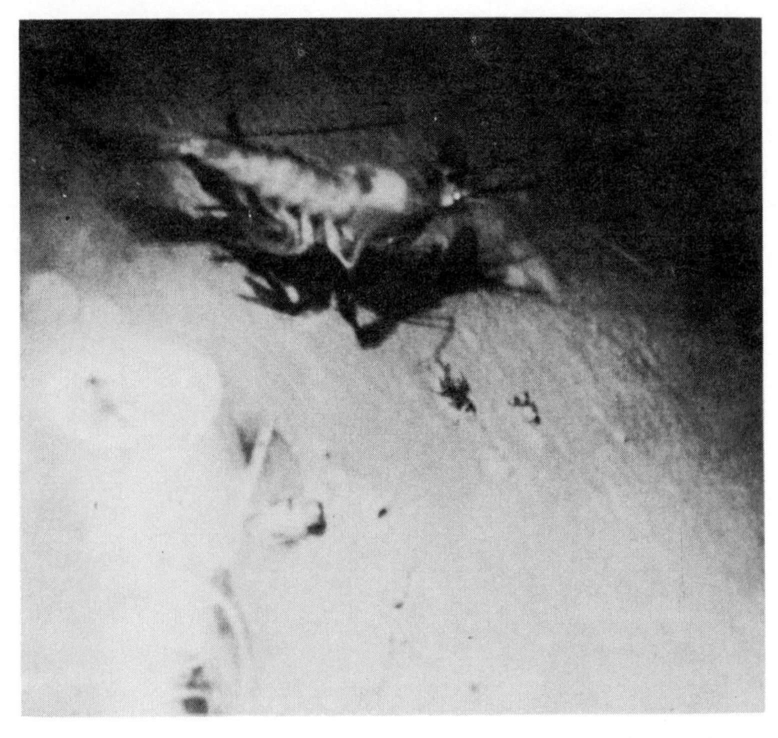

Fig. 1.25. Bottom dwelling crustacean at a depth of 4000 m off the Bahamas. (Heezen & Hollister, 1971; photo: B. Heezen)

portion of the body; it is dotted with papillae and may appear segmented. The animal lies upright in its tube in the mud and can extend its tentacles and upper portions out into the water.

It is a characteristic feature of the Pogonophora that they have no mouth, gut or anus at any stage in their development. The body wall consists of an epidermis covered by a thin cuticle. Beneath the epidermis there lies a thin muscle layer. Blood vessels run longitudinally inside the body cavity and a ventral vessel, swollen and apparently muscular, pumps blood around the body (Fig. 1.29). The blood in *Siboglinum* and some others contain haemoglobin.

The nervous system of the *S. caulleryi* consists of ganglion cells in a dorsal groove in the epidermis which runs posteriorly from a 'cerebral' ganglion at the base of the tentacle. Giant nerve fibres exist in some species, presumably playing a part in a rapid

Fig. 1.26. Acorn worm (enteropneust) moving over the surface of the sea floor at 4871 m depth (Kermadec Trench). (Heezen & Hollister, 1971; photo: B. Heezen)

withdrawal of the body back into its tube. Tentacles are innervated by an epidermal nerve net. In general these animals lack eyes but may have chemo-receptor cells. The sexes are separate with the gonads situated in the metasoma. In common with many aquatic invertebrates, sperm cells are transported in a package, a spermatophore, but the mechanism by which it is conveyed to female Pogonophora is not known. Fertilisation is thought to occur in the tube of the female and the embryos subsequently develop there. Larvae are not known in the group so it is generally assumed that the young emerge from their mother's tube and build their own tube nearby.

Because Pogonophora have no gut, nutrients from the environment must pass through their body wall to become assimilated. The epidermis lies between a basement membrane and the outer cuticle. Some species have bristles, cilia and microvilli projecting through the cuticle, which is flexible and extensible. It is constructed of fibres in criss-cross layers. Chemical analysis of the

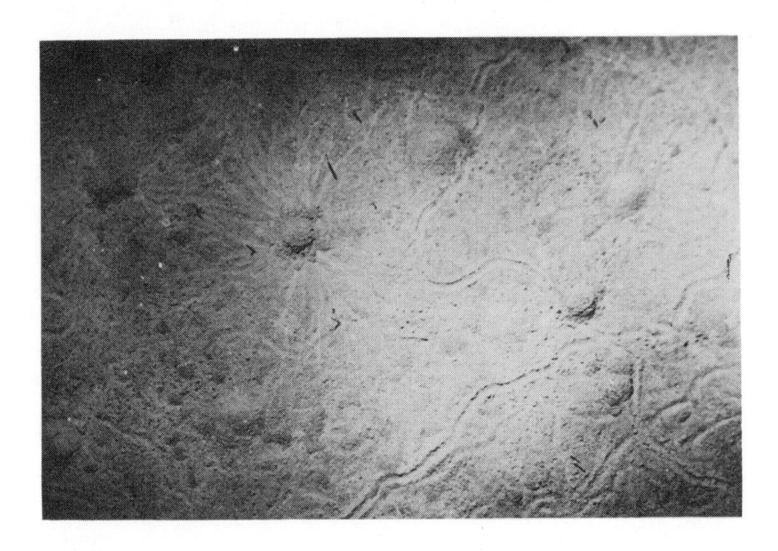

Fig. 1.27. Animal tracks on the sea floor at a depth of 7680 m; note the coconut (New Britain Trench). (Heezen & Hollister, 1971; photo R. Fisher)

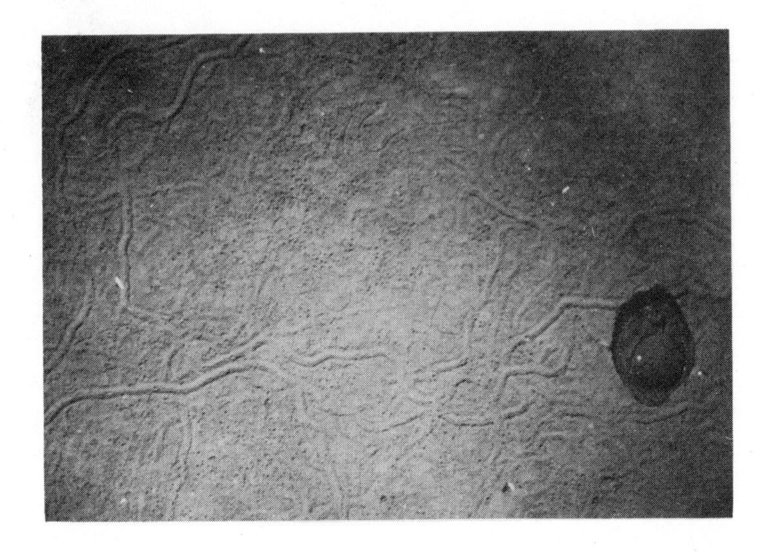

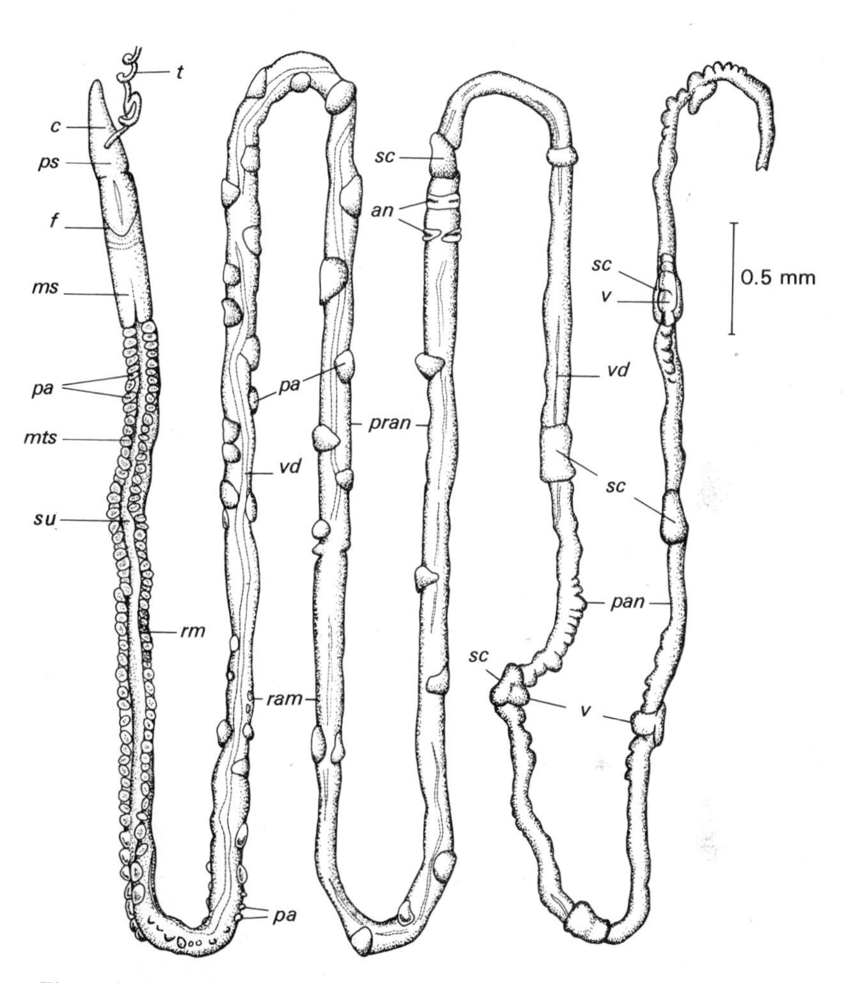

Fig. 1.28. A female *Siboglinum caulleryi*. The tentacle and hind part of the body are omitted; *an*, girdles: *c*, cephalic lobe: *f*, bridle: *ms*, mesosoma: *mts*, metasoma: *pa*, papillae: *pan*, postannular region of the metasoma: *pran*, preannular region of the metasoma: *ps*, protosoma: *ram*, nonmetameric part of the preannular region of the metasoma: *rm*, metameric part of the preannular region of the metasoma: *sc*, dorsal shields: *sv*, ventral sulcus: *t*, tentacle: *v*, postannular papillae: *vd*, dorsal blood vessel seen by transparency. (Ivanov, 1963)

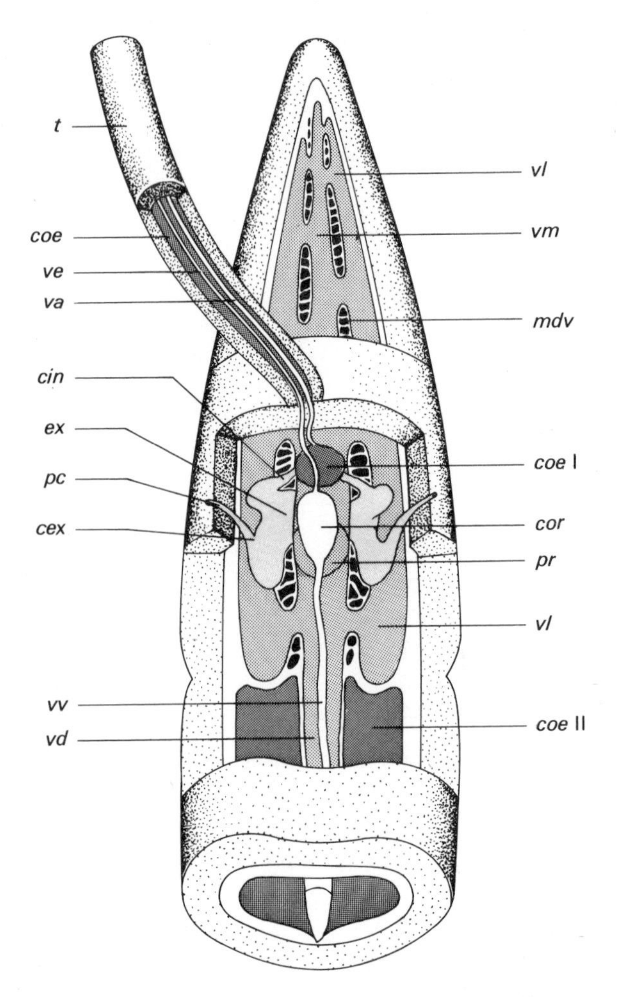

Fig. 1.29. Stereogram of the structure of the front end of the body of *Siboglinum caulleryi*, in ventral view; *cex*, distal canal of coelomoduct: *cin*, proximal canal of coelomoduct: *coe*, coelomic canal of tentacle: *coe* I, protocoele: *coe* II, mesocoele: *cor*, heart: *ex*, excretory portion of coelomoduct: *mdv*, dorso–ventral muscles: *pc*, external pore of coelomoduct: *pr*, pericardial sac: *t*, tentacle: *va*, afferent vessel of tentacle: *vd*, dorsal vessel: *ve*, efferent vessel of tentacle: *vl*, lateral cephalic vessel: *vm*, median cephalic vessel: *vv*, ventral vessel. (Ivanov, 1963)

cuticle is incomplete but it is collagen-like with outer (acid) and inner (neutral) polysaccharide layers (Gupta & Little, 1970). The epidermis is composed of unspecialised cells which appear capable of pinocytosis (Little & Gupta, 1968) and a variety of secretory cells, some of which may secrete digestive enzymes into the cavity between the cuticle and external tube. The physiology of the epidermis is currently under investigation and the results available at the time of writing are discussed in Chapter 6. It is difficult to avoid the conclusion that the epidermis is the site of an efficient nutrient-uptake mechanism, making the pogonophores the only free living macroscopic animals to feed on soluble nutrients obtained from the environment. Their ability to live at high pressure also implies special adaptations at the molecular level.

Siboglinum caulleryi has the distinction of occupying the greatest depth range of any animal, living at depths from 20–8160 m (Ivanov, 1963). Other pogonophores are much more restricted in the vertical range, and occur throughout the world's oceans. According to Southward and Southward (1967) Pogonophora typically live where there are plenty of other benthic animals, in relatively nutrient-rich areas of the ocean floor.

Deep sea micro-organisms, both pelagic and sedimentary, play an important role in the nutrition of animals and possess interesting physiological properties. Some of these are mentioned in Chapters 3 and 6.

Summary

Table 1.5 summarises the physical features of the deep sea.

Two distinctive and biologically interesting conditions prevail. Deep sea organisms live in a unique combination of high pressure with low temperature, a combination which acts at the molecular level in ways which we poorly understand. Further, deep sea organisms occupy an enormous environment in which biomass and sensory stimuli tend to be greatly dispersed or attenuated.

For all grades of organisms in the deep sea, nutrition would appear to pose special problems; and in the case of multicellular animals sensory physiology has particular interest. In the treatment which follows special emphasis is given to hydrostatic pressure, a factor which affects the molecular machinery of simple and complex organisms alike.

TABLE 1.5. *Major features of the deep sea*

		Also present in
Low temperature	Less than 10 or 4 °C (according to definition)	Shallow seas
High pressure	50–1100 atm (according to definition)	Oil-well brines Lake Baikal 100 atm
Constant temperature	Typically no more than ± 0.2 degC annually or less, below 1000 m	Ice-covered seas
Sunlight reduced or absent; other stimuli attenuated	Less than 10^{-10} surface value below 1000 m	Caves
Distance from primary production	Several km; primary production restricted to top 50 m	—
Lack of or diminished rhythms, seasonal and diurnal	—	—
Three-dimensional environment, large scale	Volume 1300×10^6 km^3 or approx. 60% earth's surface	—

It will be clear from this book that we have a great deal to learn about the physiology, or more generally the dynamic aspects, of deep sea life. In reciprocal fashion the deep sea has much to teach us about molecular mechanisms, about nutrition in a food-scarce situation and about sensory physiology, and to ignore it would result in an unsatisfactory, anthropocentric biology.

Pioneer biologists such as Regnard (1891) and Certes (1884) realised this nearly a hundred years ago, but the development of modern biology and marine science have only recently made an experimental investigation of deep sea organisms feasible, although by no means easy.

2 MOLECULAR ASPECTS OF HIGH HYDROSTATIC PRESSURE

Introduction

An understanding of how high hydrostatic pressures and low temperatures affect chemical and physiological processes is basic to an understanding of deep sea biology. In this section we shall deal with the volumetric aspects and pressure sensitivity of simple solutes and proteins. It will be argued that there are three distinct types of solute–solvent interactions which involve significant volume changes. These interactions however, are responsible for the pressure sensitivity of physiologically important equilibria and chemical reactions. The section closes with a brief account of the ways in which pressure affects the kinetics of biochemical reactions.

The effect of high pressure on the electrical conductivity of seawater

Electrical conductivity may be measured accurately and its variation with temperature, pressure and solute concentration has been studied most recently by Horne (1969). Electrical current is carried by ions whose random diffusion is biased by the electrical potential applied during the conductance measurement. A simple model of charged particles moving as if in a viscous medium accounts for the experimental results. The velocity of a spherical ion, j, acted upon by an electrical force F is:

$$V_j = \frac{F}{6\pi\eta A_j} \qquad (1)$$

where η is the viscosity and A_j the effective radius of the ion. At atmospheric pressure an increase in the temperature of seawater increases its electrical conductance but decreases its viscosity. The equivalent conductance Λ, which we may regard as proportional to V_j, thus varies inversely with viscosity η, giving the relationship

TABLE 2.1. *The constancy of the Walden product.* (Horne, 1969)

Solution	Percentage change in going from 0 to 20 °C		
	Equivalent conductance, Λ	Viscosity, η	Walden product, $\Lambda\eta$
Inf. dil. NaCl	+68	−44	6.0
0.5 M NaCl	+62	−42	3.7
Seawater salinity 35‰	+61	−49	5.1

known as Walden's rule. Table 2.1 from Horne (1969) shows how the Walden product, $\Lambda\eta$, remains approximately constant while conductivity increases by 61 per cent and viscosity decreases by 49 per cent over a 20 degC temperature rise. This suggests that the radius of the ions carrying the electrical current is unaffected by the change in temperature. At elevated pressures conductance increases in a more complicated manner and Walden's rule is not obeyed. Fig. 2.1 shows a series of curves each of which depicts a contribution to the total conductance from separate sources (Horne & Courant, 1964). Curve *B* shows the increase in conductance arising from the action of pressure on the equilibrium

$$MgSO_4 \rightleftharpoons Mg^{2+} + SO_4^{2-}$$

Other ions in seawater are either completely dissociated or contribute only a very small part to the increase in conductance. Interestingly Mg^{2+} ions probably behave in a manner similar to that envisaged in the model, being large, heavily hydrated and therefore bulky. How it is that dissociation is increased by high pressure will be dealt with shortly.

A major part of the increased conductance derives from the bulk compression of the solution producing a more concentrated solution of ions (see curve *C*). Seawater is not incompressible and contains a significant open structure, which even the presence of ions has failed to destroy. The difference in conductivity between curves *C* and *D* may be attributed to a change in the viscosity of seawater caused by pressure (i.e. a change in η in equation (1)). The viscosity change is clearly a decrease permitting a more rapid movement of ions. Exactly how pressure causes a decrease in viscosity will also be explained shortly. The discrepancy between the experimental points and curve *D* is accounted for if it is assumed

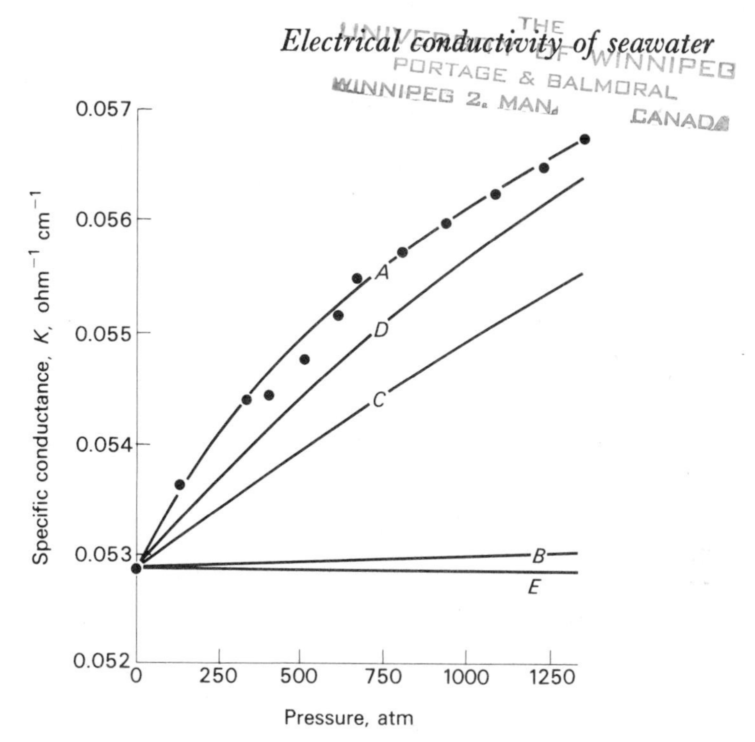

Fig. 2.1. Analysis of the electrical conduction of seawater of salinity 35‰ under pressure at 25 °C. *B*, correction for dissociation of $MgSO_4$: *C*, correction for compression of seawater volume: *D*, correction for the decrease in the viscosity of seawater: *A*, correction for the diminished effective ionic radii: *E*, uncorrected base curve. See text. (Horne & Courant, 1964)

that pressure decreases the effective, hydrated, size of ions. This would account for the more rapid rise in conductance with pressure than can be explained by the other factors. This presents us with a paradox; having taken pressure to favour dissociation (curve *B*) it is curious that pressure should apparently diminish the effective, i.e. hydrated, size of an ion. In order to explain this and the anomalous viscosity effect (curve *D*) the structure of water has to be considered.

WATER STRUCTURE

Chemists frequently describe water as anomalous because it differs from ordinary liquids such as liquid argon. To a biologist this is not particularly helpful. Water shows interesting biphasic changes in several properties which ordinary liquids do not, and water therefore has to be provided with a special structure to account for these complicated properties. A well known example is its temperature of maximum density at about 4 °C. Why should water become less dense at temperatures above and below 4 °C? Why should the density maximum shift to lower temperatures when ions are added, as if ions made water more 'normal'? A population of monomeric water molecules could hardly account for these observations. In the most general way we have to imagine individual water molecules hydrogen bonded to each other and forming ever changing clusters whose group behaviour is altered by temperature, pressure, and solutes, both charged and uncharged.

IONIC DISSOCIATION

Dissociation is favoured by increased pressure because a molecular volume decrease occurs in the process. This may be directly observed. Horne (1969) notes that a volume of NaCl equal to 13.5 ml (0.5 moles NaCl) added to water at a certain temperature to make 1 kg of solution, should yield a total volume of 987.2 ml (assuming a simple addition of volumes), but the observed volume is 983.0 ml. The presence of ions brings about a constriction of water molecules into the so-called hydration shell of the ion. According to the principle of Le Chatelier, pressure will thus favour the formation and hydration of ions, a prediction long ago substantiated experimentally. The consequence of introducing ions, electro-striction, is a decrease in the compressibility of the solution because a portion of the water is pre-compressed by electrical forces.

VISCOSITY CHANGES AT PRESSURE

The anomalous decrease in the viscosity of seawater with an increase in pressure has already been noted. Horne proposes that an ion exerts two effects on the behaviour of the surrounding water. One is the short range effect, electro-striction, already described. The other longer range effect is to disturb the structure of bulk water in a way which brings about a more open, less dense arrangement of molecules. The diagram in Fig. 2.2 shows a 'cluster' model of water structure, following earlier workers in this field. It is the

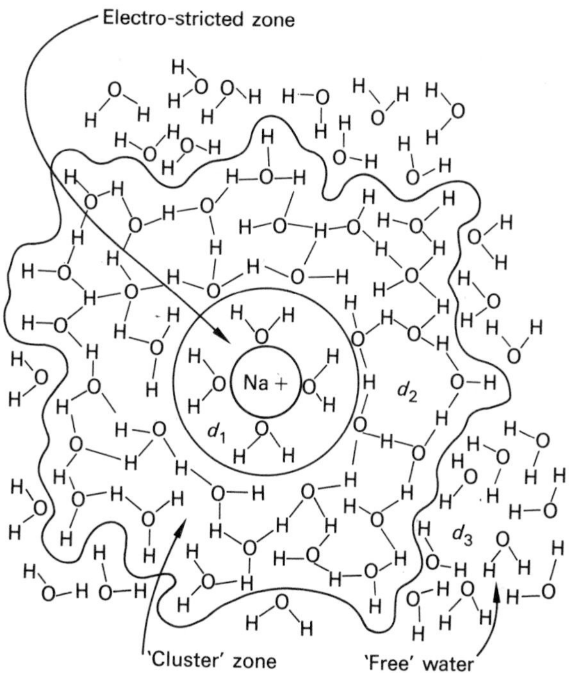

Fig. 2.2. Two-dimensional representation of the total structure-enhanced hydration atmosphere of the sodium ion. At 1 atm $(\text{density})_1 > (\text{density})_3 > (\text{density})_2$. (Horne, 1969)

layer d_2 which accounts for the extra conductivity increase in seawater at elevated pressure. Fig. 2.3 shows how pressures in the range occurring in the deep sea diminish viscosity. If the cluster zone, d_2 in Fig. 2.2, is crushed by pressure then a more freely interacting system of molecular aggregates might come into existence. Under an applied electrical force ions will move with less hindrance. At higher pressures a general compaction will take place to diminish the internal mobility of molecules, thus increasing viscosity. Normal liquids show this compaction effect at all pressures. For a thorough treatment of this subject the reader should consult Horne (1969) and other recent texts, for example Drost-Hansen (1972).

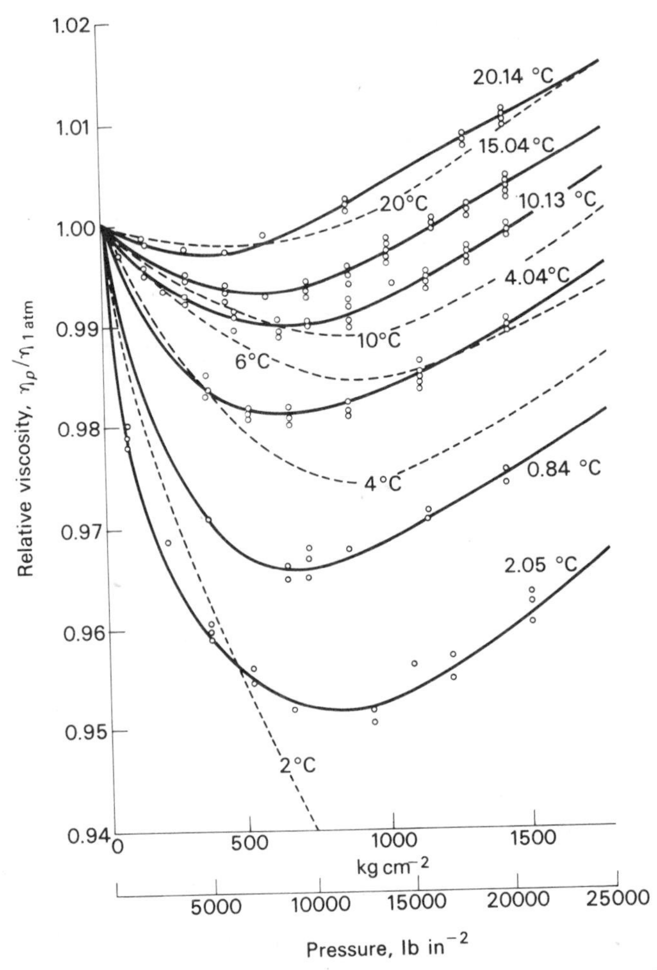

Fig. 2.3. Effect of pressure on the relative viscosity of pure water (dashed lines) and seawater (continuous lines) at 35‰ salinity and stated temperatures. $kg\,cm^{-2}$ closely approximates to atm. (Horne & Johnson, 1966)

Uncharged solutes

VOLUMETRIC ASPECTS OF UNCHARGED SOLUTES IN WATER

Uncharged solutes affect water structure causing important volumetric changes. The experiments of Masterton (1954) demonstrate a decrease in the partial molal volume of hydrocarbons when they are transferred from an oily hydrophobic phase to water. The apparatus which was used in the experiments consisted of two vertical tubes with interconnections top and bottom; the top connecting tube allowed the level of two liquids, water in one tube and a solution of hydrocarbon in the other, to equilibrate. When a valve in the bottom tube was opened a slight flow of liquid took place because of the difference between the densities of the two solutions. The method permitted the specific gravity of aqueous solutions of hydrocarbons to be determined to an accuracy of 1 part in 10^7. Molal volumes were calculated from specific gravity data by the equation:

$$\text{Apparent molal volume} = \left(1 - \frac{d}{d_0} - \frac{y}{d_0}\right)\frac{M_2}{y} \qquad (2)$$

in which M_2 is the molecular weight of the hydrocarbon, y is the concentration of hydrocarbon in g ml^{-1} and d and d_0 the densities of the test solution and pure water respectively. Fig. 2.4 shows the partial molal volumes of selected hydrocarbons in water at different temperatures. In all cases the volume occupied by an individual hydrocarbon molecule in an aqueous environment is smaller than that occupied when the same molecule is surrounded by molecules of its own type. For example, at 25 °C the molal volume of benzene in benzene is 89.5 ml while that of benzene in water is 83.1 ml. On entering an aqueous phase benzene thus undergoes a volume decrease of 6.4 ml mole^{-1}. The effect of temperature on this molal volume is interesting. In the case of benzene an increase in temperature causes a linear expansion in partial molal volume, but in the three aliphatic hydrocarbons studied a different relationship is seen. Aliphatic hydrocarbons show a larger volume decrease on dissolving in water. To some extent the volume decrease may be attributed to the internal pressures of water compressing the solutes; organic solvents generally have lower internal pressures than water. Additionally, the hydrocarbons may occupy voids in the water structure. However, the effect of temperature on partial molal volumes seems to indicate that some water-ordering takes place around the aliphatic hydrocarbons. Fig. 2.4 shows how all partial molal volumes increase steeply with rise in temperature; this is due to thermal expansion. But the collapse of the partial molal volume of methane in particular, and to a lesser extent of propane, at elevated temperatures is consistent with the presence

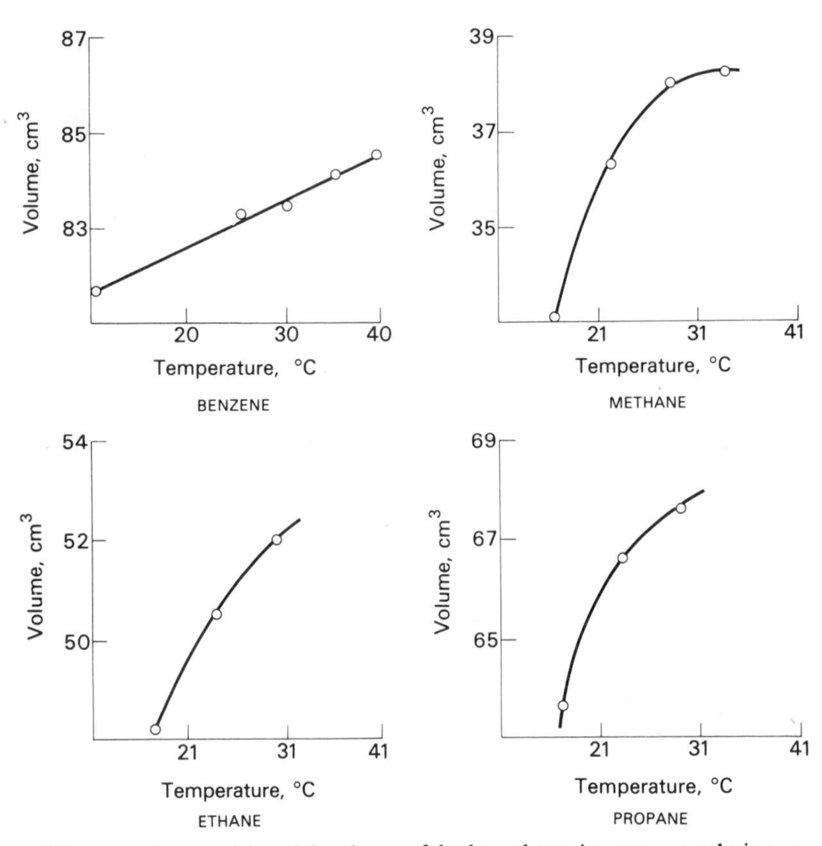

Fig. 2.4. The partial molal volume of hydrocarbons in aqueous solution at atmospheric pressure and selected temperatures. (Masterton, 1954)

of ordered water being broken down by thermal motion. Masterton (1954) points out the aliphatic compounds form stable clathrates (hydrates) whereas benzene, whose partial molal volume is linearly related to temperature, does not. The curves for methane and propane in Fig. 2.4 provide evidence for the existence of a clathrate type of structure in solution.

From this brief consideration of an apolar solute in water we have encountered three processes all of which may be involved in a change in molecular volume: the compression of the solute by the solvent, the possibility that a solute can occupy a void in a solvent and thus dissolve without adding its own volume to a system and finally the stabilising of a clathrate structure which may not exist in the absence of a solute. The relative contributions of these processes to the overall decrease in volume measured by Masterton are obscure. The important point is that a net volume decrease occurs when an apolar solute enters an aqueous phase and this is of considerable biological significance.

MICELLE FORMATION AT HIGH PRESSURE

Certain solutes may be induced to form colloidal particles or micelles when their solubility is decreased by, for example, a change in temperature. The non-ionic surface active agent PNE (polyoxyethylene nonylphenyl ether) is such a solute, forming a cloudy suspension of micelles when the temperature is raised. Suzuki and Tsuchiya (1969) have shown that pressures of less than 2000 atm increase the solubility of PNE, and thus raise the temperature at which micelles form, the so-called cloud point. At higher pressures micelle formation is favoured and the temperature of the cloud point declines (Fig. 2.5). In the first case the volume decrease brought about by the entry of the hydrophobic solute into the aqueous phase determines the pressure effect. At higher pressures the bulk compressibility of the micelles causes a greater volume decrease and so pressure favours the formation of high pressure micelles. Tuddenham and Alexander (1962) found a similar biphasic effect of pressure in the formation of micelles of charged molecules. In the case of the soaps which they studied, a transitional pressure of 750 atm, above which micelle formation is favoured by pressure, was observed. The authors invoked two factors to account for high pressure micelles; firstly, the bulk compression of the hydrocarbon micelle core; secondly, the increased dielectric constant of water which permits the charged molecules to approach each other.

We have now considered three relatively simple systems at pressure, namely, electrically conducting seawater, hydrocarbons in water and micelles. Several processes have been shown to cause volume decreases: electro-striction by ions; the movement of hydrophobic solutes into water with water ordering or hydrophobic hydration, and the bulk compression of an open molecular structure, as in a micelle or in the hypothetical cluster zone in water. We have not considered hydrogen bonds but there is some

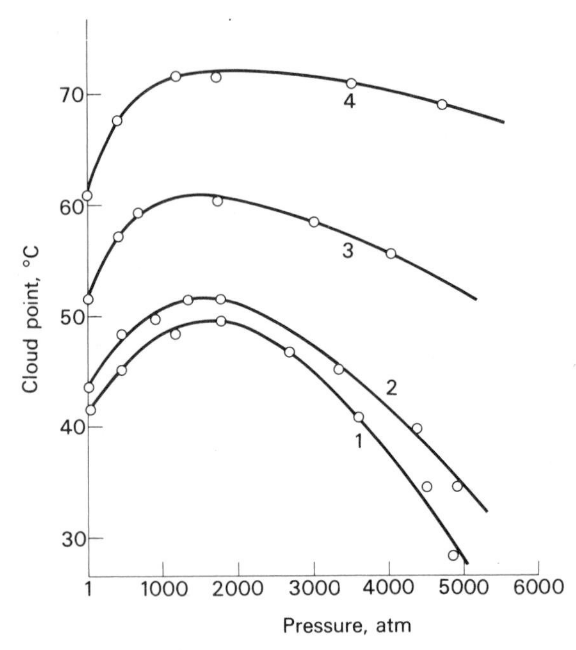

Fig. 2.5. The effect of pressure on the cloud point of PNE–water system. Percentage by weight of PNE, 1, 0.05: 2, 10.0: 3, 20.0: 4, 30.0. See text. (Suzuki & Tsuchiya, 1969)

evidence that their formation involves a small volume decrease (Suzuki, Taniguchi & Enomoto, 1972). Each of these is a major field of enquiry for physical chemists and has been dealt with here in a most elementary fashion. In the subsequent sections we will consider the extent to which these three mechanisms are responsible for the sensitivity of more complex molecules to high pressure.

Large molecules

VOLUMETRIC ASPECTS OF PROTEINS

The role of ionised and apolar groups in the conformation and biological activity of proteins is well established. We may therefore anticipate that pressure will alter the conformation of proteins by affecting their ionic and hydrophobic sidechains. Rasper and Kauzman (1962) distinguish three components in the partial molal volume of a dissolved protein. A major component is the sum of the amino acid volumes determined by the addition of bond lengths and van der Waal's radii. A secondary volume is derived from the protein's conformation. For example, solvent may be preferentially excluded from certain regions by tertiary structure. The third component of the partial volume of a protein derives from the interaction of its surface groups with solvent water. The volume changes associated with the second and third categories are of prime interest in the range of pressures found in the sea. McMeekin, Groves and Hipp (1954) have shown how the molal volume of a protein changes in response to an aqueous environment. The specific volume of anhydrous, crystalline β-lactoglobulin is 0.802, the reciprocal of its density. In a humid gaseous environment the specific volume of the crystalline protein is decreased to 0.772, an effect attributed to the presence of water molecules attached to the molecule and exerting their cohesive force (or internal pressure). In aqueous solution the apparent specific volume is 0.751. The qualifying term 'apparent' is used to emphasise that in calculating the protein's volume from density data it is assumed that the solvent remains unaffected by the protein. According to McMeekin *et al.* the difference between the hydrated crystalline specific volume (0.772) and the apparent specific volume (0.751) is a good measure of the contraction of the solvent brought about by the ionic groups on the protein. This equals $0.772 - 0.751 = -0.021$ ml water per gram of protein. Proteins thus have interesting volumetric properties* and hence pressure sensitivity. We can consider the phenomenon first in relatively simple and small molecules.

SYNTHETIC MACRO-MOLECULES

The ionisation of poly-D-glutamic acid (PDGA) and poly-L-lysine (PLL) in water causes a volume decrease involving the electrostriction of solvent. Hence the dissociation of these substances is sensitive to pressure. Suzuki and Taniguchi (1972) have described how the ionisation of these acids increases by approximately 20 per cent at 1000 atm. The salts of these acids are nearly completely ionised in solution at 1 atm and are little affected by high pressure (Fig. 2.6).

* See Zamyatnin (1972).

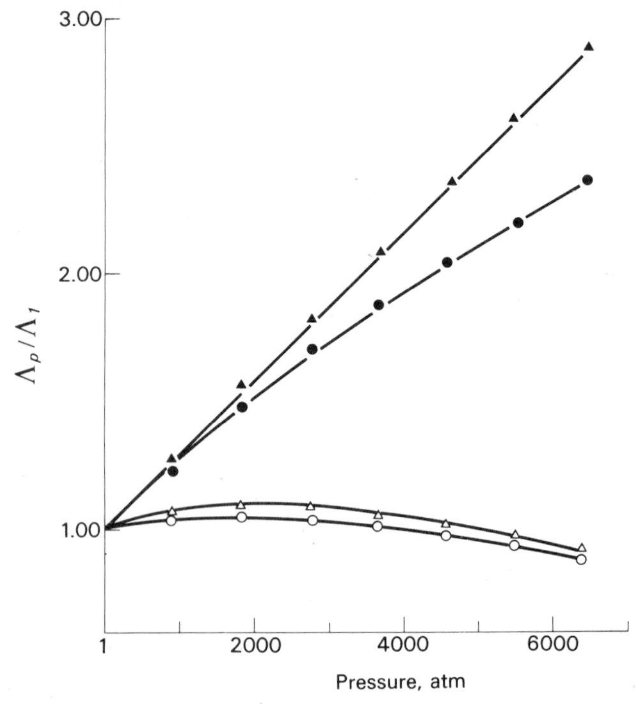

Fig. 2.6. The effect of pressure on the relative equivalent conductivities ($\Lambda p/\Lambda_1$) of poly-D-glutamic acid (●), poly-L-lysine (▲), and their salts (○, △, each conc. 5×10^{-3} M) in aqueous solutions at 30 °C. (Suzuki & Taniguchi, 1972)

The helical conformation of polyamino acids is changed when pressure alters the repulsive interactions between side groups. Makino and Noguchi (1971) have performed the converse experiment of that illustrated in Fig. 2.6 using dilatometry to measure the volume increase brought about by the addition of hydrogen ions to poly-L-glutamic acid:

$$-COO^- + H^+ \rightleftharpoons COOH \qquad (3)$$

Hydrogen ions force the equilibrium to the right, causing a molecular expansion through the release of previously bound water. Pressure would shift the equilibrium to the left.

The enzyme ribonuclease has been subjected to skilful chemical manipulation and converted to poly-L-valyl ribonuclease (PVRN-ase) by the addition of valine chains to the surface of the molecule (Kettman, Nishikawa, Morita & Becker, 1966). Thus, unlike many

$$\begin{array}{c} CH_3\ \ CH_3 \\ \diagdown\ \diagup \\ CH \\ \diagup \\ NH_2CHCOOH \end{array}$$

Valine

natural proteins, PVRN-ase carries exposed hydrophobic groups on its surface. These groups influence its solubility and volumetric properties. Whereas native RN-ase aggregates only at high temperature, PVRN-ase associates* rapidly at 30 °C. The aggregation almost certainly involves the hydrophobic methyl groups; for example, similar polyglycyl compounds show no thermal association. The association of PVRN-ase may be described as the mutual dissolution or attraction of apolar groups which, as we have seen, will involve a volume increase. The temperature sensitivity of the system is of special interest. At low temperatures the ordered water layer surrounding the apolar groups is stable and inhibits association. Note the similarity with the temperature sensitivity of the partial molal volumes of certain aliphatic hydrocarbons (Fig. 2.5). The ionic environment of PVRN-ase is also important. Association is favoured by an increased sodium chloride concentration, a 'salting out' effect (Fig. 2.7); other solutes exert a 'salting in' effect (Nishikawa, Morita & Becker, 1968). Fig. 2.8 shows how the rate of association, measured by change in optical density, is decreased by a moderate pressure. The kinetics of thermal association are discussed in a later section; the reaction scheme below merely illustrates the phenomenon.

$$\text{Sub-units} \underset{\substack{-\Delta V \\ -\Delta H}}{\overset{\substack{+\Delta H \\ +\Delta V}}{\rightleftharpoons}} \text{PVRN-ase associated by hydrophobic forces} + H_2O \text{ released} \qquad (4)$$

* The term 'associate' implies the formation of non-covalent bonds while 'aggregation' is a general term with no implications as to the type of bonds involved. Polymerisation implies the formation of covalent bonds between monomers.

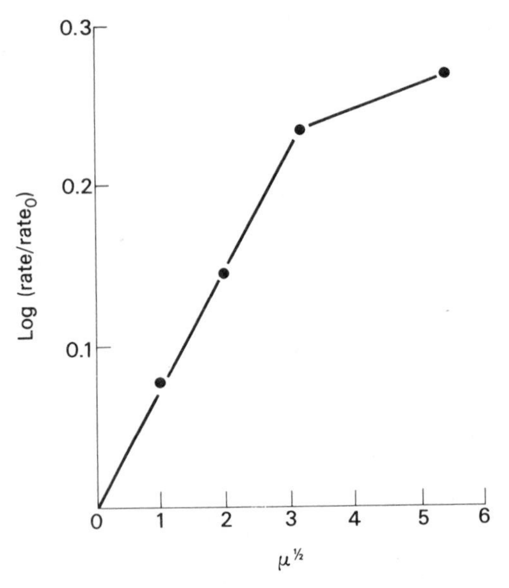

Fig. 2.7. Relative rates of association of PVRN-ase as a function of ionic strength (NaCl) pH 7.4, at atmospheric pressure. (Nishikawa, Morita & Becker, 1968)

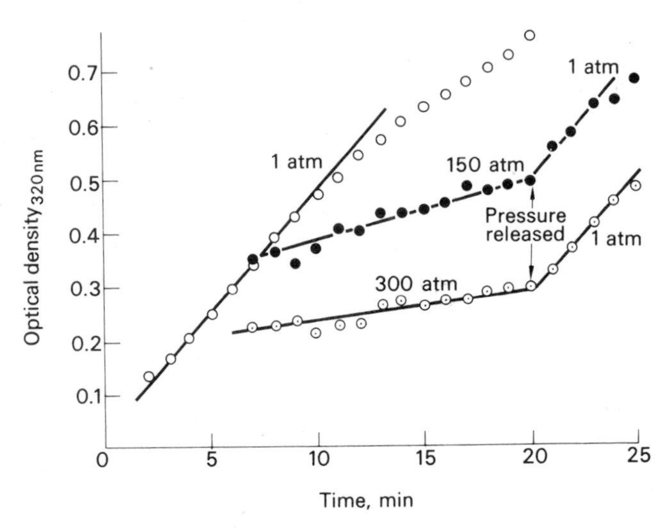

Fig. 2.8. The effect of pressure on the thermal association of PVRN-ase in 0.6 M NaCl. See text. (Kettman, Nishikawa, Morita & Becker, 1966)

UNMODIFIED PROTEINS

β-casein, a non-enzymic oligomeric protein dissociates to sub-units at temperatures below 4 °C and in nature, therefore, this protein exists in the associated form. High hydrostatic pressure dissociates the protein and Payens and Hermans (1969), who have studied the equilibrium in detail, conclude that hydrophobic interactions are probably involved in the formation of the normal associated form.

In contrast, rabbit myosin shows a reversible dissociation at moderate pressure which has been attributed to the dissociation of charged sub-units (Josephs and Harrington, 1968). At any rate no evidence for hydrophobic hydration is available and the temperature insensitivity of its equilibrium is consistent with the dissociation of ionised groups. Josephs and Harrington have studied this molecule using a complicated centrifugation technique to generate pressure and have calculated a volume increase of 380 ml mole^{-1} monomer bound for the polymerisation reaction in 0.1–0.2 M KCl at pH 8–8.5. This large volume change rendered the myosin filaments sensitive to pressures as low as 40 atm which is equivalent to 400 m depth. Tropomyosin exhibits a temperature sensitive monomer–polymer equilibrium which is unusual in shifting to the left with rise in temperature. It will be shown in a subsequent chapter how living muscle is probably affected by temperature and pressure by mechanisms other than changes in the aggregation of proteins.

The activity of enzymes is a convenient index of the integrity of their structure. Many enzymes have been exposed to pressure with interesting results. A study by Penniston (1971) provides us with some evidence that the disaggregating effect of pressure occurs in a number of multimeric enzymes. For reasons which will be discussed in a later section, the enzymes (those studied are listed in Table 2.2) were incubated at a moderate temperature with a high concentration of substrate. A pressure of up to 1500 atm was applied for 15 min after which the enzyme reaction was stopped and the products analysed. Differences were observed between those enzymes which are multimeric in structure and those which are monomeric, i.e., composed of a single unit. The former were pressure labile, their activity being reduced by increased pressure whereas enzymes composed of a single unit showed enhanced activity at elevated pressure.

TABLE 2.2. *The effect of high hydrostatic pressure on the catalytic properties of selected enzymes* (Penniston, 1971)

Enzymes and enzyme fractions	% inhibition by 1000 atm for 15 min	Enzymes	% activation by 1000 atm for 15 min
Creative kinase	16	Peroxidase	8
DPN-ase	5	Myokinase	32
Alkaline phosphatase	44	5′-Nucleotidase	107
Argininosuccinase	64	[a] Trypsin	6
Pyruvate carboxylase	122	[a] Amylase	14
ATP-ase		[a] Lysozyme	53
beef heart mitochondria	71	[a] Chymotrypsin	68
ETP_H	58		
purified mitochondrial ATP-ase	76		
rat liver mitochondria	60		
erythrocyte ghosts	49		
sarcotubular vesicles	36		
ATP-$^{32}P_i$ exchanges			
beef heart mitochondria	112		
ETP_H	95		
rat liver mitochondria	114		

Reaction conditions: high substrate concentration, temperature between 23 and 27 °C. Figures greater than 100% mean total inhibition was achieved at less than 1000 atm.
[a] Data quoted by Penniston.

Penniston, using the kinetic reasoning which will be discussed in a later section, argues that the diminished activity in the multimeric enzymes is brought about by their disaggregation at pressure. Both hydrophobic and ionic bonds are probably broken in the process. Penniston suggests that deep sea multimeric enzymes may require strong noncovalent interactions.

An early attempt by Johnson and Schlegel (1948) to detect changes in the oxyhaemoglobin–haemoglobin equilibrium in bovine red cells failed to detect any change at pressures up to 680 atm. However, Wells (1972–73) reports that 100 atm slightly increases the affinity of human haemoglobin for oxygen at concentrations of less than 20 per cent saturation. At higher oxygen concentrations 100 atm reduces the affinity of haemoglobin for oxygen. At pressures greater than 100 atm haemoglobin showed a reduced affinity for oxygen at all oxygen tensions.

The large size and multimeric structure of many respiratory pigments suggest that they may exhibit interesting pressure effects. Very high pressure treatment of haemoglobin certainly induces structural changes in the protein but interestingly an abnormal human haemoglobin, a sickle cell haemoglobin described by Murayama (1966), revealed striking changes at moderate pressure. The particular haemoglobin studied by Murayama was thought to contain a valyl–valyl (i.e. hydrophobic) bond at a specific site. As this was the only feature in which it differed from normal haemoglobin it was thought that the hydrophobic bond was responsible for the aggregation of the molecules when deoxygenated, causing a 'sickling' or kinking of the intact erythrocyte. At 200–300 atm pressure, sickled cells reverted to a normal shape, an effect which Murayama attributed to the hydration of the hypothetical hydrophobic bond. Significantly, sickle cell haemoglobin underwent aggregation at high temperature and a liquefaction at low temperature. This example is of further interest because the sickle cell condition appears to arise from a single amino acid substitution in normal haemoglobin, reminding us of the importance of amino acid sequences in the properties of proteins. Murayama and Hasegawa (1970) have recently reported that a pressure of up to only 50 atm favours the disaggregation of extracted sickle cell haemoglobin; thereafter the volume change in the aggregation reaction changes to become negative and pressure enhances aggregation. The change in the direction of the pressure effect is tentatively attributed to a phase change perhaps comparable to the formation of micelles. To date, no other respiratory pigments have been studied at moderate pressure, but the haemocyanin from *Loligo pealii* undergoes an increase in partial specific volume during aggregation. Both ionic and hydrophobic bonding appear to be involved (Van Holden, 1966). This molecule is therefore probably pressure sensitive.

CONFORMATION CHANGES IN SYNTHETIC POLYPEPTIDES

Synthetic polypeptides present the experimenter with a simplified macro-molecule which serves as a model for natural proteins. The volume changes which occur when such a polypeptide undergoes major conformational changes are particularly instructive. Makino and Noguchi (1971) measured the volume changes associated with

TABLE 2.3. *Volume changes by dilatometry of poly(S-carboxymethyl-)L-cysteine and poly-L-glutamic acid during titration.* (Makino & Noguchi, 1971)

Samples	ΔV per H^+ bound (ml $mole^{-1}$)	ΔV (transition) (ml $mole^{-1}$)	NaCl concn
		Coil to β	
PCMC	12.7	2.35	0
	11.4	1.90	0.2
		Coil to helix	
PGA	12.4	1.25	0
	11.4	1.1	0.01
	11.1	0.62	0.2

the transition of poly(*S*-carboxymethyl-)L-cysteine (PCMC) from the coiled to the β-configuration. Titration of the sodium salt of the polypeptide with hydrochloric acid yielded a volume increase of 12.7 ml per mole H^+ ions bound, corresponding to the reaction $COO^- + H^+ \rightarrow COOH$. The volume change was slightly less in the presence of sodium chloride. The change in the ionisation of the polypeptide causes a conformational change to the β-state, envisaged as an ordered arrangement of amino acid residues keeping solvent water from the 'body' of the molecule. In contrast, the ionised coiled conformation is thought to be stabilised by mutually repulsive electrostatic charges arranged in a more open solvated structure. Hence a volume increase occurs with a release of electro-stricted water molecules when the ionised groups are neutralised. Makino and Noguchi were able to follow the change in conformation by measuring the shift in the velocity of sound passing through a 15 ml solution of the polypeptide, as well as determining the volume change directly by dilatometry. The same authors also studied the synthetic polypeptide (poly-L-glutamic acid, PLGA) which reacts with hydrogen ions undergoing a transition from coil to helical conformation with a volume increase of between 12.4 ml per mole H^+ bound. Again a smaller volume change occurred in the presence of salts. The coil to helix transition involved a slightly smaller volume change in PLGA than the coil to β-configuration in PCMC (Table 2.3). This was attributed to differences in void volume, i.e. conformational differences, between the two polypeptides. Although the reaction $COO^- + H^+ \rightarrow COOH$

may be given a precise volume change, even in model polypeptides higher order complexity introduces overall volume changes which, at the present state of knowledge, are difficult to predict.

STRUCTURAL PROTEINS: ACTIN, FLAGELLIN AND TMV PROTEIN

A wide variety of eukaryotic cells from low pressure environments contain filaments which react with heavy meromyosin and are therefore demonstrably actin.

Ikkai and Ooi (1966) have studied the monomer $\rightleftharpoons$ polymer equilibrium of actin *in vitro*. The experiments involved mixing a solution of buffered globular (monomeric) actin and ATP in the presence of magnesium chloride in a dilatometer at 25 ± 0.0002 °C. The change to F-actin produced an overall volume increase. Similar mixing experiments were undertaken to show viscosity changes during the G–F (globular to filamentous) transition (Fig. 2.9). The small discrepancy between viscosity and volume need not concern us. The increase in viscosity arises from the change in shape of the molecules from globular to filamentous; the volume change was considered by Ikkai and Ooi to represent the net volume change of several reactions. The dilution of the actin and the $MgCl_2$ solutions caused a negligible change but the action of ATP involved two reactions, each with a significant volume change. A volume increase of 22 ml mole^{-1} occurs when ATP binds Mg^{2+} ions (Rainford, Noguchi & Morales, 1965), and a volume decrease of 18.6 ml mole^{-1} arises from the release of inorganic phophate ions in the hydrolysis of ATP. The resultant volume change for the G $\rightleftharpoons$ F transition was calculated at +391 ml per mole protein (mol. wt 57000). Although the shift in the monomer–polymer equilibrium involves complicating subsidiary reactions, Ikkai and Ooi concluded that the volume increase reflected the release of ordered water around polar and apolar groups at binding sites on the sub-units.

Gerber and Noguchi (1967) have studied the protein from the flagella of *Salmonella abortus-equi* (strain Sh23). The protein, called flagellin, was isolated in two forms; one consisted of a solution of flagellin in monomer form; the other consisted of 'seeds'. These are aggregates about 0.2 μm long which act as nuclei for the formation of filaments which resemble intact flagella.

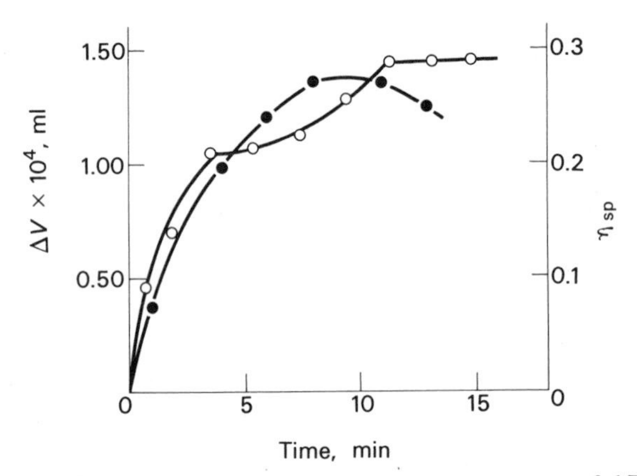

Fig. 2.9. Time course of the volume and viscosity increases of ATP–G-actin solution on polymerisation. At zero time, 15 ml of G-actin solution (1.15 mgml) in 2 mM Tris–HCl buffer (pH 8.0) containing 500 μM ATP was mixed with 3 ml of 6 mM magnesium chloride. Temperature was maintained at 25 °$\pm$0.0002 °C. ○, volume change: ●, coefficient of viscosity. (Ikkai & Ooi, 1966)

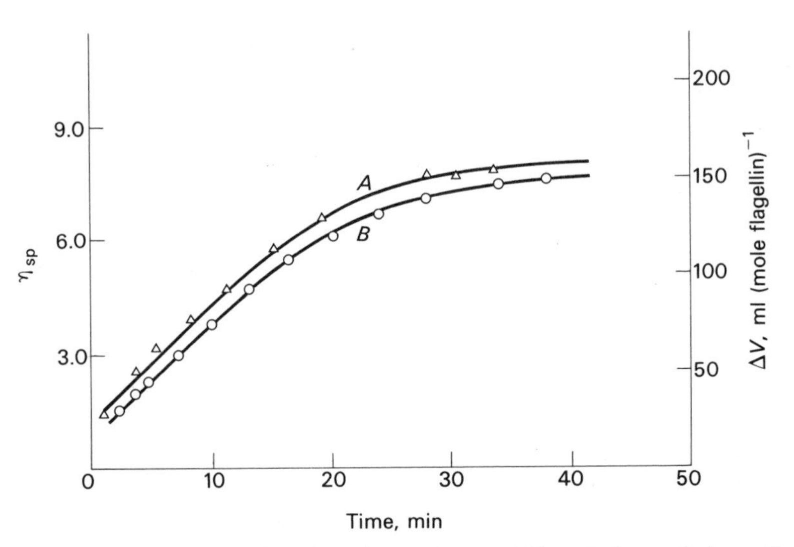

Fig. 2.10. Time course for the volume change, ΔV, and the coefficient of viscosity in the aggregation of flagellin. Curve *A*, volume change; curve *B*, coefficient of viscosity. Protein concentrations were 6.9 mg ml^{-1} in 0.15 M potassium chloride and 0.05 M phosphate buffer, pH 7.1. Monomer to seed ratio, 1:1; temperature, 22.08 °C. (Gerber & Noguchi, 1967)

The overall volume change occurring during the aggregation of sub-units was measured with a dilatometer and the change in viscosity was also followed separately. Fig. 2.10 shows how the viscosity rises as the increase in molecular volume occurs during filament formation at 22 °C. At temperatures between 25 and 31 °C a rise in temperature decreased the rate of polymerisation but increased the final volume change. At 25 °C, ΔV is 160 ml per mole aggregated; at 31 °C, ΔV is 255 ml per mole aggregated. This suggests that the mechanism of filament formation changes at higher temperatures. Note that ΔV is the difference between the partial molal volumes of monomeric flagellin and aggregated flagellin and is best accounted for in terms of change in the solvent around the molecules rather than by 'internal' changes in the molecules themselves. The specific volume of flagellin calculated from amino acid analysis is close to that determined experimentally. In view of the large number of hydrophobic residues in flagellin, Gerber and Noguchi suggested that hydrophobic bonding was a plausible mechanism for the aggregation. The associated increase in volume would then be caused by the release of water from hydrophobic groups. The curious effect of a rise in temperature diminishing the rate of formation of filaments is interesting. From studies of certain bacteria, *Spirillum serpens* and *Bacillus subtilis*, there is evidence for the synthesis of flagellin at elevated temperature but its assembly into functional flagellin is not achieved. Such a system would be interesting to study under conditions simulating the deep sea.

The protein component of the tobacco mosaic virus (TMV) has been thoroughly investigated by Lauffer (1964) and Jaenicke and Lauffer (1969). The protein aggregates into rod shaped particles resembling the intact virus when the temperature is raised from approximately 2 ° to 23 °C. The reversibility of the aggregation (polymerisation according to Lauffer) is shown in data plotted in Fig. 2.11. The monomer $\rightleftharpoons$ polymer equilibrium shifts to the right with a decrease in pH and conversely H^+ ions are liberated as the equilibrium is shifted to the left. The volume change involved is 590 ml mole^{-1} or 10^5 g protein of virus particle formed. 150 molecules of water per protein sub-unit are released during the process. It is curious that high pressure studies have not been carried out on this preparation. The large volume changes involved in this and other equilibria clearly imply a sensitivity to high

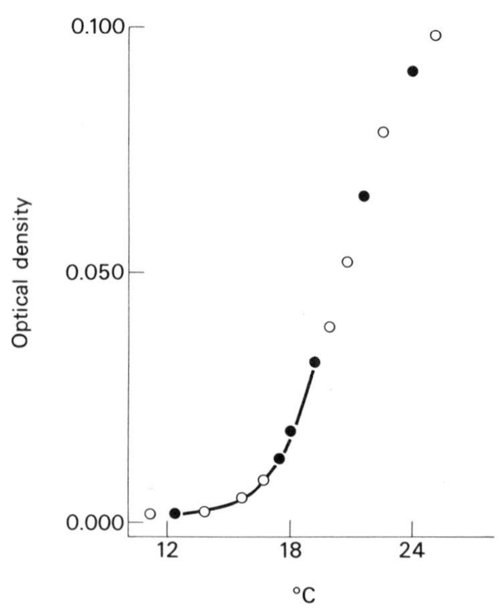

Fig. 2.11. Reversible polymerisation of tobacco mosaic virus (TMV) protein dissolved in 0.1 ionic strength phosphate buffer at pH 6.5 ○ corresponds to rising temperatures and ● to falling temperatures. (Lauffer, 1964)

pressure. In a following section (p. 59) the quantitative relationships between molecular volume change and pressure will be discussed in both equilibria and rate processes.

We conclude this section with two cautionary and complicating observations. Kauzman, Bodensky and Rasper (1962) measured the volume changes brought about by ionisation or neutralisation of various chemical groups in large proteins and compared them to the volume changes seen in simpler systems. In the case of simple carboxylic acids the combination H^+ ions, $R{-}COO^- + H^+ \rightarrow R{-}COOH$ yields a volume increase of between 10 and 14 ml per mole of H^+ ions bound. In more complicated molecules, particularly proteins, a similar volume change was detected when individual side chains underwent the same reaction. However, the dissociation of a carboxylic acid is modified by its neighbour.

Thus: $H{-}COOH \rightarrow H{-}COO^- + H^+$

$$\text{yields } -7.8 \text{ ml per mole } H^+ \text{ released} \quad (5)$$

$HO{-}COOH \rightarrow HO{-}COO^- + H^+$

$$\text{yields } -29 \text{ ml per mole } H^+ \text{ released} \quad (6)$$

We have to recognise that neighbouring groups complicate the volumetric properties of macro-molecules. Secondly, pressure itself changes the apparent molal volume of such simple molecules as glycine and alanine in aqueous solutions (Yayanos, 1972).

The preceding examples suggest that high pressure is of considerable significance in the molecular biology of deep sea organisms. The volume changes and pressure effects which have been described were largely derived from *in vitro* systems and we have eventually to tackle the functional, living state. It is conceivable that a great deal of pressure-sensitivity may disappear when an isolated protein is returned to its physiological environment. It is equally possible that the pressure sensitivity of intact organisms will be greater than that of their isolated parts through amplification processes at the kinetic and integrated levels of organisation.

Dynamic aspects: the quantitative relationship between volume change and pressure in chemical reactions

LE CHATELIER'S PRINCIPLE

This principle tells us that pressure will favour the condition which occupies least volume, and so we might expect chemical reactions which involve a volume decrease to be accelerated by high pressure. In order to relate the effect of pressure on an equilibrium, or the rate at which a chemical reaction proceeds, to volume change we have to consider some basic chemistry. This account continues at an elementary level but the reader who is conversant with reaction kinetics should refer to the treatments of Johnson, Eyring and Polissar (1954) and Laidler (1951).

The reversible reaction $A+B \underset{k_{-1}}{\overset{k_1}{\rightleftharpoons}} AB$ may be considered to consist of a forward and a reverse reaction indicated by the rate constants k_1, k_{-1}. The volume change, ΔV, which occurs on going from A+B to AB may be measured by dilatometry and the effect of pressure on the equilibrium may be predicted, according to Le Chatelier's principle. However from ΔV we cannot predict the *rate* at which A+B will convert to AB. The rate of reaction is determined by the volume change involved in passing through the activated state which subsequently decays to form products. This volume change is $\Delta V^{\ddagger}$, the activation volume.

Molecular aspects

Consider the thermal aggregation of PVRN-ase which has been mentioned on p. 49. At a suitably elevated temperature aggregation proceeds without the influence of enzymes. The rate at which it proceeds at an elevated pressure is determined by the activation volume, $\Delta V^{\ddagger}$, which may be calculated in retrospect but which in the present state of knowledge cannot be predicted. For a shift in a simple equilibrium $\Delta V^{\ddagger}$ is likely to be similar to ΔV but for many chemical reactions this no longer holds.

When an increase in temperature forces the PVRN-ase association reaction to the right we may apply the familiar Arrhenius equation which describes how a slight rise in temperature greatly increases the number of molecules passing to the activated state. The activation energy, μ, of such a thermal reaction proceeding at constant pressure is calculated by equation (7):

$$\mu = 2.3\,R(\log_{10}kT_1 - \log_{10}kT_2)/(1/T_2 - 1/T_1) \tag{7}$$

where T_1 and T_2 are selected experimental temperatures. At constant temperature but at varying pressures the number of molecules passing to the activated state will be reduced if the formation of that state involves a volume increase. The source of that volume increase will be conformational changes and the solute-solvent interactions already discussed. By analogy with equation (7) we may write:

$$\Delta V^{\ddagger} = 2.3\,RT(\log_{10}k_{P_1} - \log_{10}k_{P_2})/(P_2 - P_1) \tag{8}$$

P_1 and P_2 are experimental pressures and R is the gas constant. $\Delta V^{\ddagger}$ is the volume of activation. In the thermal aggregation of PVRN-ase $\Delta V^{\ddagger}$ is $+203$ ml mole^{-1} when the reaction proceeds in 0.6 M NaCl, and $+259$ ml mole^{-1} in 0.3 M NaCl. Fig. 2.12 shows that the $\log_{10}$ of the reaction rate plotted against pressure yields a straight line, indicating that $\Delta V^{\ddagger}$ is unaffected by pressure. The slope of the line corresponds to the magnitude of $\Delta V^{\ddagger}$. Activation volumes $\Delta V^{\ddagger}$ are thus calculated from results obtained from pressure studies. Overall volume changes, ΔV, occurring in passing from reactants to products or from one side of an equilibrium to the other, are directly observed in a dilatometer. The difference in volume between the initial state and the final product of an equilibrium may be calculated from equation (9):

$$\Delta V = 2.3\,RT(\log_{10}K_{P_1} - \log_{10}K_{P_2})/(P_2 - P_1) \tag{9}$$

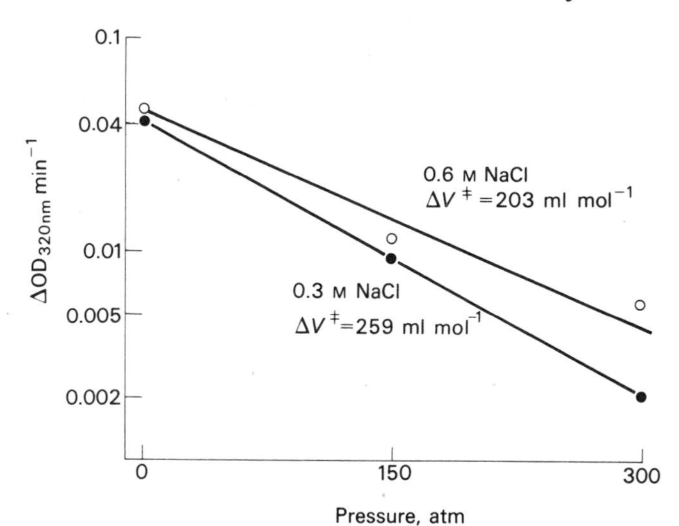

Fig. 2.12. The rate of association of PVRN-ase at selected pressures. Rate is expressed as change in optical density (OD) per min. See Fig. 2.9 and text. (Kettman, Nishikawa, Morita & Becker, 1966)

K is an equilibrium constant under given conditions: $\Delta V^{\ddagger}$ may be thought of as the ratio of the volume of the activated complex to the volume of the reactants.

ENZYME REACTIONS

Enzyme reactions are peculiarly sensitive to pressure as the rate-determining stage in the reaction may alter according to temperature, substrate concentration and other factors. And since the volume changes involved in different stages of the reaction may differ in sign as well as magnitude the net effect of pressure may change drastically according to reaction conditions. The kinetic treatment of enzymic processes at different pressures is more complicated than for reactions in which temperature is varied at constant pressure.

As is well known, an enzyme reaction may be regarded as a two stage process:

$$\mathrm{E}+\mathrm{S} \underset{k_{-1}}{\overset{k_1}{\rightleftharpoons}} \mathrm{X}$$

$$\mathrm{X} \xrightarrow{k_2} \text{products} \qquad (10)$$

Molecular aspects

Enzyme and substrate (E, S, respectively) pass through an activated state on the way to forming the enzyme substrate complex X. The rate constant for this reaction, $E+S \rightleftharpoons ES^{\ddagger} \rightarrow X$, is k_1. The reaction $E+S \rightarrow ES^{\ddagger}$ involves a volume change which may arise through conformational or solvation changes. Its activation volume is $\Delta V_1^{\ddagger}$. The formation of the enzyme–substrate complex is reversible and the rate constant for the reverse reaction is k_{-1}. The transition of the enzyme–substrate complex to products by way of a second activated stage $X^{\ddagger}$ is irreversible. Its rate constant is k_2 and its activation volume is $\Delta V_2^{\ddagger}$. Pressure alters k_1, k_{-1}, and k_2 in the manner described for non-enzymic reactions. Thus the effect of pressure would be to increase k_{-1} if the reaction $E+S \rightleftharpoons ES^{\ddagger}$ involves a volume increase k_1 would be decreased. The net effect of pressure on the overall reaction is set by the slowest stage in a manner which may be described by Michaelis and Menten equations. Laidler (1951) has demonstrated how the effect of pressure will be determined by the different rate constants according to conditions.

REACTIONS AT RELATIVELY LOW TEMPERATURES

Laidler's treatment proceeds as follows. First consider the condition in which temperature is low, as in the deep sea, and in which enzymes will be free of thermal denaturation. The Michaelis and Menten equation defining the reaction velocity when k_2 is slower than k_{-1} is

$$v = \frac{k_2 K[\mathrm{E}]\,[\mathrm{S}]}{1+K[\mathrm{S}]} \tag{11}$$

In the presence of a high concentration of substrate $K[\mathrm{S}]$ is greater than unity, so that equation (11) reduces to

$$v = k_2[\mathrm{E}] \tag{12}$$

Thus the effect of pressure is determined by the volume change in the sub-reaction X → products, that is, by the activation volume $\Delta V_2^{\ddagger}$. Under these conditions the rate of reaction is determined by the concentration of effective enzyme present.

It may be also shown that when k_2 is greater than k_{-1}, and substrate concentration high the activation volume $\Delta V_2^{\ddagger}$ will be rate-determining.

Penniston (1971), it will be recalled, used enzymes in just these conditions. Multimeric enzymes saturated with substrate were found to be pressure-labile whereas monomeric enzymes were pressure-resistant. An apparent $+\Delta V_2^\ddagger$ was attributed by Penniston to the depletion of intact multimeric enzymes, caused by their dissociation to sub-units, which led to a limitation in reaction velocity.

When substrate concentrations are low, the physiological condition, the relative velocities of k_{-1} and k_2 are important. When k_{-1} is greater than k_2, and substrate concentrations are low, equation (11) reduces to

$$v = k_2 K[\mathrm{E}]\,[\mathrm{S}] \tag{13}$$

The effect of pressure is determined by the compound volume, $\Delta V_2^\ddagger$ plus ΔV. $\Delta V_2^\ddagger$ has been defined; ΔV is the change in volume during the reaction $\mathrm{E}+\mathrm{S} \rightarrow \mathrm{X}$. Laidler uses the symbol $\Delta V^\ddagger$ for the compound volume change in the reaction $\mathrm{E}+\mathrm{S} \rightleftharpoons \mathrm{ES}^\ddagger \rightarrow \mathrm{X} \rightleftharpoons \mathrm{X}^\ddagger$.

When k_2 exceeds k_{-1} in the presence of low substrate concentrations the following equation applies:

$$v = \frac{k_1 k_2 [\mathrm{E}]\,[\mathrm{S}]}{k_2 + k_1 [\mathrm{S}]} \tag{14}$$

k_1 is rate-determining and $\Delta V_1^\ddagger$ the critical volume change. Thus by varying substrate concentration at low temperature the rate-determining stage in an enzymic reaction may change from $\Delta V^\ddagger$ (low substrate) to $\Delta V_2^\ddagger$ (high substrate) even though the relative velocities of k_{-1} and k_2 remain unchanged.

There are few examples with which to illustrate these theoretical points. Pressure accelerates the hydrolysis of starch by salivary amylase and pancreatic amylase in conditions of high substrate concentration. In these reactions $\Delta V_2^\ddagger$ is -22 ml mole^{-1} and -28 ml mole^{-1} respectively. Pancreatic lipase, in the presence of low concentrations of tributyrin, is inhibited by pressure because $\Delta V_1^\ddagger$ or $\Delta V^\ddagger$ is $+13$ ml mole^{-1}. These examples from work by Benthaus are discussed by Laidler (1951). In physiological conditions the interpretation of the effects of pressure on enzyme kinetics may be further complicated by the involvement of ligands whose interaction with an enzyme may be pressure-sensitive. These biochemical complications are dealt with in Chapter 3, p. 103.

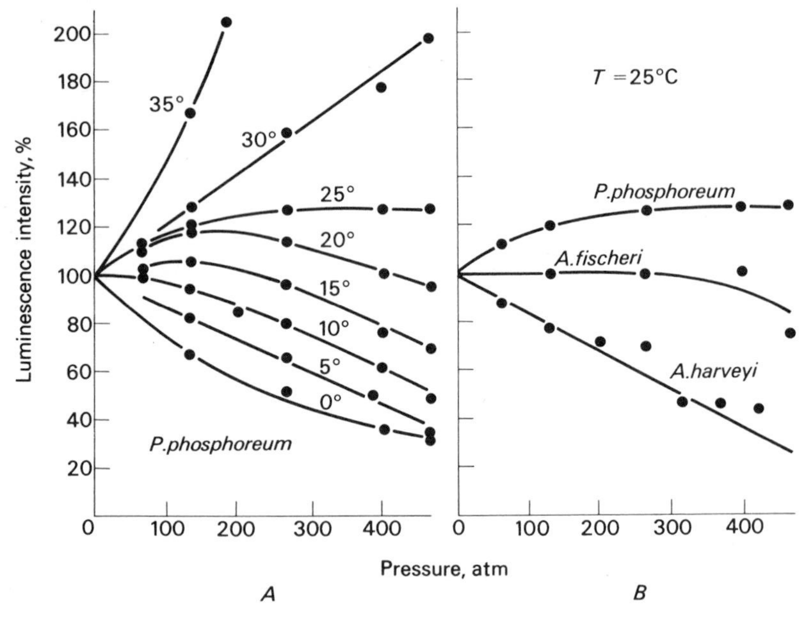

Fig. 2.13. The effect of hydrostatic pressure, *A*, on the luminescence of *Photobacterium phosphoreum* as a function of temperature, and *B*, on the luminescence of three species of bacteria at 25 °C. Intensity of luminescence at atmospheric pressure is expressed as 100 to show the percent change at selected pressures. (Brown, Johnson & Marsland, 1942)

REACTIONS AT RELATIVELY HIGH TEMPERATURES

A number of enzyme experiments have been carried out at high temperatures where pressure exerts an effect not likely to be found in the deep sea. Fig. 2.13 illustrates how pressure increases the rate of reaction at high temperature and decreases it at low temperature. The substrate concentration for the luminescent reaction shown in Fig. 2.13 is taken to be low so that at temperatures below a species specific optimum, $\Delta V^{\ddagger}$ or $\Delta V_1^{\ddagger}$ is thought to be positive. At high temperature a critical enzyme is thought to become rate-determining as it becomes progressively denatured. Pressure, opposing this, increases the rate of reaction. Note the data in Fig. 2.13 refer to relative luminescence intensity. A number of examples in which pressure accelerates reaction velocity at high temperatures have been described by Johnson, Eyring and Polissar (1954).

Summary

Pressure can only affect a system by way of change in volume. In aqueous systems in the pressure range found in the deep sea the following molecular processes involve a significant decrease in volume and are thus favoured by high pressure: (1) ionisation, (2) hydrophobic hydration, (3) compression of a bulk phase. In the case of hydrophobic hydration, lowering the temperature and decreasing the ionic strength frequently enhances the pressure effect. Complicating interactions occur in proteins, such as the creation of 'void volume' by conformation changes, which involve a molar volume increase. Neighbouring groups influence the magnitude of volume changes. The distinction between ΔV the overall volume change which occurs in a reaction and $\Delta V^{\ddagger}$ the volume change which occurs in the formation of the activated complex, is fundamental. It is the activation volume, $\Delta V^{\ddagger}$ of the rate-limiting stage which determines the rate of change at high pressure in ordinary chemical reactions.

Volume changes in reaction mechanisms render enzymic processes particularly sensitive to pressure. Further examples are discussed in the next chapter.

Pressure also affects the behaviour of aqueous solutions at interfaces, their transport properties (Drost-Hansen, 1972) and other special properties such as the equilibrium pressure of dissolved gases. The latter will be considered in its biological context in Chapter 4.

3 BIOCHEMICAL ASPECTS OF HIGH HYDROSTATIC PRESSURE

It is extremely difficult to investigate in detail the many effects which pressure exerts on organisms. First there is the dilemma familiar to biologists, that precisely controlled reaction conditions necessary for a precise interpretation may themselves introduce factors not present in nature. Additionally, in deep sea biology we have the very considerable difficulty of procuring experimental material in good condition. This section will consist of a discussion of some interesting pressure experiments on relatively convenient organisms with a limited account of some experiments on deep sea material.

The discussion will start with the question of the stability of DNA at high pressure and proceed through macromolecular synthesis to some effects of pressure on bacterial metabolism. Here we will meet deep sea metabolism for the first time. Then there follows a section on experimentally induced changes in the pressure tolerance of bacteria. The final section describes some recent work on the effect of pressure on enzymes and their control in deep sea fish.

The reader may find it useful at this stage to inspect Fig. 3.1, taken from Hammes and Wu (1971). The diagram illustrates some of the ways in which biochemical events are regulated in cells. In a preliminary fashion this section discusses the action of pressure on first, the initiating macro-molecules (shown in the lower rectangle, Fig. 3.1), and then on some metabolic paths ($A \rightarrow Z$) and then on some controlling processes in deep sea animals (the dotted lines in Fig. 3.1).

DNA

As an isolated compound *in vitro* DNA is much more resistant to very high pressures than many proteins. Perhaps this is a consequence of its hydrogen bonded, double helix structure. In any

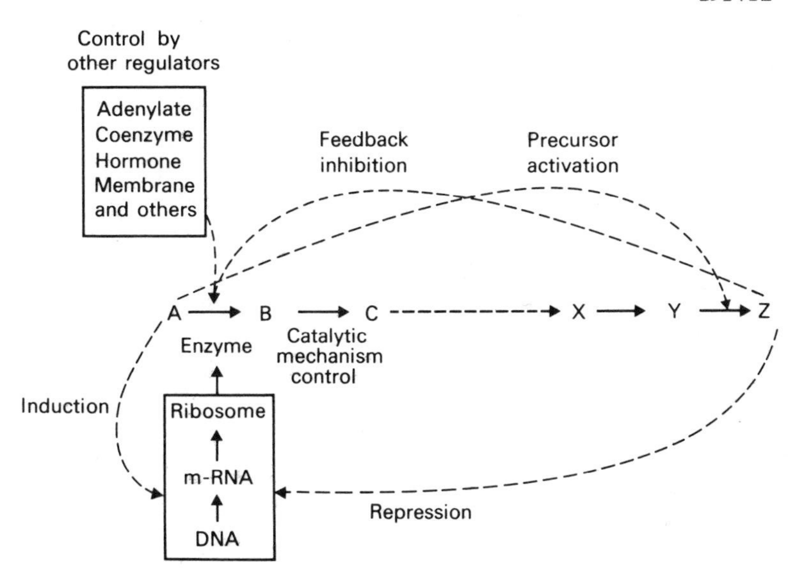

Fig. 3.1. A schematic representation of mechanism for the regulation of enzyme activity. (Hammes & Wu, 1971)

event Suzuki and Taniguchi (1972) report that DNA extracted from either salmon sperm or calf thymus glands is unaltered in its optical density at a wavelength of 260 nm after exposure to 10000 atm for an hour. The temperature range and pH range employed in the experiment were 25–40 °C and 4.8–9.0 respectively. A similar treatment of proteins leads to irreversible coagulation. Further evidence of the stability of extracted double stranded DNA comes from work by Heden and his colleagues (1964). DNA which was isolated from *Bacillus subtilis* and then exposed to several thousand atmospheres for 30 sec or to hundreds of atmospheres for longer, showed no change in its ability to transform other bacteria.

Is DNA in a living organism more susceptible to pressure? This question can be viewed from at least two aspects. Pressure may exert a direct, mutagenic effect and it may also affect DNA metabolism. More work has been done on the question of DNA metabolism than on the mutagenic effects of pressure, but both areas are relatively little explored.

In view of the variety of physical and chemical agents whose mutagenic properties have been studied it is surprising that the effect of pressure has been neglected. There is no theoretical reason for supposing that mutations will involve a large volume increase or decrease. A mutation which occurs with an activation volume increase in a high pressure environment would appear to be an unlikely event, consequently it is difficult to avoid the conclusion that pressure-stable mutations will prevail in the deep sea. McElroy (1952) has studied the pressure sensitivity of mutations which were induced with nitrogen mustard and ultraviolet irradiation in the spores of the *Neurospora crassa*. The idea was put forward that a transitional state is involved in the process of mutation. In the case of nitrogen mustard, mutations presumably involve a $+\Delta V^{\ddagger}$ since pressure reduced the mutagenic potency of the agent. Treatment of *Neurospora* spores with UV light produced mutations which were similarly suppressed by a pressure of 660 atm. However, mutations which arose after a more severe dose of UV, sufficient to kill more than 95 per cent of the spores, were enhanced by the same pressure. It therefore appears that pressure is capable of influencing the transitional state or the gene products of experimentally induced mutations. If such mutations are similar to those which occur spontaneously in nature, then presumably pressure will play at least a modifying role in the mutation process occurring in the deep sea. Moreover it is important to know if the high pressure–low temperature environment in the depths produce a characteristic mutation rate.

Two other organisms have been subjected to high pressure in an attempt to induce inherited changes. The unicellular organism *Euglena gracilis*, which normally photosynthesises, yielded colourless mutants after exposure to pressures of between 500 and 1000 atm over periods of time ranging from 20 min to 2 hours (Gross, 1965). One mutant, designated PR-1, continued to grow in nutrient medium for 14 serial transfers but failed to redevelop chlorophyll. This interesting phenomenon does not appear to have been followed up. The discovery of algal-like cells in deep water (p. 274) suggests another reason for exploring this phenomenon further. However, Gross showed that green *Euglena* were more pressure resistant than cells bleached by growth in the dark.

The littoral copepod *Tigriopus* has been exposed to several hundred atmospheres pressure at the nauplius stage with the result that females predominated in the ensuing cultures (Vacquier, 1962). The number of animals which survived the most severe

TABLE 3.1. *Effect of pressure on* E. coli*: DNA in relation to cell mass.* (ZoBell & Cobet, 1964)

Pressure (atm)	Mean cell-length (μm)	Protein per cell ($\mu g \times 10^{-7}$)	DNA ($\mu g \times 10^{-8}$ per cell)	DNA ($\mu g \times 10^{-8}$ per μm cell-length)
E. coli, strain B				
1	1.97	0.74	0.34	0.17
200	2.28	1.04	0.44	0.19
400	4.56	1.85	0.49	0.11
425	4.77	1.83	0.41	0.08
450	5.39	2.14	0.34	0.06
E. coli, strain S: forms filaments in presence of radiometric agents				
1	2.18	0.80	0.28	0.13
200	2.23	0.89	0.31	0.14
400	2.83	1.15	0.27	0.09
425	3.31	1.35	0.35	0.10
450	3.61	1.48	0.29	0.08
E. coli, strain R_4 (radiation resistant)				
1	1.76	0.067	0.29	0.16
200	1.70	0.067	0.29	0.17
400	2.42	0.094	0.37	0.15
425	2.68	0.103	0.37	0.13
450	2.81	0.109	0.28	0.09

pressure treatment (700 atm) was low, and it remains to be seen if selection or sex conversion occurred. This experiment, like the previous one, deserves to be followed up. Both the case of *Euglena* and of *Tigriopus* serve to remind us that pressure may act as a selective agent at a variety of levels.

The metabolism of DNA may be an important pressure sensitive area. One of the first investigations into the effect of pressure on the DNA metabolism in bacteria was carried out by ZoBell and Cobet (1964). The DNA was extracted for assay from cultures of the bacterium *Escherichia coli* which had been incubated at selected pressures. Growth at pressure was abnormal, resulting in long filamentous cells. We shall return to this phenomenon but for the present the data in Table 3.1, taken from ZoBell and Cobet's results, are of immediate interest. Three different strains of *E. coli* were grown anaerobically with nitrate as a hydrogen acceptor at 30 °C and at various pressures. DNA synthesis was inhibited.

Expressed as DNA per cell, little change is seen, but expressed as DNA per length of cell a substantial inhibitory effect of pressure is apparent. These results are consistent with the idea that pressure inhibits the replication of DNA which in turn inhibits cell division. Growth, however, continues.

The synthesis of DNA in a marine psychrophile *Vibrio marinus* MP4*, which was isolated from seawater collected from a depth of 1200 m off the Oregon coast, has been studied by Albright and Morita (1968). Under aerobic conditions 15 °C proved the optimum temperature for the growth of the *Vibrio*. This was the temperature selected for anaerobic incubation at high pressure with nitrate as a hydrogen acceptor. Under these conditions growth was not exponential but linear with time. DNA was extracted and assayed by the diphenylamine reaction. The synthesis of DNA in pressurised cultures was not affected by 200 atm (Fig. 3.2). 400 atm and 500 atm exerted little inhibitory effect in the first hour of incubation, but thereafter substantially reduced DNA synthesis. 1000 atm caused a prompt and total inhibition of DNA synthesis (Fig. 3.2).

At this point two obvious questions come to mind. What is the effect of pressure on other aspects of the cells' metabolism, and is there a particularly pressure sensitive target in the DNA synthesis machinery? Experiments conducted by Yayanos and Pollard (1969) provide evidence of at least one pressure sensitive site in the replication of DNA in *E. coli* under specified conditions. Their experiments involved a strain of *E. coli* (strain 15) which requires thymine and L-leucine for growth. The criterion for DNA synthesis was a conventional one, namely the amount of labelled thymine incorporated into a cell fraction which was insoluble in cold trichloracetic acid. Cells in the log phase of growth were incubated at selected temperatures in a pressure vessel. The incorporation of labelled thymine was followed by withdrawing cells from the culture at intervals, the cycle of decompression, sampling and recompression taking only 5 sec. The study was a complicated one and here we may note only some selected results. First, pressure diminished the rate at which thymine was incorporated into DNA. At a pressure of 250 atm the rate of incorporation was below that of control cultures but showed a progressive increase as the population of cells grew (Fig. 3.3). Pressures of

* Bacterium which grows well at low temperature.

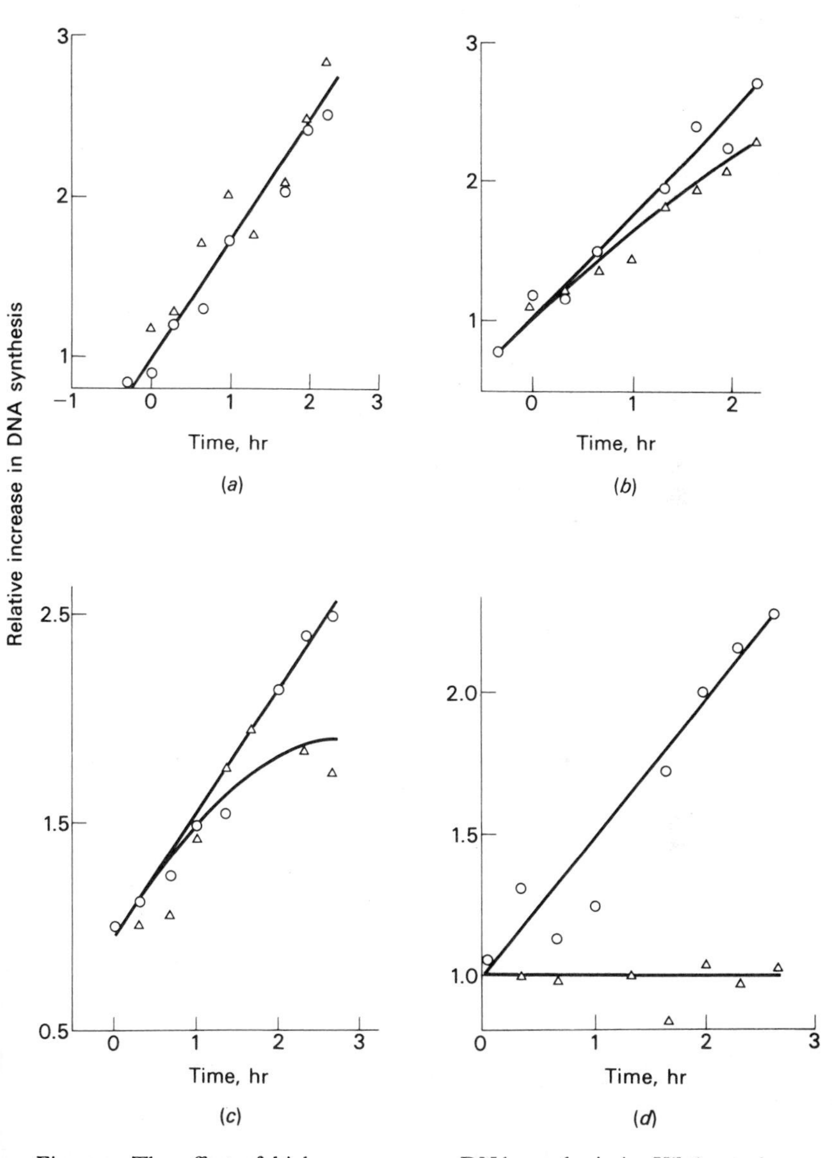

Fig. 3.2. The effect of high pressures on DNA synthesis in *Vibrio marinus* MP4, at 15 °C. (*a*) 200 atm: (*b*) 400 atm: (*c*) 500 atm: (*d*) 1000 atm. ○, control cells: △, experimental cells. Vertical scale, relative increase: horizontal scale, time in hours. (Albright & Morita, 1968)

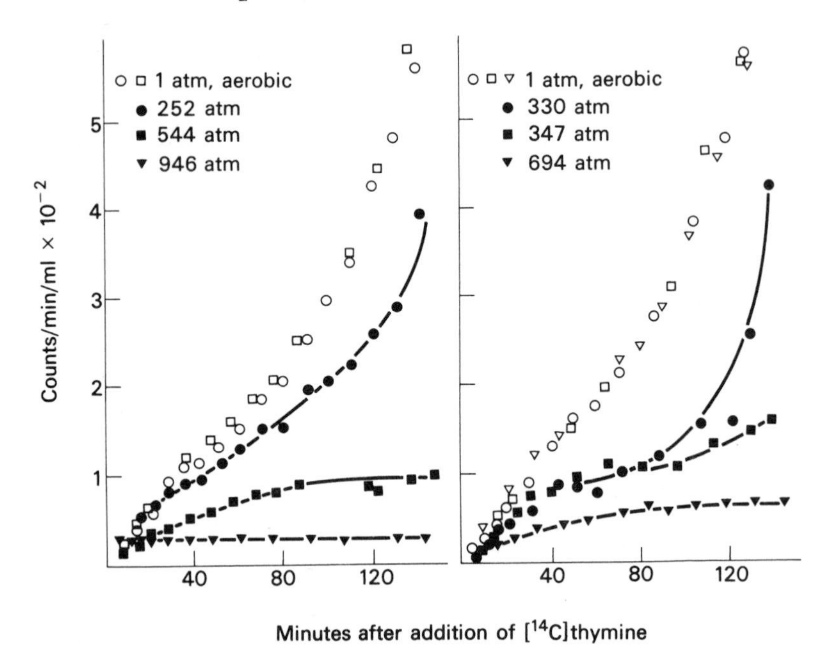

Fig. 3.3. The synthesis of DNA in *E. coli* at high hydrostatic pressures. [^{14}C]thymine incorporation at 37 °C as a function of time after addition of [^{14}C]thymine. The incubations at high pressure were begun at the following times: 252 atm, 11 min; 544 atm, 5 min; 946 atm, 4 min; 330 atm, 4 min; 347 atm, 5 min; 694 atm, 5 min. (Yayanos & Pollard, 1969)

250–450 atm showed a sudden increase in the rate of incorporation after a delay and at pressures of between 500 and 800 atm thymine incorporation ceased after a time. Immediate inhibition of thymine incorporation, that is, of DNA synthesis, was only brought about by the application of 946 atm at 37 °C, or a lower pressure at a lower temperature.

The last, extreme result was interpreted as a direct inhibition of the DNA polymerase. This cannot account for the short lived period of DNA synthesis at lower pressures. The effect of intermediate pressures was likened to the withdrawal of an essential amino acid which stops new replications but allows the completion of existing growing points. Thus moderate pressure (500 atm at 37 °C) may inhibit the initiation of growing points while a much higher pressure is required to inhibit polymerase activity.

Pressure has been shown to retard the transport of amino acids (glutamate, glycine, phenylalanine and proline) into a marine bacterium, which strengthens this interpretation (Paul & Morita, 1971).

No comparable study of DNA synthesis at high pressure has been undertaken with eukaryotes but preliminary autoradiographic experiments with sea urchin eggs shows the following; (*a*) 330 atm at 20 °C fails to stop the incorporation of [^{3}H]thymidine into the nuclei of fertilised eggs, even though the pressure inhibits pronuclear fusion: (*b*), 500 atm inhibits [^{3}H]thymidine incorporation in similar cells (Zimmerman & Silberman, 1967).

We cannot readily translate the results of these experiments to deep sea conditions but one conclusion is clear enough. DNA metabolism in prokaryote and eukaryote cells is certainly affected by deep sea pressures. The stability of DNA as a chemical substance at high pressure may not mean very much in a biological context. Compared to other biochemical events, DNA synthesis in bacteria and perhaps eukaryotes is of intermediate sensitivity to pressure under the experimental conditions employed. Furthermore, it is interesting to note that separate stages in DNA synthesis may be differently affected by pressure. Perhaps this is not surprising, but further work may well focus on specific pressure sensitive stages where adaptation will be called for in deep sea organisms.

RNA and protein synthesis

We may now turn to consider the effects of pressure on RNA and protein synthesis in some of the cells already mentioned. Returning to the growth of *E. coli* at high pressure, ZoBell and Cobet (1964) found protein to constitute a nearly constant percentage of the dry weight of the cells grown at pressures up to 450 atm (Table 3.2). DNA synthesis, compared to protein synthesis, as we have seen, is inhibited. On the other hand RNA synthesis, compared to protein synthesis, is stimulated. So the order of sensitivity to pressure in the macro-molecular synthesis in *E. coli* appears to be DNA, then protein, and RNA synthesis the least sensitive. The total biomass (growth yield) which was produced in the cultures was substantially reduced at pressure.

Pollard and Weller (1966) came to a different conclusion from a study of the differential effects of pressure on macro-molecular

TABLE 3.2. *Protein, RNA and DNA content of three strains of* Escherichia coli *at different pressures.* (ZoBell & Cobet, 1964)

Strain	Incubation pressure (atm)	Protein (% of dry wt)	RNA (% of protein)	DNA (% of protein)	DNA (% of RNA)
B	1	73.7	6.8	4.6	67.6
	200	75.6	6.5	4.2	64.6
	400	74.8	7.5	3.2	42.6
	425	71.0	8.4	2.8	33.3
	450	72.8	10.1	1.6	15.8
S	1	75.5	6.0	3.6	60.0
	200	79.8	6.3	3.5	55.6
	400	77.6	6.9	2.4	34.8
	425	77.0	8.1	2.6	32.1
	450	76.4	8.6	1.3	15.1
R_4	1	76.1	6.2	4.4	70.9
	200	80.5	6.2	4.3	69.3
	400	74.8	7.1	4.0	56.3
	425	73.3	8.0	3.6	45.0
	450	73.6	8.6	2.6	30.2

synthesis in *E. coli*, finding that the incorporation of labelled amino acids into protein was more pressure sensitive than DNA synthesis. Interestingly, these authors provided some evidence to show that pressure did not significantly diminish the volume of intact bacteria and also suggested that ribosomes may be deformed at pressure. We shall see this proved a shrewd guess.

Landau (1967), in a later study, employed the technique of enzyme induction in which cells are exposed to an inducer, isopropyl thiogalactopyranoside (IPTG). This compound stimulates the transcription processes leading to the synthesis of the enzyme β-galactosidase (Fig. 3.1). According to Landau's reasoning the three processes, induction, transcription and translation could be separately studied in the intact cell. Fig. 3.4 shows the effect of pressure on the synthesis of β-galactosidase in *E. coli* at 37 °C. Note that 670 atm exerts an immediate and complete inhibition. Lesser pressure exerts a graded effect with 265 atm having no effect. All the pressures between 265 and 670 atm shown in Fig. 3.4 exerted an immediate effect on enzyme synthesis. The results of assays at intervening time intervals fall on the dashed lines but are not plotted in the figure. Landau argues that

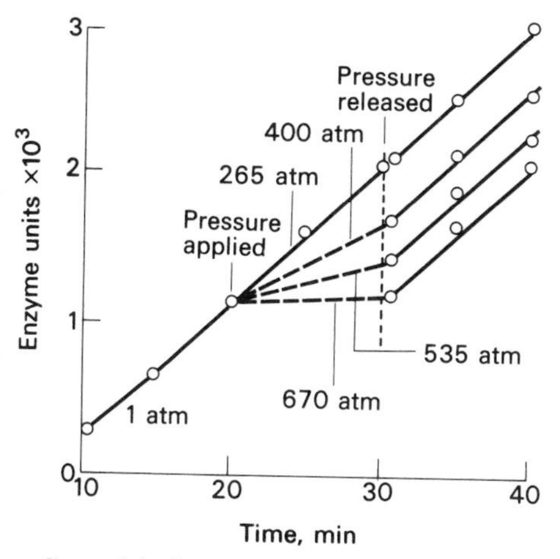

Fig. 3.4. The effect of hydrostatic pressure on β-galactosidase synthesis in *E. coli* at 37 °C. Inducer was added 20 min prior to pressure application. (Landau, 1967)

these results mean pressure inhibited a stage following the synthesis of m-RNA, that is, translation. If pressure only acted on the synthesis of m-RNA (transcription) the inhibition would be gradual and determined by the decay of the m-RNA already in existence. Landau calculated the activation volume $\Delta V^{\ddagger}$ of β-galactosidase translation as approximately 100 ml mole^{-1}, although the precise definition of this activation volume is not clear.

The complete inhibition of β-galactosidase synthesis by 670 atm is rapidly reversed (Fig. 3.4). Landau used this effect to study the action of pressure on the synthesis of the m-RNA in the synthesis of β-galactosidase, and the subsequent stability of m-RNA. Cells were exposed to IPTG for a few minutes to initiate the synthesis of m-RNA. 670 atm was then applied to inhibit the synthesis of the β-galactosidase. A second chemical agent (borate) could then be added to block m-RNA synthesis at selected times. By the ingenious combination of these two agents (Fig. 3.5), Landau was able to produce the data plotted in Fig. 3.6. Curve *A* is a measure of the m-RNA for β-galactosidase synthesised in the first four minutes at atmospheric pressure. The decay of β-galactosidase m-RNA following chemical block at 1 atm is rapid, the half life being about 2.5 min. Landau assumed a similarly rapid decay at high pressure so, after the 5 min period when m-RNA synthesis is

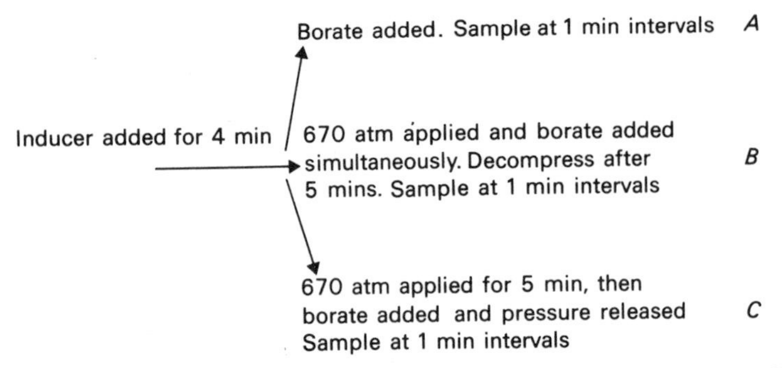

Fig. 3.5. The experiment of Landau. See Fig. 3.6 and text. (After Landau, 1967)

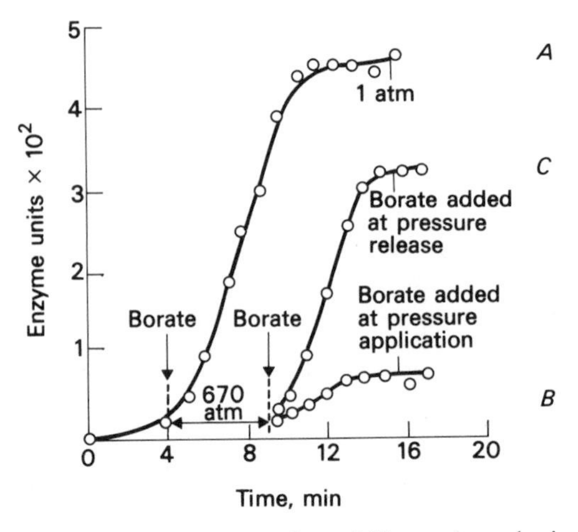

Fig. 3.6. The effect of 670 atm on the stability and synthesis of m-RNA in *E. coli*. Inducer was added at zero time. (Landau, 1967)

blocked by a chemical agent, the cells should contain only 25 per cent of the control level of β-galactosidase. Curve *B* (Fig. 3.6) shows that much less enzyme is present after exposure to pressure, so m-RNA decay has been accelerated. However, Curve *C* in the same figure shows that some synthesis of m-RNA occurs at 670 atm. Experiments in which induction was started after the application of pressure demonstrated a graded inhibitory effect up to 265 atm. At higher pressures no induction took place. The evidence favoured the interpretation that pressure inhibited the formation of the inducer–repressor complex and further suggests

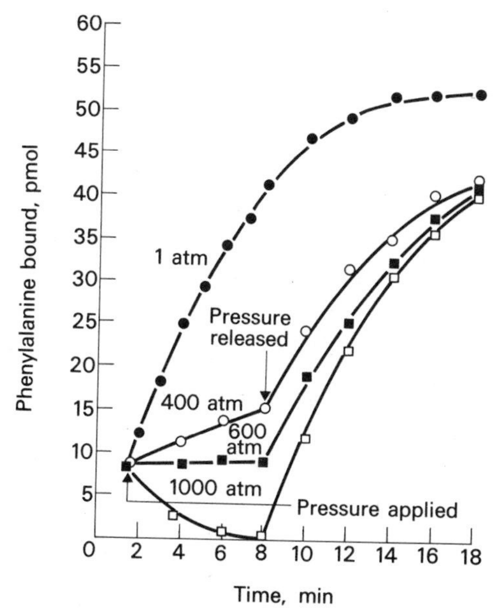

Fig. 3.7. The effect of hydrostatic pressure on the rate of binding of [^{14}C]phe-t-RNA to *E. coli* ribosomes. Reactions were pressurised after 1.5 min at 1 atm and depressurised at intervals up to 6.5 min later. Values for binding under pressure have been corrected for binding which occurs in the 35 sec period between depressurisation and reaction termination. Each point is the result of a duplicate assay. (Arnold & Albright, 1971)

that in the synthesis of induced β-galactosidase the order of decreasing sensitivity to pressure is induction, translation and transcription. Other work (Landau, 1970) indicates that translation in Hela cells is also susceptible to inhibition by pressure. Attempts have been made, using cell-free extracts, to pinpoint the inhibitory effect of pressure in the translation process (Arnold & Albright, 1971). Fig. 3.7 shows the extent to which pressure affects the binding of [^{14}C]phenylalanyl-transfer-RNA to *E. coli* ribosomes; at 20 °C 1000 atm reversibly dissociates the [^{14}C]phenylalanyl-t-RNA–poly U-ribosome complex. Similarly 1000 atm was also reported to impair the stability of the m-RNA–ribosomal complex.

Dipeptide formation (phenylalanyl phenylalanine) was found to be inhibited by high pressure in conditions in which reactants were freely available (Arnold & Albright, 1971). The volume of reaction, ΔV, of peptide bond formation has been indirectly determined by Linderstrøm-Lang and Jacobsen (1941) as approximately +15–20 ml mole^{-1} at 1 atm

$$NH_3^+—R_1CH—COOH^- + NH_3^+—R_2CH—COOH^-$$
$$\rightarrow NH_3^+—R_1CH—CO—NH—R_2CH—COOH^- + H_2O$$

ΔV was determined by dilatometry following the tryptic digestion of protamines such as clupein. The cleavage of peptide bonds in a bigger and more complicated molecule, lactoglobulin, involved a much larger volume change which was associated with conformational changes. For a more detailed discussion the reader is referred to Linderstrøm-Lang and Jacobsen (1941) and a recent paper by Yayanos (1972).

A eukaryote cell whose biochemistry has been studied at high pressure is the Protozoa *Tetràhymena pyriformis*. The sensitivity of the cell to short pulses of high pressure is demonstrated in a number of experiments carried out by Zimmerman and his co-workers. In cells whose division cycle had been synchronised by heat shocks, a two minute exposure to 670 atm reduced the subsequent uptake of labelled phenylalanine and uridine into trichloroacetic acid insoluble fractions at normal atmospheric pressure (Lowe-Jinde & Zimmerman, 1971). Significantly, Infante and Krauss (1971) report that the pressure generated in a centrifuge can lead to a dissociation of sea urchin monoribosomes to sub-units and Hermolin and Zimmerman (1969) found pressure reduced the polysome: total ribosome ratio in *Tetrahymena*. The nomogram shown in Fig. 3.8 from Hedén (1964) relates centrifugal speed, diameter of rotor, and pressure, clearly demonstrating the ease with which a few hundred atmospheres pressure can be inadvertently produced. Heden's paper should also be read for its speculations on the effects of high pressure on cellular and molecular systems.

Protein synthesis in deep sea conditions may involve more changes in the machinery of assembly than in the end product.

The volume expansion which takes place in the synthesis of a peptide bond involves work against the ambient pressure which would appear to be unavoidable at great depth. An interesting question is, does deep sea life expend a significant amount of energy in bringing about molecular volume increases? The question of the energy which is required to expand molecules against the ambient pressure is considered in the next section, which is concerned with energy metabolism at high pressure.

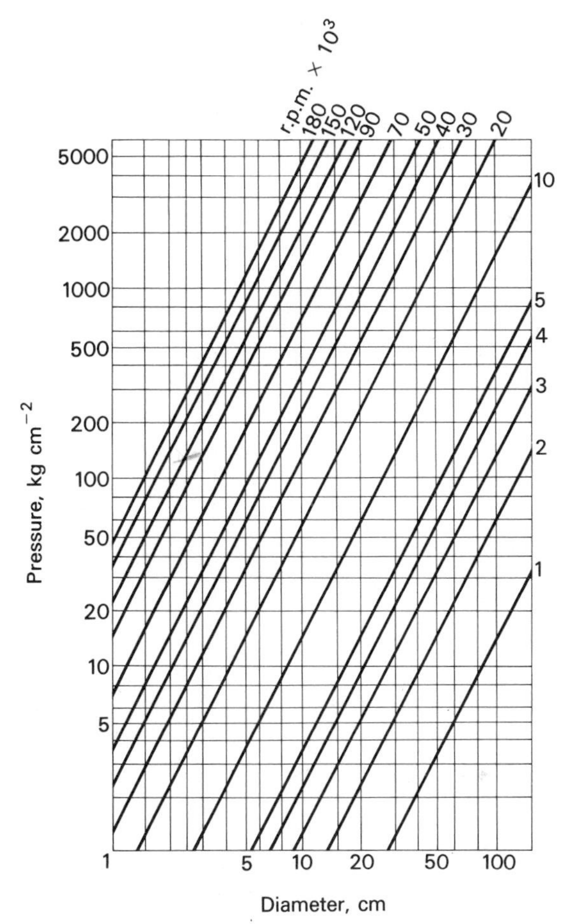

Fig. 3.8. Nomogram illustrating the centrifugally produced liquid pressures in water. Vertical scale: kg cm^{-2}, which approximates closely to atmospheres. Horizontal scale: diameter of the rotating liquid. Oblique lines: rotor speed. (Heden, 1964)

Volumetric aspects of energy metabolism

In some ingenious and original work Marquis and Fenn (1969) have given consideration to the question of the work of expansion in synthetic reactions. The system studied was not the synthesis of a macro-molecule, as might have been anticipated on the grounds of maximal molar volume changes, but the anaerobic metabolism of the bacterium *Streptococcus faecalis*.

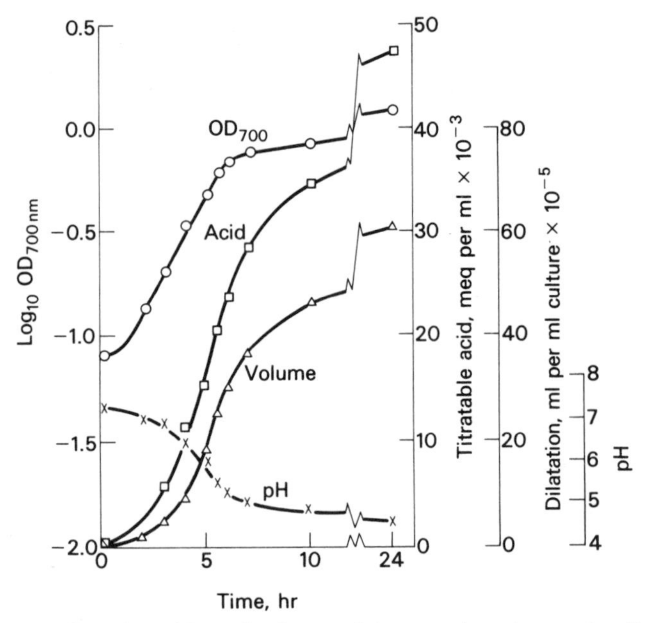

Fig. 3.9. Growth, acid production, and increase in volume of a *S. faecalis* culture in tryptone–glucose–Marmite medium. Growth temperature was 24 °C, and acid production was assessed by titrating culture samples to pH 7.2 with potassium hydroxide. The optical density (OD) of the culture was measured with 700 nm light. (Marquis & Fenn, 1969)

The experiments consisted of measurements of the molecular volume change, ΔV, involved in the growth of cell cultures and the simultaneous determination of selected metabolic products.

First *S. faecalis* was grown in 25 ml dilatometers using a medium comprising mainly tryptone, glucose and marmite (TGM). Increase in mass was followed by measuring the change in optical density of the culture; lactic acid by enzymatic assay and pH by electrode. Ammonia was determined by the Conway microdiffusion method and ATP by an assay procedure involving phosphoglycerate kinase and NADH. Fig. 3.9 shows the logarithmic phase of growth in mass plotted with total acid, pH, and the molecular volume change for the entire growth process. ΔV varied according to the buffer used in the growth medium. It was concluded that glycolysis was responsible for the volume increase but secondary

TABLE 3.3. *Volume changes of buffer reactions.* (Marquis & Fenn, 1969)

Buffer reaction	Volume change (ml per mole)	p*K* of buffer	Expected volume change for glycolysis and buffer reaction (ml per mole lactate produced)
$HPO_4^- + H^+ \rightarrow H_2PO_4^-$	+24.0	6.64	+24.1
Imidazole + H^+ → imidazole-H^+	−1.1	6.99	−1.0
Diglycine-NH_2 + H^+ → diglycine-NH_3^+	−4.4	8.25	−4.3
Protein-COO^- + H^+ → protein-COOH	+11.0 (average)	3.6 and 4.6	+11.1

reactions involving H^+ ions and buffers altered the value of ΔV considerably. It was argued that glycolysis is effectively:

$$\text{glucose} + 2 \times \text{ADP} + 2\ \text{phosphate} \rightarrow 2\ \text{lactic acid} + 2 \times \text{ATP}$$

As the overall changes in ADP, P_i and ATP are negligible compared to those in glucose and lactate, the ΔV for the simplified reaction corresponds to the change:

$$\text{glucose} \rightarrow \text{lactate}$$

The partial molar volume of glucose was determined as 111.8 ml mole^{-1} and that of lactic acid as 67.8 ml mole^{-1}. Therefore, per mole of lactic acid synthesised from glucose, a volume change of +11.8 ml mole^{-1} is to be anticipated. This agreed well with the overall volume change measured in the growing culture but unfortunately proved fortuitous. In the cultures lactic acid was mostly dissociated, with the dissociation involving a change in volume of −11.7 ml mole^{-1} lactate. The association of H^+ ions with buffer involves a volume change similar to the volume increase of growth shown in Fig. 3.9. Experiments using TGM medium and different buffers produced changes in the dilatation which were associated with lactate production. For example, phosphate buffer was associated with a larger volume increase than imidazole buffer. Table 3.3 compiled by Marquis and Fenn from previously published data, lists the volume changes of selected buffer reactions in the first column and the net volume change to be expected if lactate is dissociated in the third column.

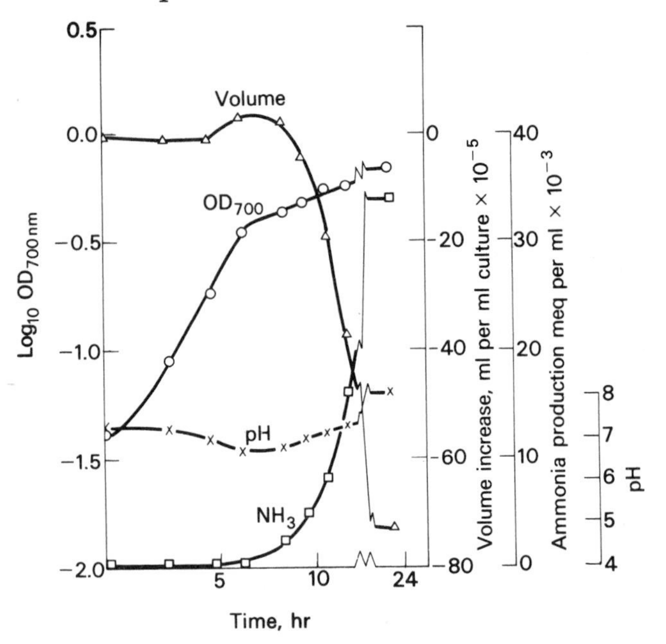

Fig. 3.10. Growth, ammonia production, and volume change of a *S. faecalis* culture in tryptone–Marmite medium with 1 g glucose and 20 g arginine per litre. Growth temperature was 24 °C. The optical density (OD) of the culture was measured with 700 nm light. (Marquis & Fenn, 1969)

Growth of *Streptococcus* on a medium which included arginine showed a spectacular volume decrease (Fig. 3.10), which coincides with the production of ammonia at the end of the log phase. A consideration of the partial molar volumes for the degradation of arginine was based on the following reaction:

$$\text{Arg} + 4H_2O + \text{ADP} + \text{phosphate} \rightarrow \text{Orn} + CO_2 + 2NH_4OH + \text{ATP}$$

As ATP did not build up in the cells it may be neglected, but CO_2 accumulated and is claimed to be taken into account in the volumetric balance sheet:

Arginine + $4H_2O$

$\bar{V} = 144.9 + \bar{V} = 18.1\ (\times 4)$ $= 217.3$ ml

Ornithine + CO_2 + $2NH_4OH$

$\bar{V} = 112.2 + \bar{V} = 33 + \bar{V} = 13.2\ (\times 2) = 171.6$ ml

-45.7 ml per mole Arginine metabolised

Per mole of ammonia synthesised, the volume change thus predicted is -22.9 ml (or $\frac{1}{2} \times 45.7$ ml mole^{-1}). The observed change in volume, expressed in relation to ammonia produced was -22.8 ml mole^{-1}. The agreement is remarkable and doubtless coincidental. It implies that the molar volume change is caused by the conversion of arginine to ornithine, CO_2 and NH_4OH, and the synthesis of all other cell constituents in some way cancel each other out.

How will the growth of *Streptococcus* be affected by high pressure? This question is answered by the work of Marquis, Brown and Fenn (1971). The cells were grown at 24 °C in various media and the total biomass was expressed in relation to the lactic acid synthesised. Media containing maltose rendered the growth of the cells less sensitive to pressure than media containing gluconate, lactose, ribose or pyruvate. In the case of pyruvate, growth failed to take place at pressures higher than 300 atm, whereas in the presence of glucose, growth of mass was only moderately inhibited by 400 atm. Energy sources did not appear a limiting factor in glucose-based growth. Indeed glycolysis was less inhibited than growth (Fig. 3.11). Direct determinations were made of the volume increase associated with growth at pressure. Two methods were used. One involved a measurement of density by determining the change in weight of a 100 ml volume of growing culture. The second method used a dilatometer (or volumeter) containing a growing culture (Fig. 3.12). Results are shown in Table 3.4. The somewhat variable volume increases which were associated with growth at 270 and 410 atm were broadly similar to those obtained for growth at 1 atm. The cultures expanded the system against the ambient pressure. Did this cost the organism much energy? Marquis, Brown and Fenn argued that such an expansion involves the additional work $P\Delta V$. Taking 15.7 ml mole^{-1} of lactate at 410 atm, and approximating to gm/cm units:

$$15.7 \text{ ml mole}^{-1} \times 410 \text{ atm} \times 1033.3 \text{ g cm}^{-2} \text{ atm}^{-1} = 6.65 \times 10^6 \text{ g cm mole}^{-1}$$

The energy released in the synthesis of a mole of lactate (12.37×10^8 g cm mole^{-1}) is very large in comparison with the extra energy required to expand the system against 400 atm applied pressure.

What does this mean in terms of the energy metabolism of the

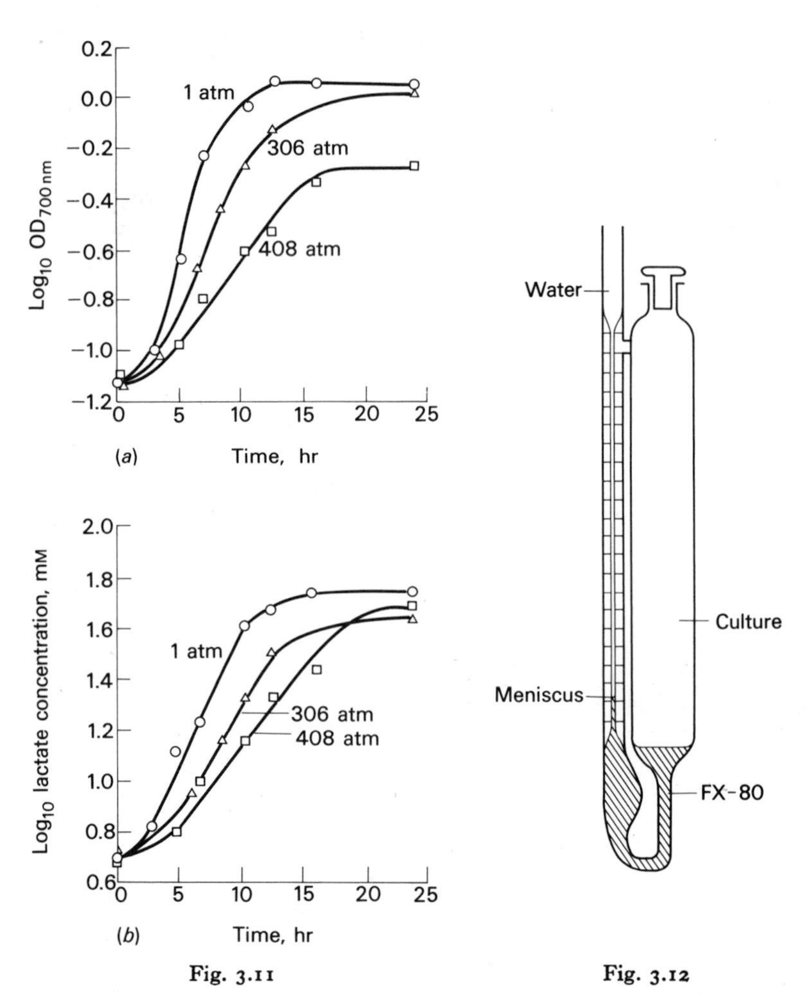

Fig. 3.11. Growth and glycolysis of *Streptococcus faecalis* under pressure in TGM medium at 24 °C. Growth in mass measured by change in optical density. (*b*) Lactate accumulation. (Marquis, Brown & Fenn, 1971)

Fig. 3.12. Volumeter for measuring changes in volume during growth under pressure. The apparatus is designed to fit into the lumen of a pressure chamber which measures 3.7 cm in diameter and 27 cm in depth. The meniscus was formed by a water–FX-80 interface; oil can be used in place of water. The capillary holds 3.17 $mm^3 cm^{-1}$, and the main chamber holds about 50 ml of culture. The stopper is held in by elastic bands not shown. (Marquis, Brown & Fenn, 1971)

TABLE 3.4. *Volume changes of* S. faecalis *cultures grown under pressure in relation to lactate production.* (Marquis, Brown & Fenn, 1971)

Pressure (atm)	Incubation time (h)	ΔV (as % culture volume)		ΔV (ml per mole of lactate)
		Volumeter	Pycnometer	
270	16	0.017	0.029	24.1[a]
270	17	0.052	0.078	12.2
270	18	0.071	0.116	14.7
410	17	0.110[b]	0.074[b]	16.6
410	24	0.045	0.083	17.7
	+21[c]	0.088[b]	—	—
410	21	0.022	0.033	9.0
	Mean	0.058	0.069	15.7

[a] Lactate assays were performed on pycnometer samples only.

[b] If the values indicated are neglected, average ΔV values of 0.041 for the volumeter method and 0.065 for the pycnometer method are obtained.

[c] In this experiment, the volumeter was kept under pressure for a second night, and a higher ΔV was obtained.

cell? ΔV, the volume of reaction, is of limited significance. $\Delta V^{\ddagger}$, the activation volume, is the important factor in determining the rate of chemical range at high pressure. We must conclude that the $\Delta V^{\ddagger}$ of some critical, rate-limiting stage in the lactate pathway must have been sufficiently small to allow metabolism to continue at high pressure. The relationship between ΔV and $\Delta V^{\ddagger}$ in a complex biochemical reaction need not be particularly close. In the simpler case of the thermal aggregation of poly-valyl-ribonuclease discussed previously one might expect a measure of agreement between ΔV and $\Delta V^{\ddagger}$. In intact cells the $\Delta V^{\ddagger}$ of a reaction may well be large and ΔV negligible. In streptococcal glycolysis the situation is rather the reverse, namely the ΔV for the overall process, including buffer reactions with protons, is sizeable but not rate determining. The energy cost of dilatation is a small component of the free energy of activation defined as

$$\Delta F^{\ddagger} = \Delta H^{\ddagger} - T\Delta S^{\ddagger} + P\Delta V^{\ddagger}$$

in which H is enthalpy, T, temperature, degrees Kelvin, S, entropy, P pressure, and V volume.

We therefore conclude that some metabolic processes involving a net volume increase will incur only a small extra energy expenditure on the part of deep sea organisms. However, although the energy cost of molecular expansion against a high ambient pressure appears small it may be of significance in an environment singularly lacking in food.

The growth and metabolic performance of bacteria, including deep sea bacteria, at high pressure

In the context of energy metabolism it is worth noting some of the effects which high hydrostatic pressure has on convenient experimental organisms.

Morita has confirmed Certes and Cochin's demonstration of 1884 (Morita, 1965; Certes & Cochin, 1884) that pressure inhibits glycolysis in yeast (Table 3.5). Morita (1957) has also used the Thunberg method of methylene blue reduction to measure the effect of pressure on the dehydrogenation of formate, malate and succinate in intact *E. coli*. Dehydrogenation of these substrates was inhibited by pressure (Table 3.6). Perhaps the results should be qualified by the fact that pressure may have altered the pH of the solution which contained phosphate buffer, but probably this would have only caused a minor change in dehydrogenase activity. The reduction of methylene blue in the absence of substrate was enhanced by pressure. Presumably, therefore, the data in Table 3.6 (Morita, 1957) show the effect of pressure on a reaction other than the reduction of methylene blue itself. Further, as the half life of succinic dehydrogenase in lyophilised cells was 4 hours at 600 atm at 30 °C (Morita & ZoBell, 1956) it is reasonable to suppose that the effect of 15 min exposure to pressure did not deplete that particular enzyme to the point where it became rate limiting.

The possibility that pressure may disrupt the structural organisation of a cell, leading to an impaired biochemical performance, has been considered by Hill and Morita (1964). The fresh water mould *Allomyces macrogynus* was subjected to high pressure, and although dehydrogenase activity in isolated mitochondria was variously inhibited, depending on the specific enzyme, mitochondria *in vivo* appeared to be unaffected by 600 atm at 26.5 °C. Tris buffer was used in these experiments so pH changes caused by pressure were therefore minimal (Distèche, 1972).

A large number of species of marine bacteria have been cultured at high pressures. One of the earliest investigations into the

TABLE 3.5. *Effect of hydrostatic pressure on the production of ethanol by* Saccharomyces cerevisiae.[a] (Morita, 1965)

Pressure (atm)	Ethanol[b] (mg per 6 ml)
1	4.12
200	3.16
400	1.77
600	0.95
800	0.41
1000	0.15

[a] The reaction mixture contained equal volumes of 0.278 M glucose in 0.067 M phosphate buffer, pH 5.6 and washed cells in 0.067 M phosphate buffer, pH 5.6. The cells were harvested from an 18-hour culture, washed twice in 0.067 M phosphate buffer, pH 5.6, and adjusted to give a transmittance of 30% at 600 nm in a spectrophotometer (Bausch and Lomb Spectronic 20). The reaction mixture was incubated for 3 hours at 27 °C. The values corrected for controls.

[b] Milligrams of ethanol produced per 6 ml of reaction mixture.

TABLE 3.6. *Dehydrogenase activity in* E. coli *at high hydrostatic pressure. The relative rate of methylene blue reduction calculated over a period of* 9–12 *min.* (Morita, 1957)

Pressure	Malate	Succinate	Formate
1	1.27	1.93	2.67
200	1.26	1.70	2.60
600	1.20	1.00	2.43
1000	0.17	0.04	0.01

susceptibility of marine bacteria to pressure was carried out by ZoBell and Johnson (1949). Table 3.7 records the growth obtained after prolonged incubation of marine bacteria at different temperatures and pressures. The culture was made up of glucose and other nutrients in filtered seawater. Bacteria from shallow depths, such as the luminescent *Achromobacter fischeri* and *Photobacterium splendidum* produced less turbid cultures than those bacteria isolated from a depth of several thousand metres, for example *Bacillus submarinus* and *B. thalassokoites*. The bacteria, including those from the deep sea, had been in culture at atmospheric pressure for several years. ZoBell and Johnson (1949) regarded this as evidence that the pressure tolerance shown by the deep sea cultures

TABLE 3.7. *Relative turbidity caused by the multiplication of marine bacteria in nutrient broth for six days at 20 °C, four days at 30 °C, or one day at 40 °C at different hydrostatic pressures.* (ZoBell & Johnson, 1949)

All cultures listed below except those marked with an asterisk, which failed to grow at 40 °C, showed four plus (++++) growth in the controls incubated at normal pressure

Culture	300 atm			400 atm		
	20 °C	30 °C	40 °C	20 °C	30 °C	40 °C
Achromobacter fischeri	++++	++	*	++	—	*
A. harveyi	++++	++++	*	+	++++	—
A. thalassius	—	++++	++	—	—	+++
Bacillus abysseus	+	++++	++++	—	++++	++++
B. borborokoites	++	++++	++++	—	++++	++++
B. cirroflagellosus	++	+++	*	—	++	*
B. submarinus	++	++++	++++	+	++++	++++
B. thalassokoites	+++	++++	++++	++	++++	++++
Flavobacterium okeanokoites	++++	++++	++++	++++	++++	++++
F. uliginosum	++++	++++	++++	++++	++++	++++
Micrococcus infimus	+	++++	*	—	+	*
Photobacterium splendidum	++++	++++	*	++	+++	*
Pseudomonas pleomorpha	++	++++	*	++	+++	*
P. vadosa	++	++++	++++	++	++++	++++
P. xanthochrus	++++	++++	*	++++	+++	+
Vibrio hyphalus	—	—	*	—	—	—
Mixed microflora from mud	++++	++++	++++	++++	++++	++++

TABLE 3.7 (*cont.*)

	500 atm			600 atm		
	20 °C	30 °C	40 °C	20 °C	30 °C	40 °C
Achromobacter fischeri	—	—	*	—	—	*
A. harveyi	—	—	—	—	—	—
A. thalassius	—	—	—	—	—	—
Bacillus abysseus	—	+	++++	—	—	++++
B. borborokoites	—	++	++++	—	—	++++
B. cirroflagellosus	—	—	*	—	—	*
B. submarinus	—	++++	++++	—	++++	++++
B. thalassokoites	—	++++	++++	—	++++	++++
Flavobacterium okeanokoites	++	++++	++++	—	—	++++
F. uliginosum	—	++++	+	—	—	—
Micrococcus infimus	—	—	*	—	—	*
Photobacterium splendidum	—	+	*	—	—	*
Pseudomonas pleomorpha	—	++	*	—	—	*
P. vadosa	—	++++	++++	—	—	++++
P. xanthochrus	++	++	++	+	+	—
Vibrio hyphalus	—	—	*	—	—	*
Mixed microflora from mud	++++	++++	++++	++++	++++	++++

was genetically determined. This is a reasonable interpretation of the data but neither of the deep sea species grew particularly well at the lowest temperature and highest pressures used, 20 °C and 600 atm. The data in Table 3.7 seem to indicate that growth at deleterious high pressures is favoured by high temperatures. There would be little selective advantage in genetic adaptation to high temperature and pressure in the ocean depths.

From these growth experiments ZoBell and Johnson (1949) proposed the term barophilic to describe bacteria which grew well or preferentially at high pressure. This term is now widely used and, it should be noted, says nothing about the temperature characteristics of organisms. Barophilic bacteria need not come from, or be potential inhabitants of, the deep sea. They may derive from oil wells where both pressure and temperature are high. However, Quigley and Colwell (1968) have demonstrated that certain bacteria isolated from deep sea sediments are not only barotolerant but taxonomically distinct from other marine bacteria as judged by a multitude of physiological tests.

To set alongside the demonstration that certain deep sea bacteria grow well at high pressure, we may briefly consider some further data, Table 3.8, from the work of ZoBell and Oppenheimer (1950) which records bacterial growth on a peptone/yeast extract/seawater medium at an initial pH of 7.5. The species which grew well at high pressure were not isolated from great depths. Their barophilic property may disappear at deep sea temperatures and even if it does not they would appear to be fortuitously rather than adaptively pressure-tolerant. Growth at pressure in a number of cases produced filamentous cells as noted on p. 69 (see also Oppenheimer & ZoBell, 1952).

A number of interesting observations were made by ZoBell and Budge (1965) on the ability of marine bacteria to grow and reduce nitrate (NO_3^-) to nitrite at high pressure. *Pseudomonas perfectomarinus* grew slowly at 600 atm at 25 °C on a broadly based nutrient medium. Washed cells continued to reduce NO_3^- in a glucose containing medium at a diminished rate at high pressure and at 10 °C (Fig. 3.13), and at a very greatly diminished rate at 2 °C, a typical deep sea temperature. The apparent initiation of nitrate reduction after 16 hours (Fig. 3.13*B*) is extremely interesting. On the whole, one gains the impression that the ability to reduce NO_3^- is not a particularly pressure-sensitive component of metabo-

TABLE 3.8. *Multiplication as indicated by turbidity* (– *or* + *to* + + + +) *or death* (d) *of marine bacteria in nutrient broth after a* 96-*hour incubation at* 30 °*C at different hydrostatic pressures.* (ZoBell & Oppenheimer, 1950)

Name of organism	Hydrostatic pressure (atm)			
	1	200	400	600
Achromobacter stationis	+ + + +	+ + + +	+ + + +	–
A. thalassius	+ + + +	–d	–d	–d
Actinomyces marinolimosus	+ + + +	+ + + +	–	–
A. halotrichis	+ + + +	–	–	–
Bacillus epiphytus	+ + + +	+ + +	+ +	–d
B. borborokoites	+ + + +	+ + + +	+ + + +	+ + +
Flavobacterium neptunium	+ + + +	+	–	–
Micrococcus euryhalis	+ + + +	–d	–d	–d
M. sedentarius	+ + + +	+ + + +	+ + + +	–d
M. aquivivus	+ + + +	+ + + +	+ + + +	+ + + +
Pseudomonas obscura	+ + + +	+ + + +	–	–d
P. hypothermis	+ + + +	–d	–d	–d
P. perfectomarinus	+ + + +	+ + + +	+ + + +	+ + + +
Serratia marinorubra	+ + + +	+ + + +	+ +	–
Vibrio algosus	+ + + +	+ + + +	+ + +	+ +
V. marinoflavus	+ + + +	–	–d	–d
V. ponticus	+ + + +	+ + +	+ +	–d

lism. Table 3.9 summarises the growth performance of some thirty more species of marine bacteria at pressure. Note that eight species, designated by collection number, maintained good growth at 1000 atm. These were isolated from depths greater than 7000 m. Note also the high temperature used in the experiments.

The phosphatase activity of intact marine bacteria has been measured by Morita and Howe (1957) at high pressure using *p*-nitrophenylphosphate at a concentration of 0.4 per cent as a convenient artificial substrate (Table 3.10). The removal of the phosphate group is accompanied by a change in colour which allows the reaction to be followed colorimetrically at pressure.

Two species of bacteria, B13A and B13B2 which were isolated from deep sea muds during the Mid Pacific Expedition (Morita & ZoBell, 1955) from an ambient pressure of nearly 500 atm, showed enhanced phosphatase activity at pressures up to 1000 atm, at 30 °C. Other marine bacteria showed diminished or unaltered activity at high pressure. As the experimental temperature was

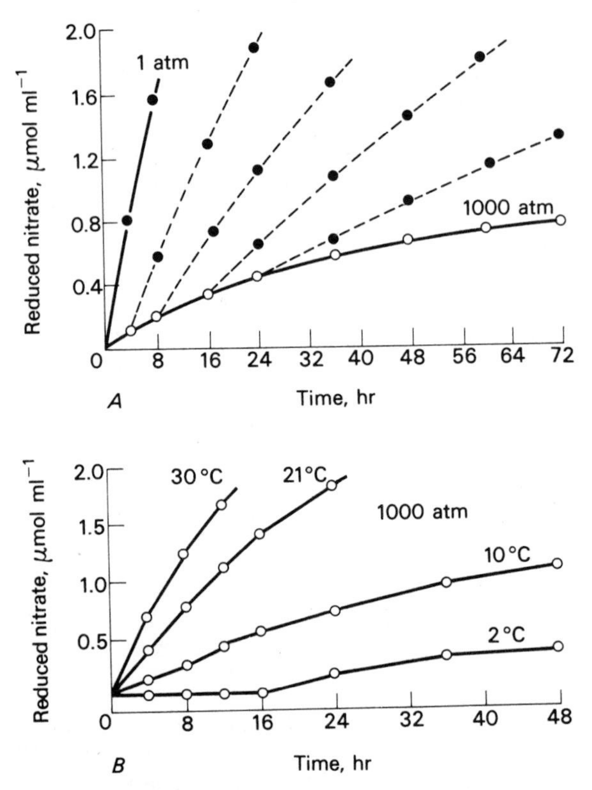

Fig. 3.13. *A*. Amounts of nitrate reduced by washed cells of *Pseudomonas perfectomarinus* during different periods of incubation at 10 °C. ●, results at 1 atm; ○, results at 1000 atm. The broken lines represent results at 1 atm after holding the cells for 4 to 24 hr at 1000 atm. Temperature: 10 °C.

B. Nitrate reduced by washed cells of *Pseudomonas perfectomarinus* during incubation at different temperatures at 1000 atm. (*A* & *B*, ZoBell & Budge, 1965)

much higher than that found in the deep sea it is once again not possible to conclude that the phosphatase activity measured in B13A and B13B2 resembles *in situ* activity. Fortunately, deep sea temperatures were used to incubate other deep sea bacteria obtained during the *Galathea* deep sea expedition. During the expedition ZoBell and Morita (1957) obtained sedimentary bacteria at several stations from depths greater than 10000 m. The oceanographic sampling methods which were used are mentioned in Chapter 7.

TABLE 3.9. *Relative amounts*[a] *of bacterial biomass detected by light transmittance measurements in cultures after a 7-day incubation at 25 °C in nutrient seawater medium as influenced by hydrostatic pressure.* (ZoBell & Budge, 1965)

	Hydrostatic pressure (atm)							
Culture	1	100	200	300	400	600	800	1000
Achromobacter stationis	81	84	82	79	51	d	d	d
A. aquamarinus	93	92	86	57	36	0	d	d
A. halotrichis	46	41	24	14	0	0	d	d
A. thalassius	96	87	48	31	d	d	d	d
Bacillus abysseus	85	84	87	72	56	0	0	0
B. borborokoites	92	93	89	71	65	43	0	d
B. filicolonicus	89	82	64	54	38	d	d	d
B. submarinus	81	71	63	49	39	0	d	d
Bacillus G-14	76	80	77	74	68	65	47	32
Bacillus G-17	86	83	85	81	72	53	36	28
Micrococcus euryhalis	82	44	28	0	d	d	d	d
M. infimus	91	81	82	56	0	d	d	d
M. sedentarius	94	81	85	76	65	d	d	d
Micrococcus G-12	70	73	68	64	51	47	24	18
Micrococcus G-13	78	77	72	70	49	41	30	23
Pseudomonas azotogena	90	81	50	23	d	d	d	d
P. hypothermis	97	72	38	22	d	d	d	d
P. marinopersica	73	58	41	0	d	d	d	d
P. oceanica	80	82	78	75	68	d	d	d
P. perfectomarinus	99	94	90	82	67	62	0	d
Pseudomonas G-23	68	65	67	58	48	41	35	33
Pseudomonas G-26	71	66	70	62	63	58	31	24
Sarcina pelagia	85	74	58	lost	41	0	d	d
Serratia marinorubra	83	86	80	71	52	d	d	d
Vibrio algosus	80	74	72	48	36	22	d	d
V. hyphalus	88	72	68	53	d	d	d	d
V. marinagilis	72	46	37	28	0	0	d	d
V. ponticus	65	68	58	lost	27	d	d	d
Vibrio G-19	63	63	61	52	50	41	29	16
Vibrio G-30	67	64	64	56	42	34	28	19

[a] 0 = negligible biomass or no growth; 99 = maximum biomass detected; d = dead, as indicated by failure of cultures to grow at 1 atm when decompressed

Cell counts were carried out employing the minimum dilution technique (Chapter 6) which yields the most probable number and requires a growth performance from the cells. The nutrient medium which was used consisted of peptone, soluble starch, yeast extract and ferric phosphate. The initial pH was 7.7. Bacteria which reduced sulphate were grown on a different medium.

TABLE 3.10. *Micromoles of phosphate (p-nitrophenol) liberated from p-nitrophenyl phosphate by various marine bacteria at various hydrostatic pressures* (Morita & Howe, 1967)

	Hydrostatic pressure			
Culture	1 atm (μM ± 5%)	200 atm (μM ± 5%)	600 atm (μM ± 5%)	1000 atm (μM ± 5%)
Bacillus borborokoites	0.058	0.048	0.076	0.076
Flavobacterium marinotypicum	2.320	3.240	2.600	1.760
Micrococcus aquivivus	0.010	0.008	0.010	0.007
M. euryhalis	0.004	0.007	0.004	0.004
Pseudomonas azotogena	0.110	0.093	0.124	0.520
P. perfectomarinus	0.040	0.118	0.044	0.054
P. xanthochrus	0.370	0.320	0.370	0.350
Number B13A	0.600	0.700	0.960	1.000
Number B13B2	2.160	2.200	3.300	3.200
Mixture	0.053	0.034	0.034	0.033

The reaction mixture contained equal volumes of *p*-nitrophenyl phosphate, cell suspension (Klett turbidimetric reading of 450, blue filter), and M/15 phosphate buffer containing 3.5 g NaCl l^{-1}. Incubation was for 1 hour at 30 °C at various pressures. Values are the average of quadruplicate runs.

Appropriate assays for starch hydrolysis, nitrate reduction and sulphate reduction were carried out after a minimum of two weeks incubation. A variety of physiological types proved to be barophilic under the growth conditions used (Table 3.11). A few grew preferentially at low pressure. This suggests they were surviving but not thriving at great depth rather in the way that thermophiles are to be found in cold environments. ZoBell and Morita's data in Table 3.11 are of some historical significance being the first laboratory demonstration of a physiological process which takes place at the greatest ocean depths. There is little for this author to add to Table 3.11; it demonstrates and partly defines one of the central problems in deep sea biology.

Shortly after the *Galathea* expedition, Russian marine microbiologists substantiated ZoBell and Morita's demonstration of the existence of barophilic, deep sea bacteria. The metabolism of glucose by a deep sea bacterium able to grow at high pressure has been described by Chumak (1959), and Chumak and Blokhina (1964). The bacterium, initially called Strain 187 but later *Pseudomonas desmolyticum*, was isolated by Kriss from Pacific

TABLE 3.11. *Bacteria detected in wet sediment in various ocean trenches and deeps.* (ZoBell & Morita, 1957)

A. Most probable number (MPN) of different physiological types of bacteria detected per g of wet sediment from the Philippine Trench in selective media incubated at different hydrostatic pressures at 3 to 5 °C

	Galathea station							
	No. 418		No. 419		No. 420		No. 424	
Latitude	10° 13′ N		10° 19′ N		10° 24′ N		10° 28′ N	
Longitude	126° 43′ E		126° 39′ E		126° 40′ E		126° 39′ E	
Water depth	10190 m		10210 m		10160 m		10120 m	
	Incubation pressure (atm)							
	1	1000	1	1000	1	1000	1	1000
Total aerobes	10^3	10^6	10^3	10^5	10^4	10^5	10^4	10^6
Total anaerobes	10^3	10^5	10^4	10^5	10^3	10^5	10^4	10^5
Starch hydrolysers	10^2	10^3	10	10^2	10	10^2	10^2	10^3
Nitrate reducers	10^2	10^5	10^2	10^4	10^3	10^5	10^2	10^5
Ammonifiers	10^3	10^5	10^3	10^4	10^3	10^5	10^3	10^5
Sulphate reducers	0	10^2	0	10	0	10^2	0	0

B. MPN of bacteria detected per g of wet sediment from the Kermadec–Tonga Trench in selective media incubated at different hydrostatic pressures at 3 to 5 °C

	Galathea station							
	No. 649		No. 650		No. 658		No. 686	
Latitude	35° 15′ S		32° 20′ S		35° 51′ S		28° 30′ S	
Longitude	178° 40′ W		176° 54′ W		178° 31′ W		176° 53′ W	
Water depth	8500 m		6790 m		7900 m		9800 m	
	Incubation pressure (atm)							
	1	850	1	700	1	800	1	1000
Total aerobes	10^5	10^6	10^6	10^6	10^4	10^6	10^6	10^5
Total anaerobes	10^4	10^5	10^6	10^5	10^5	10^5		
Starch hydrolysers	0	0	10	10				
Sulphate reducers	0	0	0	0				

TABLE 3.11 (*cont.*)

C. MPN of bacteria detected per g of wet sediment from deeps in Indian Ocean in selective media incubated at different hydrostatic pressures

	Galathea station		
	Pressure (atm)	No. 463 (Soenda Deep): 10° 16′ S, 109° 51′ E: water depth 7020 m	No. 492 (Weber Deep): 5° 31′ S, 131° 01′ E: water depth 7250 m
Total aerobes	1	690000	810000
	700	1050000	2300000
	1000	4800	17000
Starch hydrolysers	1	100	10
	700	1000	1000
	1000	0	0
Nitrate reducers at	1	10000	1000
	700	10000	10000
	1000	10	10
Sulphate reducers at	1	0	0
	700	100	10
	1000	0	0

sediments and shown by him to grow both at 550 atm and at atmospheric pressure. Glucose consumption was measured in cultures incubated at 27–28 °C (a surprising choice of temperature) using a glucose–peptone medium. At atmospheric pressure the bacteria are typical rods of about 3–4 μm long and 0.5–0.7 μm in diameter. At pressures between 550 and 750 atm and at 27–28 °C rods of between 200 and 300 μm in length appeared. Evidently pressure inhibits the formation of septa in these cells as it does in *E. coli*. Under these conditions the consumption of glucose at 1 atm in a 72 hour old culture was approximately 0.8 mg glucose per 10^6 cells in ordinary medium and 1.6 mg glucose per 10^6 cells in medium neutralised by the addition of calcium carbonate. At 550–750 atm the corresponding figures for glucose consumption were 1.7 and 2.8. Thus in this barotolerant bacterium, grown at 27–28 °C, pressure causes glucose consumption per cell to increase, and also inhibits carbon dioxide formation. Organic acids are preferred to carbon dioxide as the end-products of the metabolism of this cell at high pressure, as shown by the data in Table 3.12.

TABLE 3.12. Pseudomonas desmolyticum *isolated from Pacific sediments, incubated in ammonium phosphate medium.* (After Chumak & Blokhina, 1964)

Amount of glucose consumed during 7 days incubation (mg per 100 ml of culture) at 27–28 °C		Acidity (ml of 0.1 N KOH per g glucose consumed)		Volatile acids (ml of 0.1 N KOH per g glucose consumed)	
1 atm	500 atm	1 atm	500 atm	1 atm	500 atm
311.3	105.0	53.1	104.2	28.4	54.4

Note that the figures refer to volume of media and not cells, which accounts for the apparently enhanced glucose metabolism at high pressure.

It is difficult to comment on these results but one interesting feature is the enormous tolerance of the bacterium in question. It appears able to grow and metabolise to some extent at least, over the temperature range of 28° to deep sea temperatures in the region of 4 °C, and over a pressure range of 750 atm to 1 atm.

The hydrolysis of starch is carried out by many marine bacteria, including those which live at considerable depth. ZoBell and Hittle (1969) showed that two types of bacteria (*Bacillus* G-14, and *B.* G-17), which were isolated from deep sea mud at a depth exceeding 9000 m, were able to grow at 1000 atm in a suitable broth containing starch and peptones (Table 3.13). *Bacillus* G-14 cultures hydrolysed starch at a low rate at 25 °C which was little affected by pressure. This is fairly good evidence that starch hydrolysis could be part of this cell's normal metabolism at depth. The availability of starch is, of course, somewhat limited at depth, being mainly synthesised by photosynthetic plants.

Amylases from bacteria are interesting. They are relatively little affected by pressure and are of low molecular weight, being composed of a single monomeric unit. As enzymes which normally function outside the cell they may be required to work with a minimum of cofactors and without the organisation of a cell membrane. ZoBell and Hittle (1969) point out the considerable resistance to pressure shown by α-amylase.

TABLE 3.13. *Relative amounts (per cent) of bacterial biomass produced in nutrient seawater broth during 7 days incubation at different pressures, expressed on a basis of the biomass produced by each species at 1 atm being 100 per cent.* (ZoBell & Hittle, 1969)

	Hydrostatic pressure (atm)							
Species	100	200	300	400	500	600	800	1000
Bacillus abysseus	92	84	73	39	8	0	0	0
Bacillus G-14	100	100	97	89	85	74	61	42
Bacillus G-17	98	96	94	83	70	57	43	30
B. borborokoites	98	91	76	62	51	37	0	S
B. filicolonicus	89	79	60	42	11	S	S	S
B. imomarinus	63	48	16	0	0	0	0	S
B. submarinus	87	74	58	35	6	0	S	S
Flavobacterium neptunium	56	38	14	0	0	0	S	S
F. halohydrium	73	52	44	23	0	0	0	0
Pseudomonas pleomorpha	94	87	68	46	17	0	0	0
P. ichthyodermis	81	76	52	27	0	S	S	S
P. marinoglutinosus	86	80	63	38	12	0	0	0
P. marinopersica	79	53	28	6	0	0	0	0
P. membranula	68	42	20	0	S	S	S	S
P. perfectomarinus	95	92	83	67	59	31	0	S
Vibrio algosus	93	88	68	43	30	18	S	S
V. marinagilis	69	51	31	0	0	0	S	S
V. marinopraesens	96	90	74	56	14	0	0	S
V. ponticus	97	90	62	37	0	S	S	S

S = Sterilised, no viable bacteria.

Some cases of experimentally induced changes in the tolerance of bacteria to pressure

The possibility of adapting bacteria to deep sea pressures is an exciting one which has yet to be taken up experimentally. Chapter 7 describes some of the requisite high pressure techniques. The results of preliminary or fortuitous adaptation experiments suggest an intensive effort would be rewarding. For instance, in a study of the effects of pressure on the morphology of *E. coli* during anaerobic growth, ZoBell and Cobet (1962) found *E. coli* grew well at 400 atm at 40 °C after an initial decline in cell number. At 20 °C growth was stationary for over three days before the culture succumbed (Fig. 3.14). The 400 atm curve in Fig. 3.14(*a*) may be interpreted in one of two ways. Either certain types of cells with pre-adapted resistance to pressure gradually increased their

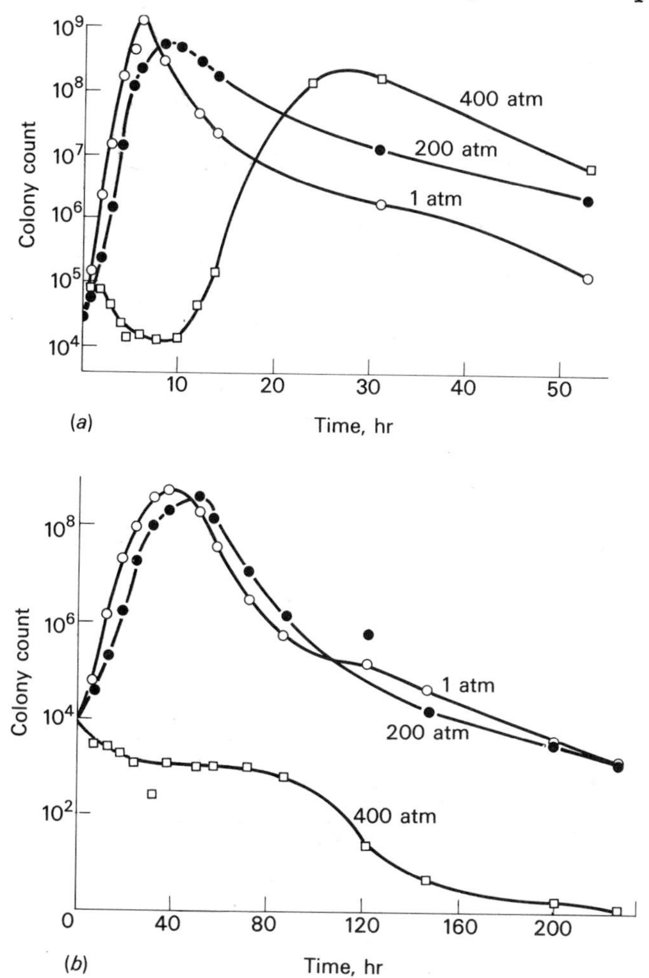

Fig. 3.14. Effect of pressure on the reproduction of *Escherichia coli* in nutrient medium (*a*) at 40 °C and (*b*) at 20 °C, as indicated by colony counts on EMB Agar. (ZoBell & Cobet, 1962)

number while others died, or metabolic changes occurred in the population as a whole which re-equipped the cells to grow at pressure. These processes are, of course, not mutually exclusive.

The effect of an initial and then a second period of compression on the growth of *Vibro marinus* and *E. coli* has been measured by Albright (1969). Four features of cultures in the middle of log

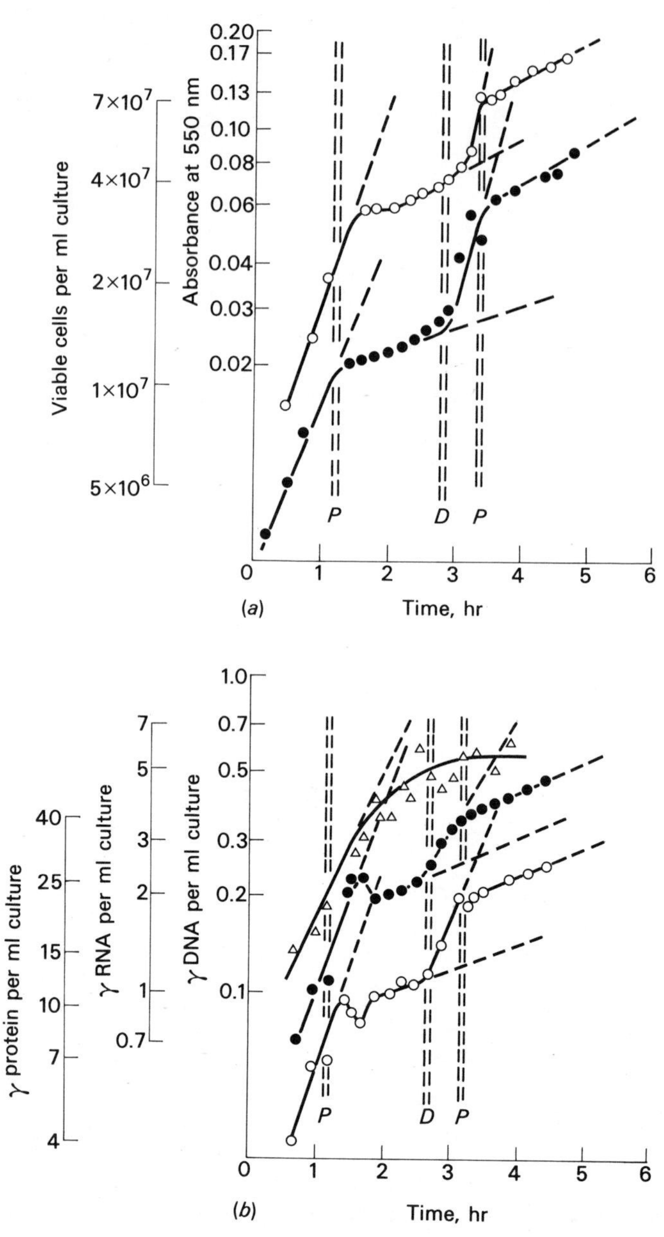

Fig. 3.15. The effect of the alternate application of hydrostatic pressures of 1 and 544 atm upon the growth of cells. (*a*) *E. coli*: ●, cell division, ○, absorption, 550 nm; (*b*) *E. coli*: ○, absorption at 550 nm, ●, RNA, △, DNA. (*c*) *Vibrio marinus*: symbols as in (*a*); (*d*) *Vibrio marinus*: symbols as in (*b*). *P*, pressure; *D*, decompression at atmospheric pressure. (Albright, 1969)

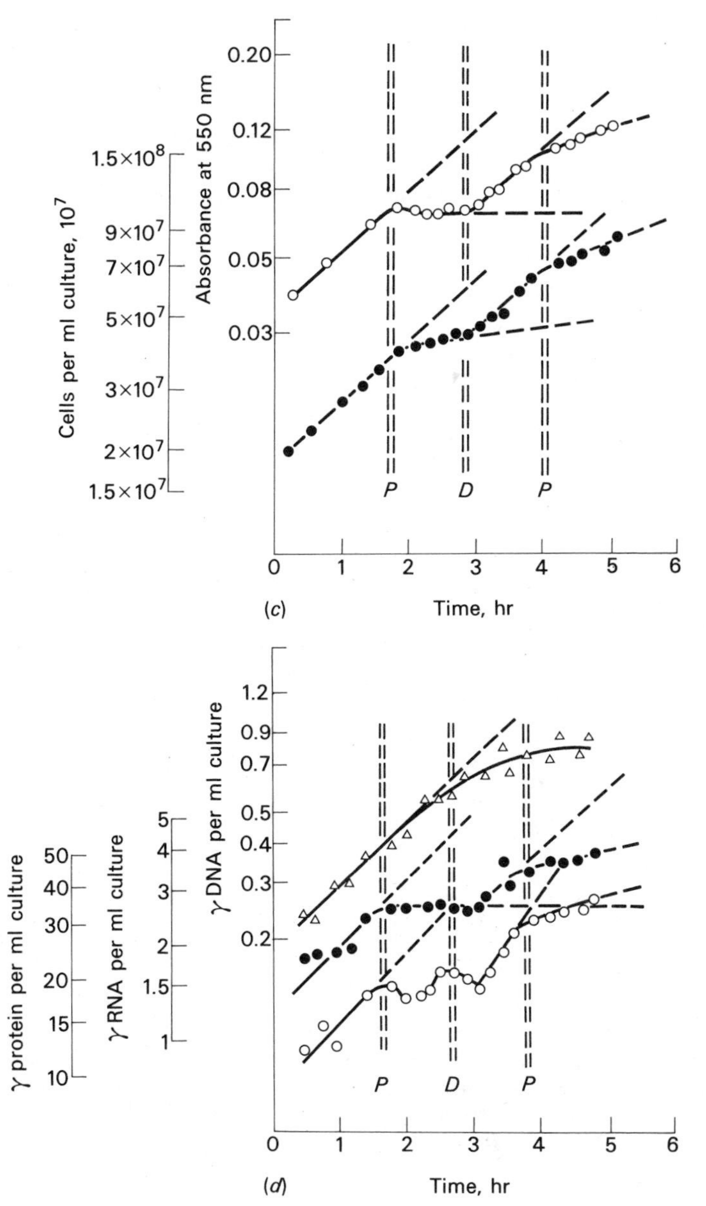
0.20
0.12
0.08
0.05
0.03
Absorbance at 550 nm
1.5×10^8
9×10^7
7×10^7
5×10^7
3×10^7
2×10^7
1.5×10^7
Cells per ml culture, 10^7
P
D
P
0
1
2
3
4
5
6
(c)
Time, hr
1.2
0.9
0.7
0.5
0.4
0.3
0.2
γ DNA per ml culture
5
4
3
2
1.5
1
γ RNA per ml culture
50
40
30
20
15
10
γ protein per ml culture
P
D
P
0
1
2
3
4
5
6
(d)
Time, hr

phase were monitored; cell number (by counts), optical density at a wavelength of 550 nm and RNA and DNA, by chemical assay on withdrawn samples. Growth in number and mass at 544 atm at 33 °C showed some recovery during the initial application of pressure. The synthesis of DNA and RNA in both cells showed a delayed decline at high pressure and then a subsequent recovery on decompression (Fig. 3.15).

During a second application of pressure the growth in number and total mass showed a lesser inhibition. Similarly RNA and DNA synthesis showed a smoother response to a second application of pressure (Fig. 3.15). There is little doubt that the cells which were subjected to pressure for a second time differed from cells at an earlier stage in the development of the culture. It is not clear from the data if this can be entirely attributed to the first pressure treatment. There may be a change in pressure sensitivity of the cells as the culture approaches stationary phase. In any event the growth response of the culture to the application of a mildly inhibitory pressure is not an immediate change in rate of growth. The curious oscillation in the RNA and DNA levels suggests a fluctuating control system settling down to a new steady rate. The apparent recovery of the rate of cell division in Fig. 3.15 reinforces the significance of the recovery of growth in Fig. 3.14. Some recovery in the rate of cell division of a eukaryote cell at high pressure is described in Fig. 4.9, Chapter 4.

Streptococcus faecalis has also been subjected to adaptation experiments. It will be recalled how different nutrients affected its growth performance at high pressure (Marquis, Brown & Fenn, 1971). When grown on ribose, *Streptococcus* is more sensitive to pressure than when grown on glucose. In the presence of glucose and $CaCl_2$ or $MgCl_2$, growth continues at pressures as high as 700 atm. Growth curves for *Streptococcus* which were reported by Marquis and ZoBell (1971) show interesting differences when $CaCl_2$ or $MgCl_2$ is added to a medium containing tryptone and glucose. Fig. 3.16 shows a significant shift in the steady state biomass in the stationary phase at 408 atm, 25 °C in the presence of $CaCl_2$. The rate of growth in the earlier log phase is little affected by the presence of $CaCl_2$. Growth at 304 atm was also increased when $MgCl_2$ was added to a different medium, one containing ribose instead of glucose. Growth rate and the final growth yield are both increased and so is the quantity of lactic acid produced. It appears that growth may be sustained in acid conditions at pressure in the presence of $MgCl_2$. These experiments do not lend themselves to precise interpretation but they illustrate the importance of specific ions in relation to pressure tolerance. Other ions such as Mn^{2+}, Sr^{2+}, K^+, and Na^+, in conjunction with the anions, Cl^-, SO_4^{2-}, Br^- had little effect on the growth of *Streptococcus* at pressure.

It is important to distinguish between the results in Figs. 3.14 and 3.15 in which bacterial growth changes in fixed conditions and those in Fig. 3.16 which shows differences in nutrient requirements at pressure. In the former case the cells change; in the latter case the medium is changed.

Effects of pressure on the control of enzymic processes in fish

A number of biochemical studies of the enzymes extracted from the deep sea fish, *Coryphaenoides* were made during the Alpha Helix expedition to the Galapagos Islands. The species is not reported but a typical example of the genus, generally known as rat tail fish, is shown in Fig. 1.21.

Current concepts of enzyme regulation hold that, in nature, enzyme activity is regulated very much more by feedforward (activation) and feedback (inhibition) mechanisms than by thermodynamic parameters of the main reaction path. Controlling agents in well studied biochemical systems include the specific substrate and product, specific ions and coenzymes as well as other 'regulator' molecules which may have diverse widespread metabolic roles. The interaction between the enzyme and the regulator (ligand) involves a process of association which, as we have seen, is likely to involve a volume change irrespective of the kind of bonding involved. Whatever the details of the enzyme–regulator interaction, control of the rate of reaction is achieved by increasing or decreasing the concentration of *effective* enzyme. Thus reaction velocity is probably determined by the rate at which the effective enzyme–substrate complex is formed. In natural conditions the substrate concentration is normally low, and in the deep sea the temperature is also low.

Before proceeding further we should note that reactions may also be regulated by the rate of synthesis or breakdown of enzymes

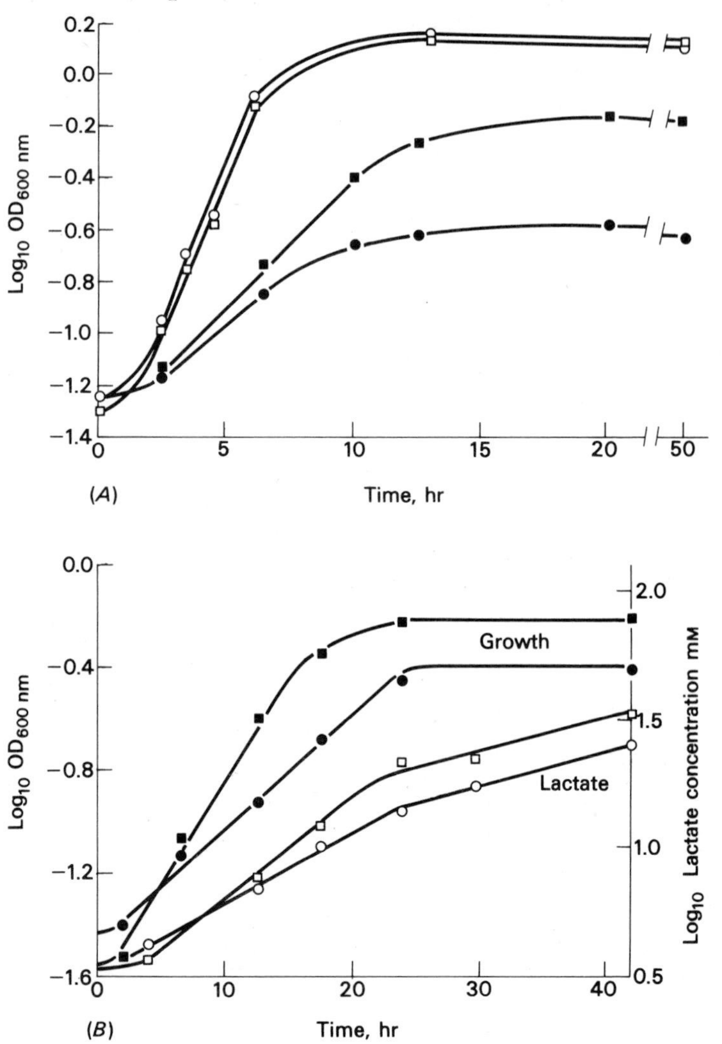

Fig. 3.16*A*. Effects of calcium chloride on growth of *S. faecalis* at one atmosphere and at 408 atm. The experimental temperature was 25 °C. The plotted points indicate growth at 1 atm in TGM medium (○) or TGM medium plus 50 mM calcium chloride (□) and at 408 atm in TGM medium (●) or TGM medium plus 50 mM calcium chloride (■).

B. Effects of magnesium chloride on growth and lactate production of *S. faecalis* in TRM medium at 340 atm. The experimental temperature was 25 °C. The plotted points indicate growth (●) and lactate production (○) in TRM medium and growth (■) and lactate production (□) in TRM medium plus 50 mM magnesium chloride. (Marquis & ZoBell, 1971)

but in the experiments which will be described in the following section this much slower and coarser control mechanism does not appear to be of immediate consequence. It is the instantaneous, fine control of enzymic activity we are concerned with here.

In many of the experiments an attempt has been made to isolate particular enzymes and to measure their activity at high pressure *in vitro*, under physiological conditions in which the activity of the enzyme is determined by the binding of a ligand. Fig. 3.17 shows the area of intermediate metabolism which has been studied in *Coryphaenoides*, namely glycolysis and gluconeogenesis. The enzymes, which were isolated from various tissues, were as follows: phosphofructokinase, PFK; fructose diphosphatase, FDP-ase; pyruvate kinase, PK: and lactate dehydrogenase, LDH. We will consider the results of experiments on PK and FDP-ase mainly, as these illustrate the kinetic approach adopted during the expedition.

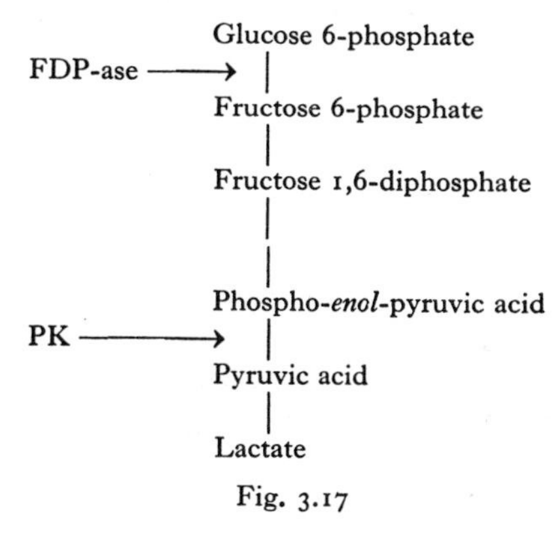

Fig. 3.17

Coryphaenoides were caught at depths in the region of 2000 m by a method developed by Phleger (Chapter 7). It is worth noting that although the animals were moribund on arrival at the surface, having experienced a rise in temperature and a decrease in pressure, some exhibited heart movements.

PYRUVATE KINASE (PK) FROM FISH MUSCLE

PK is involved in the ATP-producing reaction in which phospho-*enol*-pyruvate (PEP) is converted to pyruvate. It plays an essential role in muscle glycolysis. FDP (see Fig. 3.17) and Mg^{2+} ions activate the enzyme while ATP inhibits it.

Biochemical aspects

PK was extracted from the epaxial muscles of the trout, *Salmo gairdneri* and from *Coryphaenoides*. Electrophoresis demonstrated that the PK from the rat tail migrated towards the cathode as a single entity. PK from the trout migrated towards the anode as a single major band, although under certain conditions sub-fractions were evident. This difference in electrophonetic behaviour is interesting and unexplained.

Using a high pressure cell in a spectrophotometer it was shown that both *Salmo* and *Coryphaenoides* PK have a reduced rate of reaction at high pressure in the presence of saturating levels of PEP and ADP, a phosphate acceptor (Fig. 3.18). This demonstrates that the activation volume, $\Delta V^{\ddagger}$, for the rate limiting step is positive, in contrast to the $-\Delta V^{\ddagger}$ seen in FDP-ase (p. 114), but this observation has only limited relevance to the problem of enzyme function in the deep sea.

The pressure sensitivity of a number of controlling factors in the PK from *Coryphaenoides* was examined by Mustafa, Moon and Hochachka (1971) and Moon, Mustafa and Hochachka (1971*a*).

Mg^{2+} ions

Salmo PK is sensitive to pressure at limiting concentrations of Mg^{2+} ions and in the presence of saturating levels of other agents. *Coryphaenoides* PK shows a slight difference. The K_a for Mg^{2+} ions in the rat tail PK is unaffected by pressures up to 400 atm. The interpolated K_a for Mg^{2+} ions in the enzyme from *Salmo* at the same pressure is shown in Fig. 3.19. The data (inset, Fig. 3.19*B*) suggest a small difference between the K_a for Mg^{2+} ions in the rat tail and the trout PK at 400 atm, but hardly conclusively. The inset to Fig. 3.19*B* should be carefully noted. The significance of the difference between the points at 400 atm is not quantified by the authors. At the rat tails' normal pressure, 200 atm, there is no apparent effect of pressure on the K_a for Mg^{2+} in either enzyme.

ADP

At all ADP concentrations high pressure inhibits the rate of reaction of the PK from *Coryphaenoides* but the K_m for ADP is altered by a negligible amount. The K_m for ADP in the *Salmo* PK is increased at pressure, implying the affinity of the enzyme for ADP

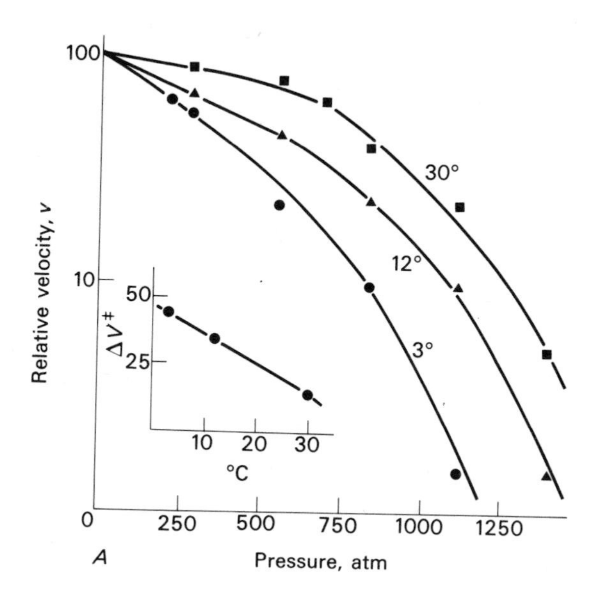

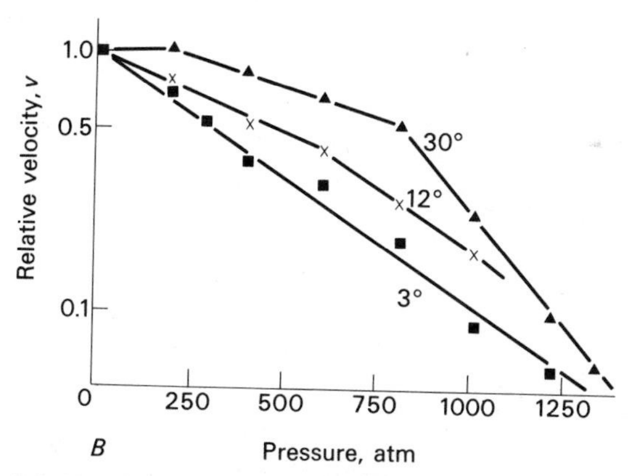

Fig. 3.18. Inhibition of muscle pyruvate kinase from (*A*) *Coryphaenoides* and (*B*) *Salmo*. Phospho-*enol*-pyruvate and ADP present in saturating concentrations. Vertical scale, log relative velocity of pyruvate kinase; horizontal scale, pressure. Inset in (*A*) shows $\Delta V^{\ddagger}$ calculated for 1–534 atm. See text. (*A* after Mustafa, Moon & Hochachka, 1971; *B* after Moon, Mustafa & Hochachka, 1971*a*)

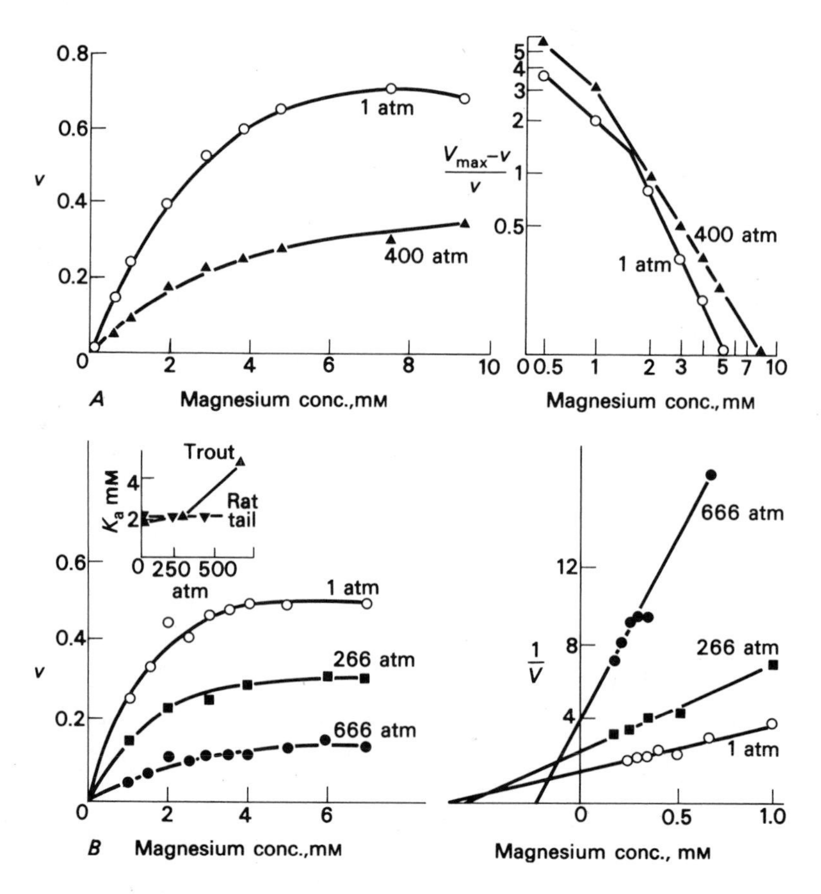

Fig. 3.19. The effect of pressure on muscle pyruvate kinase at 3 °C in the presence of low concentrations of Mg^{2+}, *A*, *Coryphaenoides*; *B*, *Salmo*. Note that the authors compare data interpolated from a Hill plot (*A*, right-hand graph) with that from a Lineweaver–Burk plot (*B*, right-hand graph). Inset in *B* shows the K_a for Mg^{2+} for *Salmo* (trout) and *Coryphaenoides* (rat tail). Note the interpolated difference between the two sets of data at 200 and 400 atm. (*A* after Mustafa, Moon & Hochachka, 1971; *B* after Moon, Mustafa & Hochachka, 1971*a*)

is greatly reduced by high pressure (Fig. 3.20) (Mustafa, Moon & Hochachka, 1971, and Moon, Mustafa & Hochachka, 1971*a*). Again the difference between the K_m values is not quantified, but appears small at pressures similar to those normally experienced by *Coryphaenoides*.

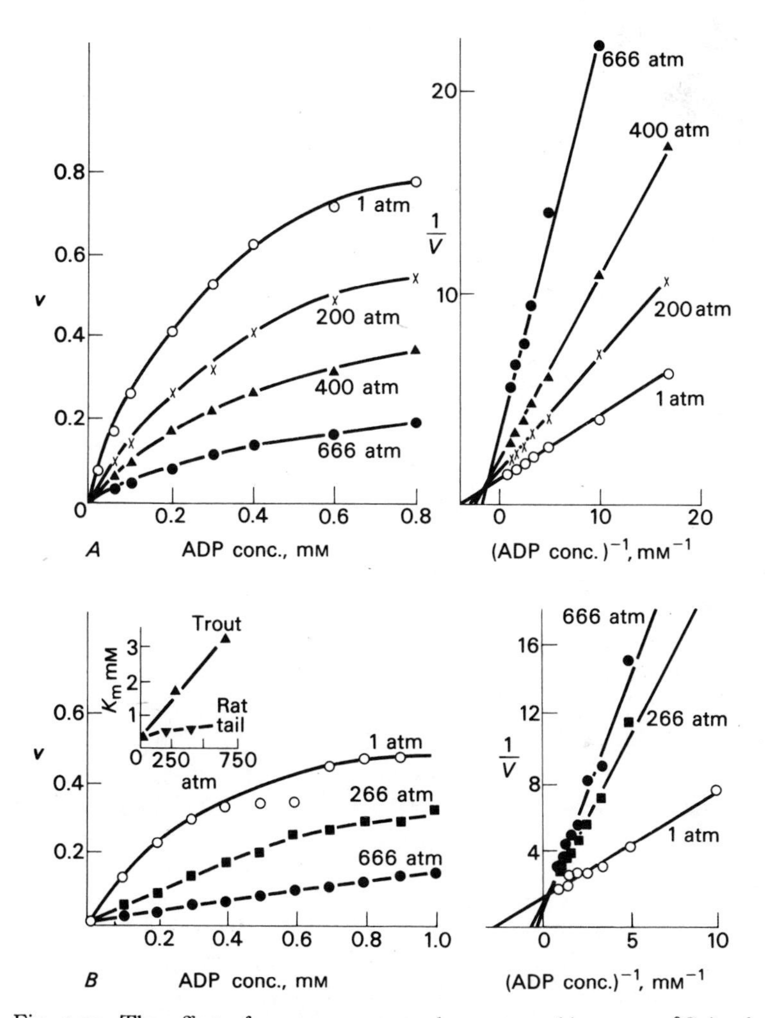

Fig. 3.20. The effect of pressure on muscle pyruvate kinase at 3 °C in the presence of selected concentrations of ADP. *A*, *Coryphaenoides* (rat tail); *B*, *Salmo* (trout). Note each K_m for ADP which is obtained from a Lineweaver–Burk plot, is shown in the inset in *B*. (*A* after Mustafa, Moon & Hochachka, 1971; *B* after Moon, Mustafa & Hochachka, 1971*a*)

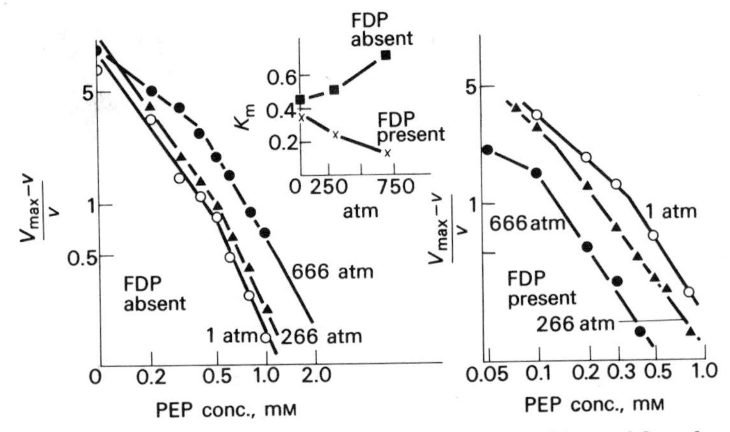

Fig. 3.21. The effect of pressure on the muscle pyruvate kinase of *Coryphaenoides* in the presence of selected concentrations of phospho-*enol*-pyruvate (PEP), both with and without fructose diphosphate (FDP) on left, and in presence of 10 μM FDP, on right. The data are presented as Hill plots. (After Mustafa, Moon & Hochachka, 1971)

Substrate

The situation in the case of PEP–enzyme interactions at high pressure is complicated. The condition in which FDP is also present as an activator is particularly interesting. *Coryphaenoides* PK then shows an increased affinity for PEP at high pressure, a shift which might tend to offset the $+\Delta V^{\ddagger}$ of the basic PK reaction. The reverse is true in the absence of FDP (Fig. 3.21). In the case of *Salmo* PK, a small but biphasic effect of pressure was found on the K_m for PEP in the presence of FDP. PK from a vertically migrating fish, a sea bass, showed no effect of pressure on the K_m for PEP in the presence of FDP. Finally the interactions between ATP (feedback inhibitor) and FDP (feedforward activator) were studied at high pressure. A possibly important distinction is seen between the shallow water and deep sea PK. In *Coryphaenoides*, FDP and ATP interact at atmospheric pressure and in specific concentrations of reactants produce a low level of PK activity. It will be seen from Table 3.14 that FDP offsets the inhibitory effect of ATP. The same is true for *Salmo* PK to a lesser extent (not shown in the table). At 200 atm or higher, however, *Coryphaenoides* PK activity is increased and FDP now has a dominant effect. *Salmo* PK activity is not increased, in fact it

TABLE 3.14. *FDP–ATP interactions in the regulation of pyruvate kinase activity in the rat tail. Conditions of assay: 0.5 mM PEP, 0.8 mM ADP, 8 mM, 80 mM* K^+*, excess coupling reagents, and constant amounts of PK, ATP and FDP added as shown below.* (Mustafa *et al.*, 1971)

Conditions	Activity as per cent of rates at 0.5 mM PEP		
	1 atm	200 atm	534 atm
0.01 mM FDP	169	225	240
1.0 mM ATP	94	110	125
0.01 mM FDP 1.0 mM ATP	122	274	260
4.0 mM ATP	60	71	80
0.01 mM FDP 4.0 mM ATP	99	213	210

declines. Mustafa *et al.* (1971) suggest that the dominance of FDP over ATP at high pressure is a feature which shows that the deep sea PK is adapted to function at high pressure.

The present author accepts this as an interesting observation and feels the differences in reactant concentration should not be overlooked; 0.01 mM FDP + 4.0 mM ATP in the *Coryphaenoides* experiments compared to 0.1 mM FDP + 3.0 mM ATP in the *Salmo* experiments.

Table 3.15 summarises the effects of pressure on muscle PK and its regulators from three species of fish as interpreted by Moon *et al.* (1971*a*).

A further kinetic study on *Coryphenoides* enzymes warns against premature and simplified conclusions. Lactate dehydrogenases (which catalyse the reaction pyruvate $\rightleftharpoons$ lactate) were extracted from heart muscle, liver and expaxial muscle (Moon, Mustafa & Hochachka, 1971*b*). The K_m for pyruvate was determined with the reaction conditions indicated in Fig. 3.22. A pressure of 533 atm more than trebles the K_m for pyruvate in the enzyme from skeletal muscle but is without effect on the homologous enzymes from heart and liver. The lactate dehydrogenase in skeletal muscle presumably differs from that of the same enzyme in other tissue and reminds us that the state of an enzyme *in situ* has considerable importance in determining its sensitivity to pressure.

TABLE 3.15. *Comparison of various catalytic and regulatory properties of muscle pyruvate kinase (PK) from the deep sea* Coryphaenoides *sp., the mid-water sea bass,* Ectreposebastes imus, *and* Salmo gairdneri *the rainbow trout.* (Moon, Mustafa & Hochachka, 1971 *a*)

	Effect of high hydrostatic pressure		
	Coryphaenoides sp.	*Ectreposebastes imus*	*Salmo gairdneri*
Selected enzyme characteristics			
$\Delta V^{\ddagger}$ (at 3 °C)	44	29	47
K_m (PEP)	Small increase	No effect	U-shaped K_m-pressure curve
K_m (PEP) in the presence of FDP	Small decrease	No effect	U-shaped K_m-pressure curve
K_m (ADP)	Small increase	No effect	Large increase
K_a (Mg^{2+})	No effect	(No effect?)	Large increase
K_i (ATP)	No effect	No effect	Substantial increase
FDP reversal of ATP inhibition	Large increase	No effect	No effect
General enzyme characteristics			
$\Delta V^{\ddagger}$ at high temperature	Reduced	Reduced	Reduced
$\Delta V^{\ddagger}$ at 3 °C	pH independent	pH independent	pH independent
Activation energy	Increased	Increased	Increased
Maximum catalytic rate	Decreased	Decreased	Decreased

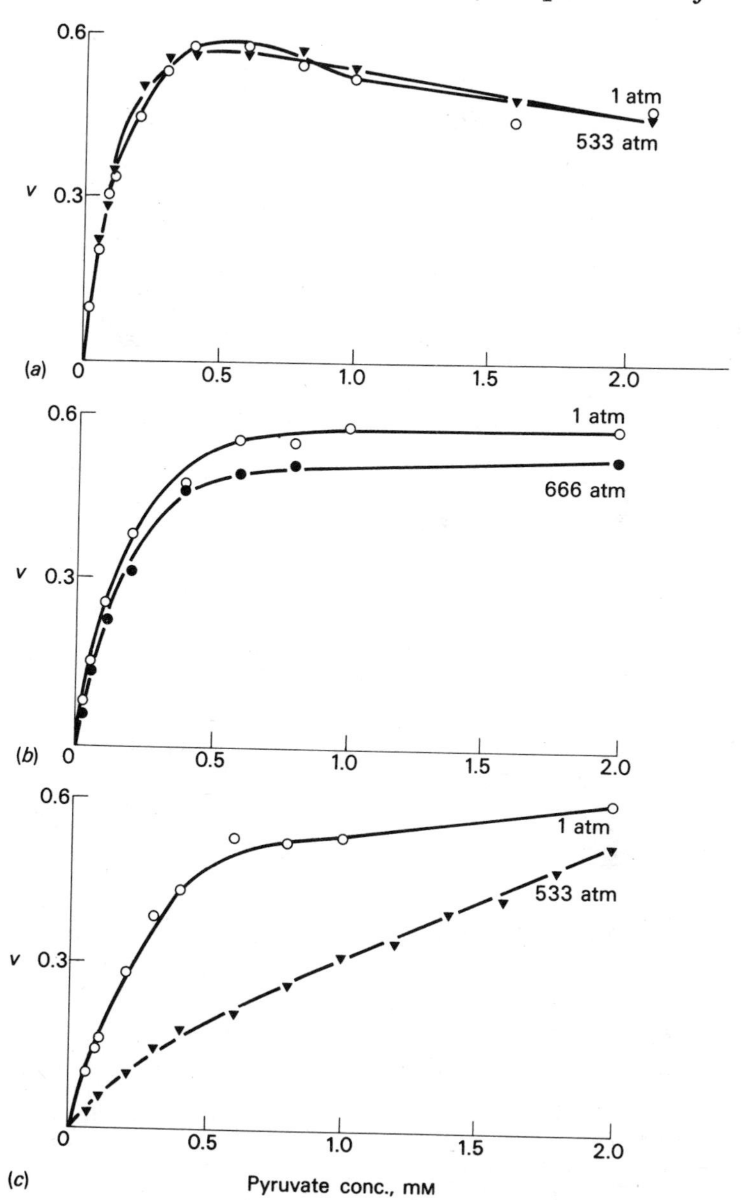

Fig. 3.22. The effect of pressure on lactate dehydrogenase from (*a*) heart tissue, (*b*) liver tissue and (*c*) muscle tissue of *Coryphaenoides*, at 3 °C in the presence of selected concentrations of pyruvate. Relative reaction velocity is plotted on the vertical scale. (After Moon, Mustafa & Hochachka, 1971*b*)

FRUCTOSE DIPHOSPHATASE (FDP-ASE) FROM THE LIVER OF *CORYPHAENOIDES*

FDP-ase was extracted from the livers of freshly caught specimens and proved amenable to study by conventional techniques. Its role in the gluconeogenic pathway should be noted (Fig. 3.17). In the mammalian liver the activity of this enzyme is stimulated by high levels of ATP and low levels of AMP. Partially purified FDP-ase was assayed against FDP by determining the inorganic phosphate liberated after a period of incubation at high pressure (Hochachka, Behrisch & Marcus, 1971, Hochachka, Schneider & Moon, 1971).

Three controlling factors were considered to be of importance; namely the concentration of substrate, magnesium ions, and AMP.

Substrate

At pH 8.75, and at saturating concentrations of Mg^{2+} ions, the activity of *Coryphaenoides*' FDP-ase is influenced by substrate concentration in the manner shown in Fig. 3.23. At atmospheric pressure an increase in the substrate concentration increases the reaction velocity, passing through a maximum at 0.2 mM FDP. At substrate concentrations higher than this, high pressures significantly increase the reaction velocity still further. Fig. 3.24 shows how different pressures affect the reaction velocity at different substrate concentrations. Clearly the reaction is insensitive to pressure only at very low substrate concentrations. The FDP-ase from the liver of the rainbow trout, *Salmo gairdneri* provides an interesting comparison. Under the same conditions (pH 8.75, Mg^{2+} ions 2–5 mM) pressure inhibits the reaction velocity markedly at low substrate (0.05 mM) concentrations (Figs. 3.23 and 3.24).

The data indicate that in the *Salmo* enzyme the K_m for FDP is increased by pressure. The authors assert that the K_m for FDP coincides with the binding constant and therefore may be regarded as a measure of the binding between enzyme and substrate. Thus pressure diminishes the affinity of the *Salmo* enzyme for its substrate. When this is a rate limiting factor, as it is at low substrate concentrations, pressure depresses the reaction velocity (Fig. 3.24). In *Coryphaenoides* FDP-ase the effect of pressure on the K_m for FDP is much less (Fig. 3.23), and as we see in Fig. 3.24, reaction velocity is little affected by pressure at low substrate

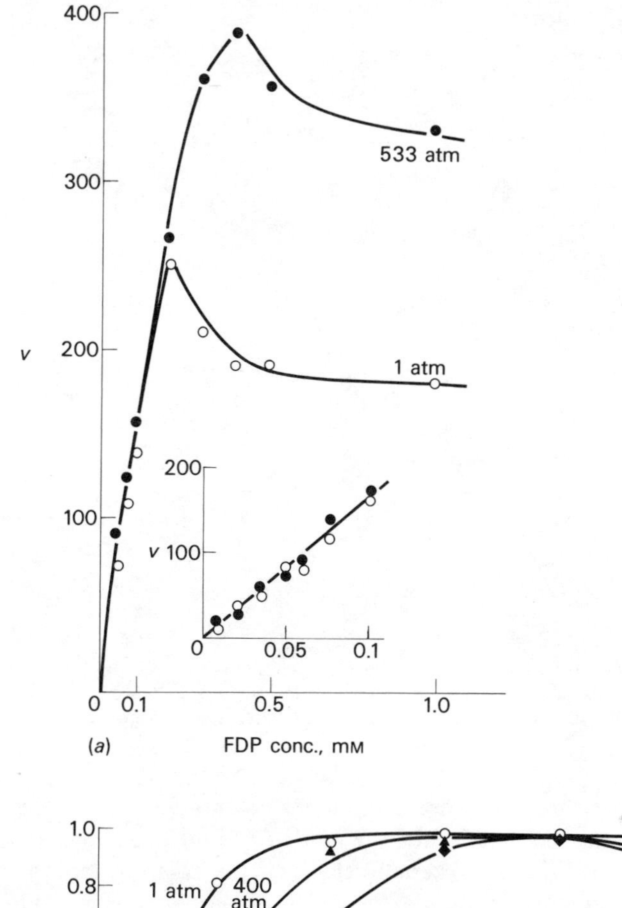

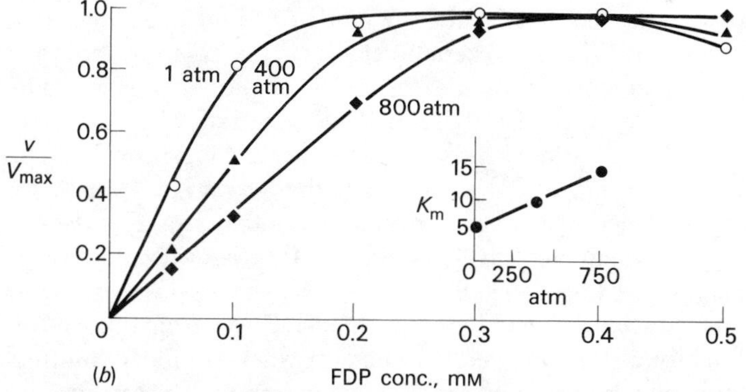

Fig. 3.23. The effect of pressure on the FDP-ase from the liver of (*a*) *Coryphaenoides* and (*b*) *Salmo*, at 2 °C, and at selected concentrations of substrate. Vertical scales are (*a*) relative velocity, (*b*) relative velocity as a fraction of the maximum velocity; horizontal scale, pressure. Inset in (*a*) shows v at low substrate concentrations. ((*a*) After Hochachka, Schneider & Moon, 1971; (*b*) after Hochachka, Behrisch & Marcus, 1971)

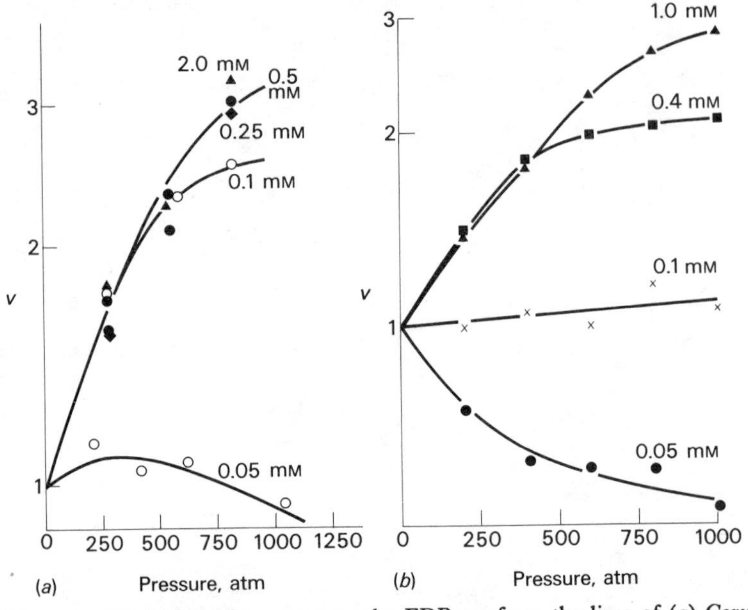

Fig. 3.24. The effect of pressure on the FDP-ase from the liver of (*a*) *Coryphaenoides* at 2 °C and (*b*) *Salmo* at 3 °C. Vertical scale, log relative velocity; horizontal scale, pressure. See text. (After Hochachka, Behrisch & Marcus, 1971)

concentrations thought to prevail in nature. A numerical comparison between the respective curves (*a*) and (*b*) in Fig. 3.24 shows that *Coryphaenoides* FDP-ase reaction velocity is increased by less than 10 per cent at 400 atm in the presence of 0.05 mM FDP. The reaction velocity of the FDP-ase in *Salmo* is reduced by more than 75 per cent at 400 atm with the same substrate concentration. The apparent cause is a difference in $\Delta V^{\ddagger}$, the activation volume for the binding reaction between enzyme and substrate. The data in Fig. 3.23 are evidence for a significant positive volume change occurring when the *Salmo* enzyme and substrate interact at low substrate concentrations. Perhaps the FDP-ase from *Coryphaenoides* involves different volume changes when similarly binding. If so, this may be a site of adaptive change to high pressure.

It must also be pointed out that at a substrate concentration of 0.1 mM the *Salmo* FDP-ase appears to be little affected by pressure, the enzyme–substrate affinity and the $\Delta V^{\ddagger}$ of the main reaction presumably cancelling each other out (Fig. 3.24).

Mg^{2+} ions and AMP

Similar kinetic measurements were carried out in conditions in which Mg^{2+} ions and AMP concentrations were varied in the presence of saturating concentrations of other ingredients. At low Mg^{2+} ion concentrations a pressure of 533 atm has no effect on the reaction velocity of *Coryphaenoides* FDP-ase but at higher concentrations pressure increases the rate of reaction (Fig. 3.25(*a*)). *Salmo* FDP-ase is sensitive to pressure at low Mg^{2+} ion concentrations (Fig. 3.25(*b*)). A numerical comparison, ignoring the 1 degC difference in temperature between the two experiments is as follows; at 533 atm the interpolated K_a for Mg^{2+} ions in the trout FDP-ase is 0.30 mM and at atmospheric pressure the K_a is 0.15 mM (Fig. 3.25(*b*)). For the rat tail FDP-ase the authors state that the corresponding figures are 0.083 mM [0.83 mM?] and 0.045 mM [0.45 mM?]. Thus the K_a for Mg^{2+} ions in the trout is doubled at 533 atm, but in the rat tails it is slightly less than doubled. The authors do not quantify the significance of these differences and in Table 3.16 describe them as 'large'.

AMP is assumed to be the main physiological controlling factor over the FDP-ase in *Coryphaenoides* as is the case in shallow water or low pressure vertebrates. Low concentrations of AMP activate, while higher concentrations inhibit, FDP-ase. Fig. 3.26(*a*) shows the effect of AMP on the activity of FDP-ase from *Coryphaenoides* at 1 atm and at high pressure. *Salmo* FDP-ase is also sensitive to AMP concentration (Fig. 3.26(*b*)) 400 atm diminishes the extent of the biphasic action of AMP and higher pressures eliminate the AMP activation of the enzyme. Hochachka, Schneider and Moon (1971) state the K_i for AMP in the rat tail FDP is less affected by pressure than is the trout FDP-ase. This is not altogether apparent from the data, comparing the effect of 200 atm on both enzymes for example. Table 3.16, which will be considered shortly in more detail, summarises the authors' interpretation of the results of experiments on the two FDP-ases.

Tables 3.15 and 3.16 list some properties of the two enzymes we have considered, PK and FDP-ase, which are similar in the two species. Such properties are referred to as 'general properties' by the authors and this is probably a good term. Establishing these similarities certainly focusses interest on the differences listed in the same tables. Note that all the experiments on the FDP-ase

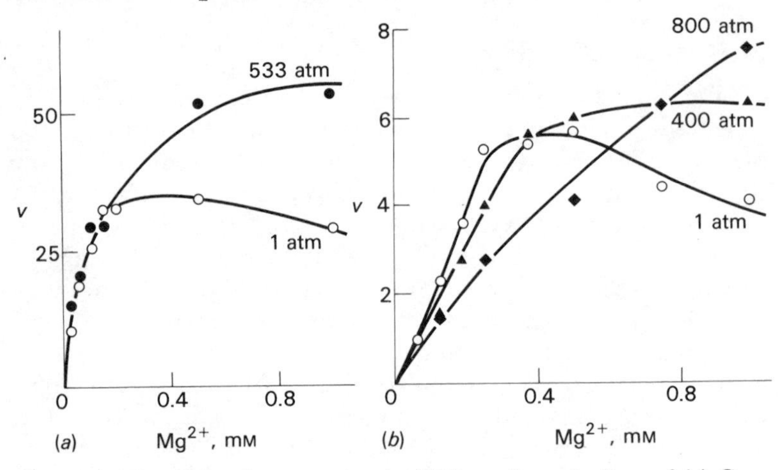

Fig. 3.25. The effect of pressure on the FDP-ase from the liver of (*a*) *Coryphaenoides* (2 °C) and (*b*) *Salmo* (3 °C) in selected concentrations of Mg^{2+} ions, at pH 8.75, and using 0.5 mM FDP. Vertical scale, relative velocity; horizontal scale, mM Mg^{2+}. ((*a*) After Hochachka, Schneider & Moon, 1971; (*b*) after Hochachka, Behrisch & Marcus, 1971)

TABLE 3.16. *Summary and comparison of various properties of liver FDP-ase from* Salmo gairdneri *and the deep sea* Coryphaenoides *sp.* (Hochachka, Schneider & Moon, 1971)

	Effect of high pressure	
Enzyme characteristics	*Salmo gairdneri*	*Coryphaenoides* sp.
General properties		
V_{max} at pH 8.75	Increase	Increase
V_{max} at pH 7.5	No effect	No effect
pH profile	Alkaline pH optimum	Alkaline pH optimum
Native enzyme	Inactivated	Inactivated
Selected properties		
K_m (FDP)	Increase	No effect
K_a (Mg)	Large increase	Small increase
K_i (AMP)	Decrease	No effect

were carried out at pH 8.75 which is close to the optimum for both enzymes at high pressure in the presence of saturating concentrations of Mg^{2+} ions and substrate at low temperature.

Another similarity between the two types of FDP-ase is their rapid inactivation at high pressure (Fig. 3.27). Under the experi-

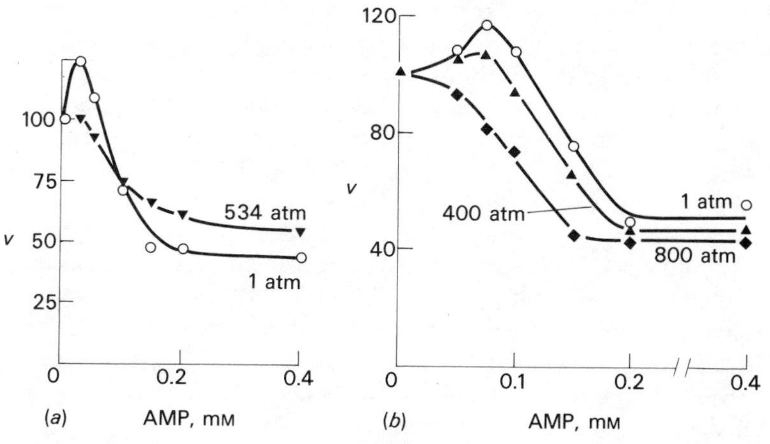

Fig. 3.26. The effect of pressure on the FDP-ase from the liver of (*a*) *Coryphaenoides* and (*b*) *Salmo* at 3 °C in selected AMP concentrations. Vertical scale, relative reaction velocity; horizontal scale, mM AMP. Experiments at pH 8.75, 2.5 mM Mg^{2+}, 0.5 mM FDP. ((*a*) After Hochachka, Behrisch & Marcus, 1971; (*b*) after Hochachka, Schneider & Moon, 1971)

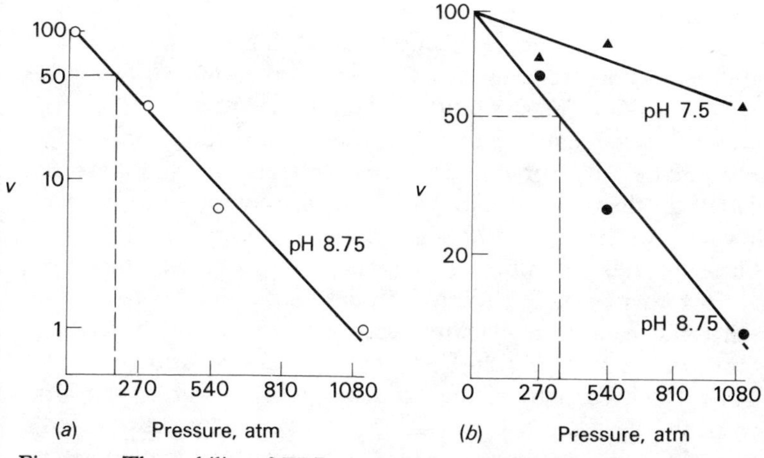

Fig. 3.27. The stability of FDP-ase from (*a*) *Coryphaenoides* and (*b*) *Salmo* at high pressure. Each enzyme was incubated for 1 hour at pH 8.75 (or 7.5) and subsequently assayed for activity against FDP at 3 °C and atmospheric pressure. Vertical scale, log relative reaction velocity; horizontal scale, incubation pressure. ((*a*) After Hochachka, Behrisch & Marcus, 1971; (*b*) after Hochachka, Schneider & Moon, 1971)

mental conditions used, the FDP-ase from *Coryphaenoides* when in the native condition is 50 per cent inactivated after 1 hour at 200 atm whereas the FDP-ase from *Salmo* required 360 atm for an hour to bring about the same inactivation; this suggests the stability of the native enzyme at high pressure is not important in nature.

We have now to consider the significance of the differences between the enzymes listed in Tables 3.15 and 3.16. However, first we should consider the normal pressure which is experienced by the *Coryphaenoides*. Although this is not known for certain it will probably be in the region of two hundred atmospheres. The pressure range to which individuals are exposed is probably small although the young may develop in shallow water. It is unlikely that the sea bass experiences as much pressure as the rat tail, and we can certainly regard the trout *Salmo* as a shallow water animal.

The pressures referred to in the tables differ according to the experiment. This author questions the conclusions concerning the K_a for Mg^{2+} ions and the K_i for AMP in the FDP-ases (Table 3.16). However the pressure sensitivity of the K_m for FDP in *Salmo* contrasts with its insensitivity to pressure in *Coryphaenoides*.

In Table 3.15 the alleged differences in the K_a for Mg^{2+} ions in the rat tail and trout PK's is questionable, but the significance of the difference between the K_m for FDP in the same two enzymes seems plausible. The similarity in the general enzyme characteristics certainly strengthens the significance of the differences listed. It would be surprising if the other differences listed were not related to the pressure in the animal's normal environment. However, we have to remember that the interpretation of these results depends on a number of assumptions, some of which have already been mentioned. Anothér assumption which has been made is that the controlling factors used in the experiments actually operate in the deep sea fish. Deep sea enzymes may utilise different regulatory ions as part of their adaptation to pressure. Deterioration in the deep sea material during recovery may prove to be a major difficulty in demonstrating adaptive kinetic parameters. Some caution has to be observed in comparing results of this type because teleologically satisfying comparisons are seized upon and inconvenient data dropped. Consider the insensitivity of the trout FDP-ase to pressure at 0.1 mM substrate concentration for example. However, some of the results of the kinetic studies seem broadly consistent with the notion that deep sea life requires adaptation of the regulatory mechanisms to high pressure. Such adaptation would confer a degree of freedom from the effects of pressure which is likely to be advantageous for free swimming animals. This prompts two further lines of enquiry (*a*) What is the molecular basis of the adaptation to pressure? How, for example, might the interaction between an enzyme and its regulator be rendered insensitive to pressure? How important is the multimeric structure of enzymes, discussed in Chapter 2? (*b*) May we anticipate a completely pressure-insensitive metabolism appearing in animals at some moderate depth or are the changes which are required for adaptation to pressure of such a nature that grades of pressure sensitivity may exist at all pressures? The apparent insensitivity of the mysterious sea bass to quite high pressure is intriguing (Table 3.15).

Concerning the first question we may speculate on how an enzyme–ion (or other molecule) interaction might alter in its volumetric properties and hence sensitivity to pressure. Assume, along with authors of Table 3.15, that the alleged increase in the K_a for Mg^{2+} ions in trout PK indicates a decreased affinity for Mg^{2+} ions by the enzyme at high pressure. It is plausible to regard the Mg^{2+}–(enzyme–substrate) interaction as an ionic bond. Adaptation on the part of the deep sea enzyme cannot be expected to alter the basic volumetric effect of ionic bond formation, so have we to anticipate some compensatory volume change which would have the effect of cancelling the $+\Delta V$ of ionic bond formation? We may readily imagine, but only imagine, the arrangement below:

$$E^{2-} + Mg^{2+} \rightleftharpoons EMg^{0}$$

where E is an enzyme and EMg^0 indicates some other molecular change.

At this stage it is worth noting the views of Simpson (1964), who argues that natural selection acts on the whole phenotype. Natural selection can single out genes only if they are phenotypically distinct from others; generally selection acts on complexes of phenotypic characters. The adaptation of the intermediate metabolism of mobile animals such as fish to deep sea conditions may require more than adaptation to high pressure and low temperatures. But if we restrict our view of the deep sea to these two parameters it is reasonable to expect that adaptation should require metabolism to function normally at high pressure, and if the animal undergoes vertical excursions metabolism should be controlled independently of the changes in pressure. This independence is likely to be evolved from the molecular machinery of shallow water animals through successive changes at the level of regulatory processes rather than by a complete redesigning of enzymes. According to Simpson's view of evolution at the molecular level, 'redesigning' is unlikely to be a common occurrence. On the present extremely limited evidence, adaptation of regulatory processes to high pressures seems to involve small changes in reversible binding reactions, in which non-covalent bonds are involved. Such adaptations may require only a few changes in the amino acid sequences of specific binding sites, as implied by Hochachka, Schneider and Moon (1971). An example of the profound effect which even a single amino acid substitution may have on the pressure sensitivity of a protein is seen in the case of sickle cell haemoglobin, described in Chapter 2. It will be recalled that the introduction of a single valine group into the haemoglobin molecule appears to bring about the phenomenon of sickling which is associated with a characteristic response to high pressure.

Some conclusions

Consistent with some of the ideas outlined in Chapter 2 we see in the present chapter how pressure affects biochemical systems in subtle ways. Thus deep sea pressures have little direct effect on DNA but upset its metabolism. The tendency for pressure to dissociate sub-units and ligands seems especially important. The extra energy cost of molecular expansion in the deep sea seems small.

4 PHYSIOLOGICAL ASPECTS OF HIGH HYDROSTATIC PRESSURE

Having considered in Chapters 2 and 3 the ways in which deep sea pressures affect some chemical and biochemical aspects of organisms we now turn to the question of how pressure affects integrated physiological process and intact organisms. As before, the discussion will largely be confined to the ways in which pressure affects systems which normally work at atmospheric pressure in the hope that this will guide our thinking about adaptation to high pressures. Only a few observations have been carried out on the activity of deep sea animals and their tissues at high pressure.

Four interrelated areas of physiology are considered in turn: the mechanical activities of cells and respiration, membrane processes, muscular contraction and the locomotor activity of intact animals.

Cellular activities

BULK PROPERTIES OF CYTOPLASM

Mechanical work is carried out by relatively undifferentiated eukaryotic cells (amoeboid movement, cytoplasmic streaming) as well as by cells possessing elaborate structures, such as cilia and mitotic figures. Both differentiated and undifferentiated cells are highly susceptible to high hydrostatic pressure.

The mechanical activity of undifferentiated cytoplasm requires a certain degree of structure. One way of measuring the coherence of the cytoplasmic structure necessary for the performance of mechanical work is to determine the viscosity or 'gel strength' of either the contents of a suitable cell or selected regions of its cytoplasm. When such measurements are made on cells simultaneously exposed to several hundred atmospheres pressure it is invariably found that the cytoplasm is liquefied. Brown (1934*a*) constructed the first apparatus to carry out measurements of cytoplasmic gel strength at high pressure. He used a centrifuge

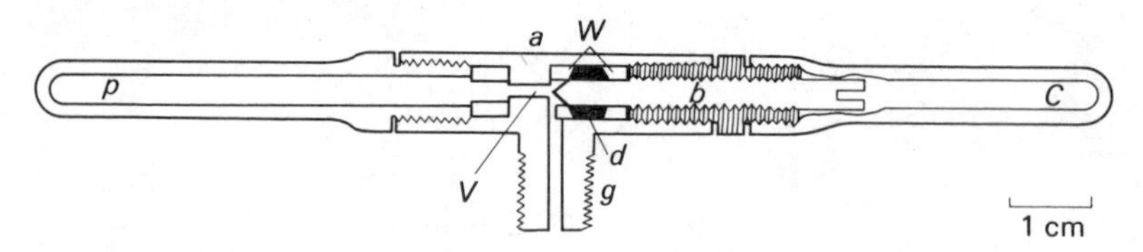

Fig. 4.1. The pressure-centrifuge. *a*, central T-shaped portion; *b*, needle valve; *C*, control chamber; *d*, rawhide packing; *g*, pump and centrifuge connection; *p*, pressure chamber; *V*, valve chamber; *W*, metal washers. (After Brown, 1934*a*)

with centrifuge tubes containing cells under high pressure (Fig. 4.1). The eggs of the sea urchin *Arbacia punctulata* are a favourite cell for such measurement as they contain convenient red pigment granules whose movement in a centrifugal field provides a measure of viscosity. Assuming Stokes' Law holds, the rate of movement of the pigment granules through the cytoplasm under an applied centrifugal force should obey equation (1):

$$v = 2\mathbf{g}\frac{(\sigma-\rho)}{9\eta}a^2 \qquad (1)$$

in which v is the granule velocity, η the viscosity of the medium through which the granules move, **g** the acceleration due to gravity, a the radius of rotation of the centrifuged cells, σ the density of the granule and ρ the density of the medium. Further, Brown assumed all the terms remain constant except η, thus $v \propto 1/\eta$. It follows that η varies directly as the time elapsed during which the granule moves a standard distance. Thus relative viscosity is measured by the time taken to sediment the pigment granules a standard distance towards the 'heavy' end of the cell. Pressurised and unpressurised specimens may be subjected to forces as high as 68000 **g**. At the higher centrifuge speeds an extra hydrostatic pressure of the order of 100 atm would be generated (Fig. 3.8). Fig. 4.2*A* shows a centrifuged *Amoeba* and Fig. 4.2*B* an *Arbacia* egg. Fig. 4.3 shows the gel strength of *Arbacia* eggs at different pressures determined by Zimmerman, Landau and Marsland (1957). The decrease in gel strength brought about by high pressure is rapidly reversed after decompression.

Quantitatively similar results have been obtained from pressure-centrifuge experiments with *Amoeba lixula*, the eggs of *Chaetopterus* (a polychaete worm) and the cells of the plant *Elodea*. A

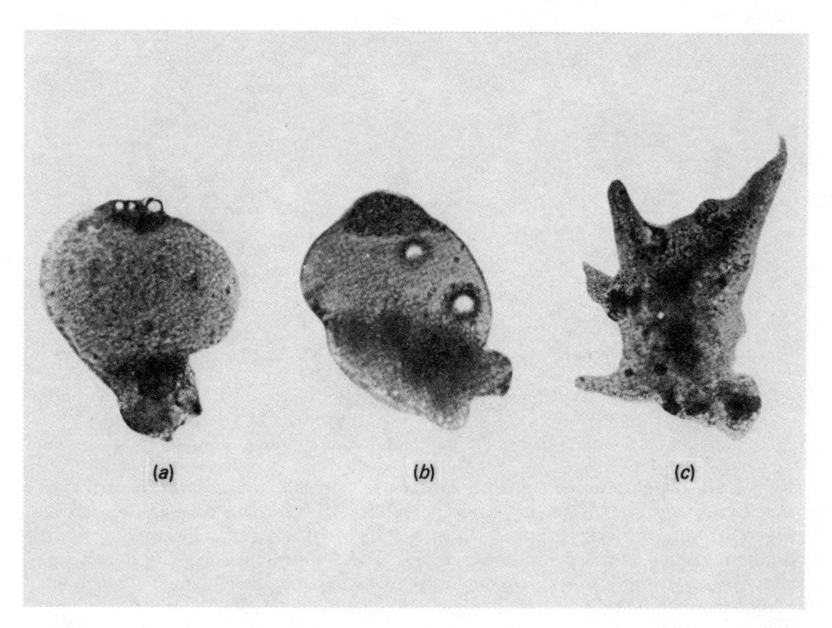

Fig. 4.2*A*. Displacement of granular elements of the plasmagel of *Amoeba proteus* pressure-centrifuged at different temperatures. Specimens heat-fixed immediately following centrifugation. (*a*) Typical specimen, centrifuged for 15 sec under a pressure of 400 atm at 25 °C. Oil cap above, zone of heavy granules below, and developing hyaline zone between. Presence of numerous fine granules in 'hyaline zone' indicates that this specimen has not quite reached the standard displacement end-point. Note oil droplets which sometimes exude from the oil cap in heat-fixed specimens. (*b*) Typical specimen, centrifuged for 15 sec at 400 atm at 15 °C. Absence of granules in hyaline zone indicates that this specimen has reached the standard end-point. Oil droplets (out of focus) are not in the cell. Width of heavy granule zone varies considerably from specimen to specimen, depending on the granular content. (*c*) Control specimen, centrifuged simultaneously with specimen (*b*), but at atmospheric pressure. (Landau, Zimmerman & Marsland, 1954)

pressure increment of 67 atm diminishes gel strength by 25 per cent. Qualitative observations of many different shallow water cells also support the notion that pressure reversibly liquefies eukaryote cytoplasm. Fig. 4.4 illustrates the interesting comparison which can be made with a myosin gel prepared from rabbit muscle. The liquefying effect of high pressure on bulk cytoplasm is also similar to its effect on methyl cellulose in which, it will be recalled, pressure dissociates hydrophobic bonds. The fact that low temperatures reinforce the pressure effect in cytoplasmic gels

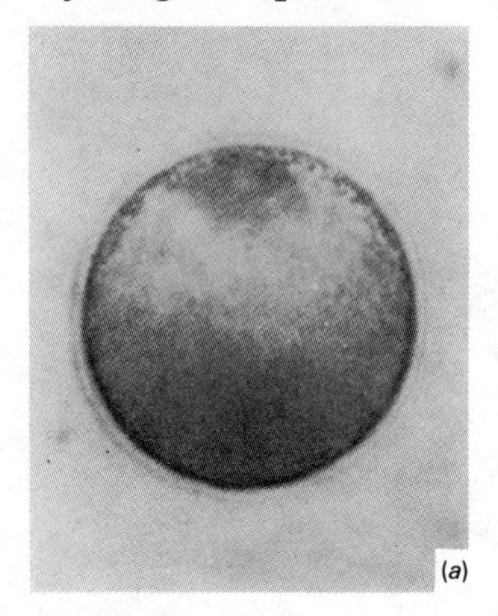

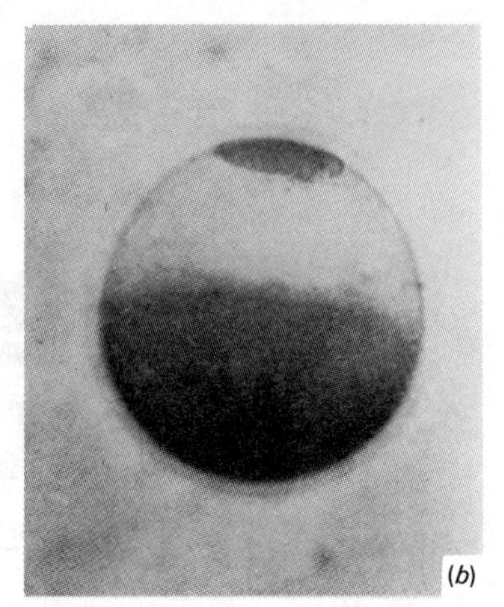

Fig. 4.2*B*. Photographs showing two eggs (*Arbacia punctulata*) centrifuged simultaneously for 185 sec at 41000 **g** and 20 °C. The control egg (*a*), centrifuged at atmospheric pressure, displays little, if any, displacement of the pigment bodies of the cytoplasmic cortex. The experimental egg (*b*), which was centrifuged at high pressure (530 atm), shows a displacement of virtually all of the pigment bodies from the cortex of the centripetal half of the egg. This represents the standard displacement end-point used in the experiments. (Zimmerman, Landau & Marsland, 1957)

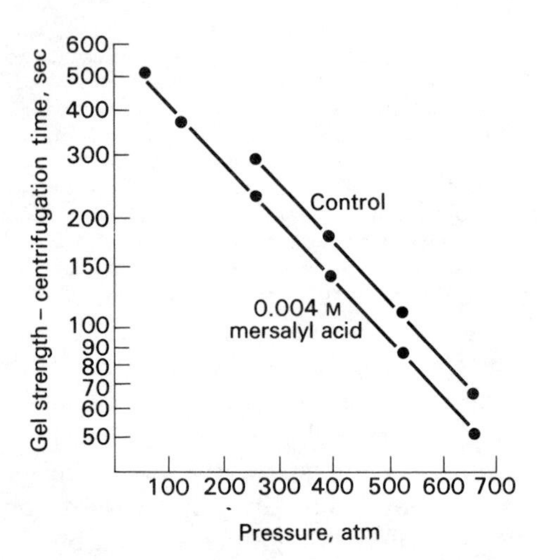

Fig. 4.3. The effect of high hydrostatic pressure on the gel strength of the cytoplasm of *Arbacia punctulata* eggs at 26 °C just before cleavage. A centrifugal force of 27800 **g** was used to determine gel strength; see text. (After Zimmerman, Landau & Marsland, 1957)

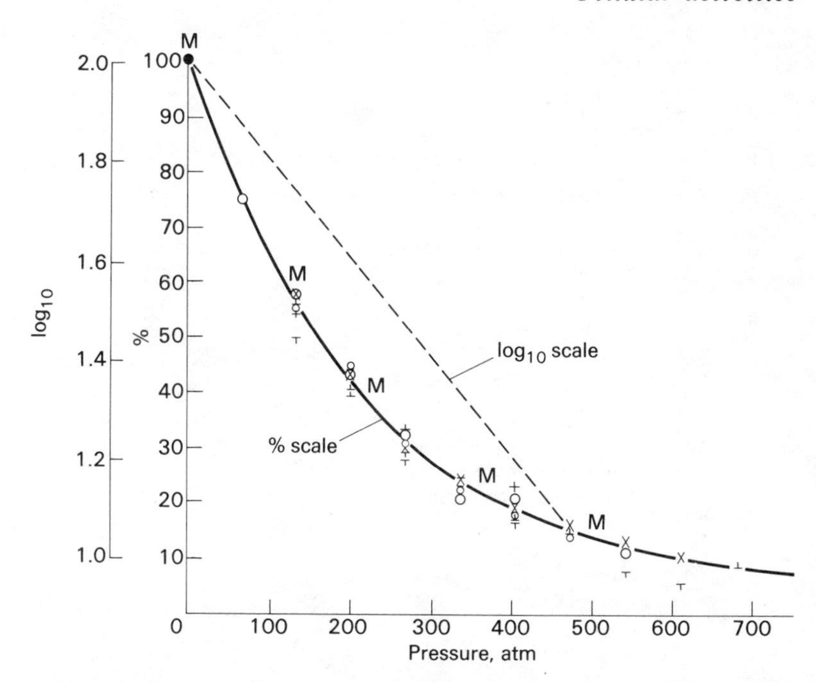

Fig. 4.4. Changes in relative viscosity of protoplasm and in rates of some biological processes; relative rates are plotted on the vertical axis. ●, arbitrary point, all other values are relative to this; ⊤, gel value, *Amoeba*; ⊥, gel value, unfertilised *Arbacia* egg; ×, gel value, cleaving *Arbacia* egg; +, rate of cleavage, *Arbacia* egg; ●, gell value, *Elodea* cells; ○, rate of streaming, *Elodea* cell; M, gel value, myosin gel (rabbit). pH = 6.5, temperature 23–24 °C. (Marsland, 1942)

suggests that hydrophobic bonds are important in determining their viscosity. It therefore seems reasonable to interpret the data in Figs. 4.3 and 4.4 in terms of the dissociation of inter- or intra-molecular bonds, particularly hydrophobic bonds.

The mechanical activity of partially liquefied cytoplasm is also diminished by pressure. The rate of cytoplasmic streaming in *Elodea*, for example, declines in proportion to the weakening of the gel (Fig. 4.4) and the ability of cells to constrict themselves into two ceases at a characteristic gel strength which may be attained with different combinations of pressure and temperature (Fig. 4.5 and Table 4.1). Both pressure effects, like the changes in viscosity, are readily reversible.

The most pressure resistant eukaryote cell yet discovered is the

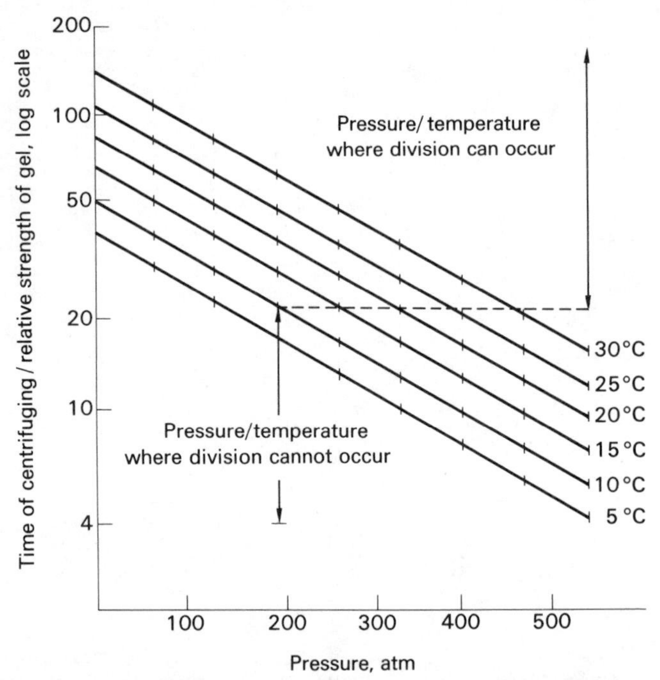

Fig. 4.5. Family of curves showing the effects of pressure, at different temperatures, on the structural strength of the gelated cortical cytoplasm of the *Arbacia* egg, and the relation of these data to the capacity of the egg to perform the work of cleavage. Any combination of temperature and pressure which reduces the cortical gel strength beyond a critical value (about 20% relative to the atmospheric value at 23 °C) is just adequate to block division. The centrifuge times (see text) are expressed directly in seconds, and the degree of variation is indicated by the length of the markers. (Marsland, 1970)

egg of the parasitic nematode worm *Ascaris megalocepha* var. *univalens*, which was observed to cleave at a normal rate at 800 atm at an unstated temperature (Pease & Marsland, 1939). Some delay of the cleavage furrow was obtained if pressure was applied before the cleavage furrow appeared.

It will be clear from Fig. 4.5 that high pressure and low temperature act together to inhibit cleavage. The simplest explanation of these phenomena is that pressure, in dissociating a gel structure, uncouples the mechanically active units in cytoplasm. It is likely, of course, that a full explanation will also have to take into account the direct effect of pressure on contraction as well as more distant

TABLE 4.1. *Pressure required to block division.* (After Marsland, 1970)

Kind of eggs	Pressure (atm)
Echinodermata	
Arbacia punctulata	330–400
Arbacia lixula (*pustulosa*)	330
Echinarachnius parma	330
Paracentrotus lividus	330
Psammechinus microtuberculosis	330
Sphaerechinus granularis	400
Nematoda	
Ascaris megalocephala	> 800
Annelida	
Chaetopterus pergamentaceus	233–267
Mollusca	
Cumingia tellenoides	267–330
Planorbis sp.	267
Solen siliqua	233
Insecta	
Drosophila melanogaster (pole cell division)	330–400
Tunicata	
Ciona intestinalis	200
Vertebrata	
Fundulus heteroclitus	330–400
Rana pipiens	330–400

energy-providing and regulating processes. In the context of adaptation to the deep sea the significance of high pressure at the structural level of cytoplasmic organisation cannot be overstated, as the following examples demonstrate.

PRESSURE-LABILE ULTRASTRUCTURE

Amoeba cytoplasm shows changes in ultrastructure when fixed at pressure after a 20 min exposure to 533 atm at 15 °C. Under these conditions Landau and Thibodeau (1962) have demonstrated the continued existence of an apparently normal plasmalemma but a disappearance of Golgi bodies and pinocytosis channels. Neither of these structures are likely to be immediately involved in the gross mechanical activity of the cell.

Physiological aspects

Other cytoplasmic elements more intimately associated with mechanical activity have proved to be pressure-labile (Zimmerman, 1971). The mitotic apparatus which is formed cyclically to distribute the chromosomes to daughter cells in the division of many eukaryote cells, is an important example. Studies by Pease (1946) and by Marsland and Zimmerman (1965) have shown that pressure disorganises the apparatus both *in situ* and when extracted from the cell. In *Arbacia* eggs the mitotic apparatus consists of bundles of radiating microtubules and other finer elements. The microtubules appear early in mitosis, as if polymerised from initiating centres, and there is evidence that the microtubules exist in dynamic equilibrium with monomeric sub-units. The action of pressure in disorganising the microtubules may therefore be to affect an equilibrium, as in the examples of simpler monomer–polymer equilibria (Chapter 2):

$$\text{monomer} \underset{-\Delta V}{\overset{+\Delta V}{\rightleftharpoons}} \text{polymer} + \text{release of 'bound' water.}$$

In the intact cell, pressure may also upset more complicated dynamic equilibria involving synthetic and breakdown reactions. The mitotic apparatus of deep sea cells is likely to be an interesting area for future work.

The elegant array of microtubules comprising the axopods of the heliozoan species *Actinophrys sol* and *Actinosphaerium nucleofilum* is caused to collapse by high hydrostatic pressure. Light microscope observations of *Actinophrys* in a pressure vessel showed the initial effect of a moderate pressure to be a shortening of the axopods (Kitching, 1957*a*). At higher pressures of between 267 and 330 atm, at room temperature, the axopods were seen to coalesce into beads of cytoplasm strung along a fine thread. At even higher pressure the thread bent and the beads aggregated near the cell body. After decompression the axopods re-established approximately their former length within a few minutes. At lower temperatures the axopods were even more sensitive to pressure. The fine thread or axoneme seen at pressure with the light microscope was later shown to consist of an orderly array of microtubules (Kitching, 1964). Fixation of pressurised *Actinosphaerium* which contains similar microtubules revealed the fibrillar products of disintegrated axonemes in otherwise normally structured cytoplasm (Fig. 4.6) (Tilney, Hiramoto & Marsland, 1966).

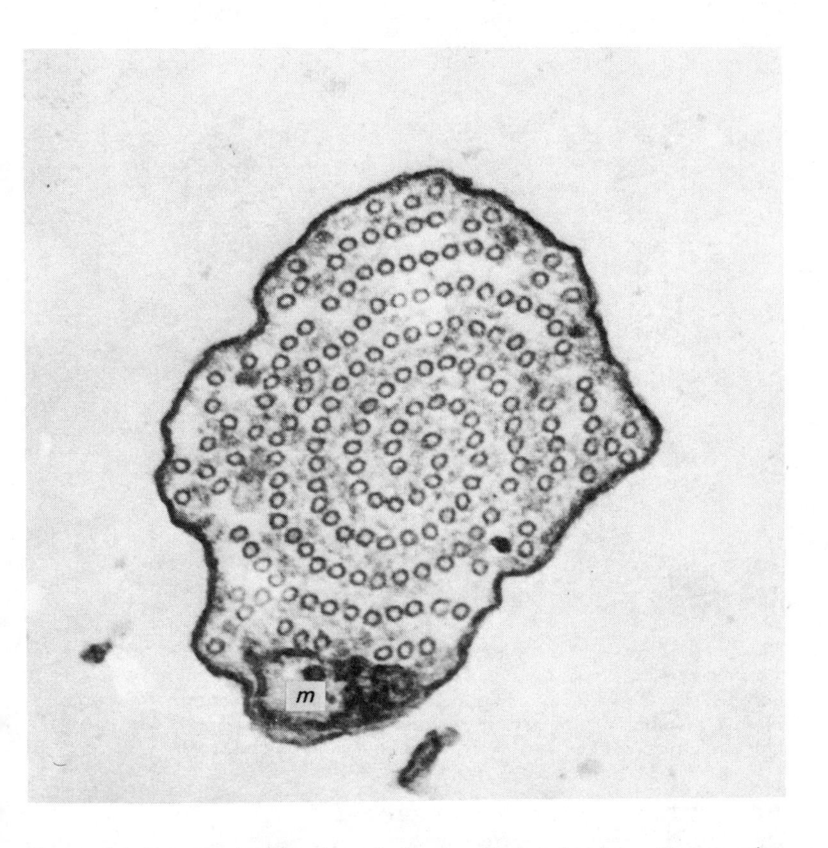

Fig. 4.6*A*. Pressure-labile microtubules in *Actinosphaerium*. Cross-section through an axopodium of an unpressurised control specimen. The microtubules, which form the axoneme, run parallel to each other and are arranged in an interlocking double coil. An amorphous material is present between the rows of the double coil. A mitochondrion (*m*) is present peripheral to the axoneme but beneath the plasma membrane which is a unit membrane. Fixation, 1% glutaraldehyde.

Not all microtubules are disrupted by high pressure. An interesting mixture of pressure-labile and pressure-resistant microtubules exist in cilia (see p. 133).

It is perhaps significant that both the mitotic apparatus and the heliozoan axopods are intracellular skeletal structures associated with shearing forces and both have the ability to collapse and reform in their natural life.

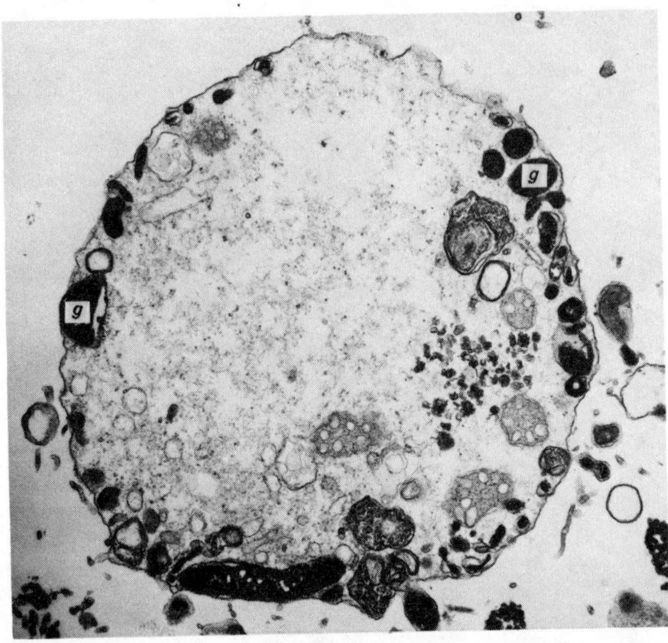

Fig. 4.6*B*. Cross-section through one of the short axopodia present after 3 min at 400 atm. Note the finely fibrillar material in the central region normally occupied by the axoneme. Peripheral to this material, beneath the unbroken plasma membrane, are several electron-opaque granules (*g*). Fixation, 1% glutaraldehyde. (*A* and *B*, Tilney, Hiramoto & Marsland, 1966)

PINOCYTOSIS AND SIMILAR PROCESSES

Many different cells engulf small particles or droplets into vacuoles which are subsequently conveyed to the cell interior. In *Amoeba proteus* pinocytosis channels can be induced by chemicals and are caused to disappear by 200 atm pressure. The dynamic nature of these structures is illustrated by the fact that they reappear some 20 seconds after decompression (Zimmerman & Rustad, 1965). Exopinocytosis (secretion) has not been studied at pressure. Contractile vacuoles which may be similar to secretory vesicles are affected by pressure, at least in certain Protozoa. Kitching (1957*b*) has measured the rate of discharge of the anterior contractile vacuole of *Paramecium caudatum* and found 330 atm stopped its activity altogether. *Vorticella* sp. at 533 atm also failed to discharge its vacuole which continued to swell. It is not surprising that such complicated organelles as contractile vacuoles are affected by pressure.

CILIA AND THEIR CONTROL

The structure of cilia exhibits both pressure-sensitive and pressure-resistant elements. When *Tetrahymena pyriformis* is subjected to an abrupt compression to 500 atm its cilia continue to beat for some time but forward locomotion dependent on coordinated beating ceases in about 5 min (Kennedy & Zimmerman, 1970). Cells which were fixed after 2 min exposure to the same pressure show partly disorganised basal bodies and are lacking the parts of the central pair of ciliary tubules which lie close to the basal body.

Pressure causes complicated changes in the frequency of beating of the lateral cilia in the excised gill of the mussel *Mytilus edulis* (Pease & Kitching, 1939). A transient acceleration following compression is seen over a wide pressure range; at pressures higher than 330 atm the frequency of beating declines to below the control level after an initial acceleration (Fig. 4.7). Temperature changes caused by the compression were examined in detail and found to be insignificant. Although Johnson, Eyring and Polissar (1954) have discussed this pressure effect in complex kinetic terms, exactly what controls the ciliary frequency in this preparation is not known; nervous control seems to play no part. The cilia were kept active in the excised gill by the presence of veratrine which accelerates the ciliary frequency in intact gills. Transient changes in the heart rate of certain animals have been seen in experiments involving step-wise compression, so some rhythmic processes may possess common kinetic properties which render them highly susceptible to abrupt changes in pressure.

In contrast to *Mytilus* cilia, the compound cilia or membranelles on the giant ciliate *Stentor* beat more slowly at pressure and do not show transient frequency changes (Sleigh, 1962). A reduction of 20 per cent in the ciliary frequency is brought about by approximately 400 atm. Ponat (1967) has observed the activity of *Mytilus* gill cilia during exposures to pressure and reports that activity at high pressure was prolonged in the presence of a low magnesium concentration, or an increased calcium or potassium concentration, in both normal and half-strength seawater. Generally, pressure resistance is higher in normal seawater than in dilute seawater. Ciliary activity in marine bivalves has provided criteria for some interesting dose-response curves. Excised gill tissue from *Ostrea edulis* and *Mytilus edulis* (shallow water, temperate sea) are com-

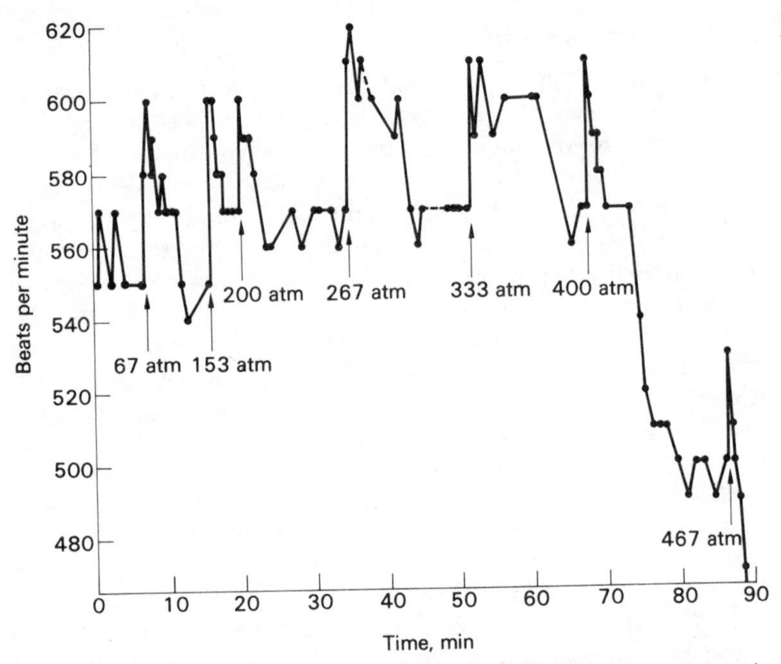

Fig. 4.7. Influence of successive increases in hydrostatic pressure (given in atm) on the frequency of beating of gill cilia of the mussel *Mytilus edulis* L. at room temperature, measured stroboscopically. (After Pease & Kitching, 1939)

pared with similar tissue from *Modiolus auriculatus* and *Chama cornucopia* (shallow water, tropical sea) in Fig. 4.8 which shows the dose-response curves obtained from pressure exposures lasting 6 hours at 22–25 °C. The gill cilia in bivalves from cooler waters are more pressure resistant at relatively high experimental temperatures than are the cilia from warm water bivalves.

Pressure exerts interesting effects on the higher order coordination of cilia in some Protozoa. Ciliary coordination in the Protozoa exhibits a wide variety of responses to pressure (Kitching, 1957*b*). Ebbecke (1935*c*) saw that *Paramecium* was not stimulated by pressure but exhibited a slowing and increased thigmotaxis. In *Spirostomum ambiguum* pressure interferes with avoidance reactions (Kitching, 1969) and the cell has interesting contractile properties which are also upset by pressure. These may be briefly summarised as follows. The whole cell may be stimulated to contract by a flow of electrical current lasting 2 sec. 3 V were

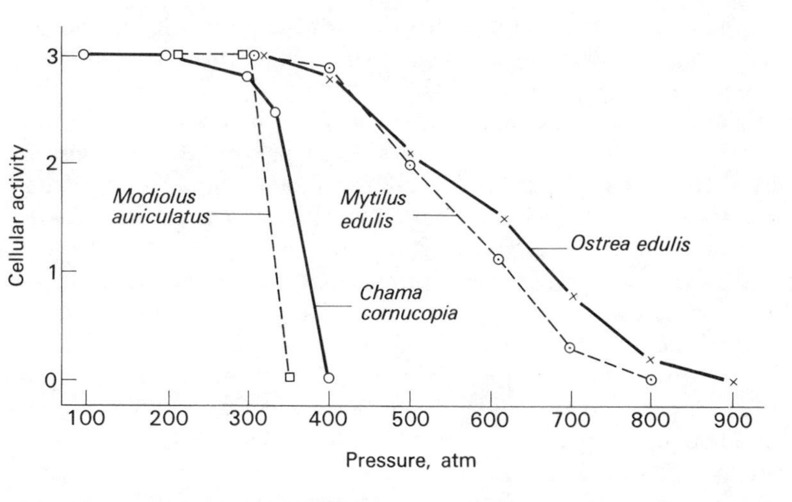

Fig. 4.8. Ciliary activity in selected bivalves scored in arbitrary units (vertical axis) following 6 hours exposure to selected pressures (horizontal axis). (Schlieper, Flügel & Theede, 1967)

normally required to elicit a response in most of the cells in Kitching's apparatus. At pressures of 267–333 atm the same electrical stimulus proved ineffective, a voltage of 10 V or more being required to achieve a comparable response. Thus pressure diminishes the cell's excitability. Higher pressures alone were able to cause whole body contractions; 400 atm or less had no effect on body length whereas pressures of more than 530 atm caused the body to contract by 10–15 per cent of the relaxed length. The relationship of this contraction to that seen in undifferentiated cells needs investigating.

The avoidance reaction is normally seen when the animal swims into an object, whereupon it reverses its cilia, swims back a short distance, and then continues its forward swimming at an angle to its previous path. Electrical and chemical stimuli also cause the cilia to reverse. At pressures in excess of 267 atm *Spirostomum* fails to perform its avoidance reaction in the face of mechanical stimulation, but pressure appears to have no effect on the ciliary reversal induced by a strong and localised concentration of NaCl. Kitching concluded from these observations that pressure affects the tactile receptor whose normal functioning is a prerequisite for ciliary reversal.

Physiological aspects

Spirostomum is also interesting in showing a remarkable change in its tolerance to pressure. Young cultures normally contain cells which divide rapidly and which contain few particles of the calcium salt hydroxyapatite which accumulates markedly in slowly dividing cells in old cultures. In some experiments reported by Bien (1967) 'young' cells were caused to contract by 300 atm whereas 'old' calcified cells showed no change in length at 500 atm and only slight contraction at 1000 atm.

Tetrahymena also shows a diminished ability to carry out its avoidance reaction at high pressure. The sudden application of 267 atm or 530 atm causes *Tetrahymena*, which may be observed swimming in a hanging drop, to circle round the edge. Normally the cells perform an avoidance reaction on colliding with the edge of a hanging drop and as a result become evenly distributed. The circling behaviour ('Red Indian effect') lasts a few minutes and has not been conclusively attributed to pressure *per se*. It is possible, although unlikely, that the abrupt and slight temperature increase caused by compression impairs avoidance. *Tetrahymena* also shows interesting changes in its responses to electrical stimulation at pressure and these have been studied by Murakami and Zimmerman (1970). Cells which normally swim towards a cathode 4 V negative with respect to a nearby anode become attracted to the anode at a pressure of 530 atm.

DYNAMIC ASPECTS AND RESPIRATION

The liquefaction of cytoplasm and the cessation of cell movements at pressure may be partly explained as a shift in an equilibrium of inert molecules constituting a gel, but there are a number of observations which force us to consider additional ways in which pressure affects the gel strength and contractility of cells.

The initial effect of a moderate pressure applied to human cells in tissue culture is the formation of so-called blebs which are small, localised pseudopodia (Landau, 1961). Very soon these disappear as the cytoplasm liquefies and rounds up. When pressure is released an overall contraction takes place before a normal cell shape is resumed. Thus a moderate pressure can stimulate transient contractile activity, and prolonged pressure can cause the build up of some contraction-inducing substance.

Moderate pressures inhibit the preparation for, as distinct from

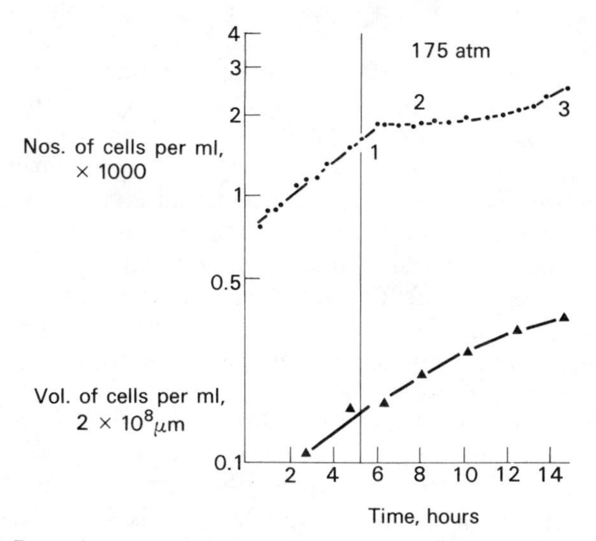

Fig. 4.9. Rate of multiplication, of *Tetrahymena pyriformis* at 175 atm, which was applied after 5½ hours. The lower curve shows volume of cells per ml. Three stages in the response to pressure, 1, 2 and 3, are considered in the text. (Macdonald, 1967)

the execution of, division in the ciliate *Tetrahymena pyriformis*, in the eggs of sea urchins and *Ascaris megalocephala* (Macdonald, 1967, Marsland, 1938, Pease & Marsland, 1939). Fig. 4.9 shows how 175 atm causes successively 1 no effect, 2 a severe inhibition and 3 a diminishing inhibition of the rate of cell division in *Tetrahymena*. The absence of any immediate pressure effect at 175 atm is thought to be due to the pressure being insufficient to weaken the contractile force generated in cells whose cleavage furrow was formed before pressure was applied. The severe inhibition which subsequently sets in (2) is due to pressure inhibiting the preparation for division and not the contractile or cleavage process itself. *Tetrahymena* may be separately observed to cleave at 175 atm in an optical pressure vessel. The gradual recovery of division at high pressure (3) may be regarded as a consequence of internal adjustments in the sequence of reactions involved in preparing the cell for division.

The length of the axopods of *Actinosphaerium* subjected to a constant high pressure undergo a partial extension after an initial shortening, which also suggests some adjustment is taking place in reactions which sustain the normal steady state length of the axopods (Kitching, 1970).

These are three examples of pressure affecting the mechanical activities of cells in a complex physiological way and there are at least four components of cellular metabolism in which pressure may do this: (i) The synthesis, assembly and breakdown of the contractile elements may be differentially affected by pressure as suggested in the case of *Tetrahymena.* (ii) Pressure may also affect energy-supplying reactions which precede the contractile, energy-releasing reaction. (iii) The latter is itself the third potentially pressure-sensitive target. (iv) Pressure may exert indirect effects by upsetting other parts of cell metabolism, particularly processes of ionic regulation.

It is well known that ATP is involved in such cytoplasmic activities as streaming and cleavage. Landau and Peabody (1963) have exposed cells to pressure in an attempt to correlate the cessation of contractile activity with ATP levels. To do this it was necessary to inactivate pressurised cells very rapidly and this was achieved by plunging the pressure vessel containing the cells into a bath at −60 °C. Human cells maintained in tissue culture were used. In the case of FL amnion cells whose normal culture generation time is 17 hours, ATP levels increased to a new steady state at a pressure of 660 atm (Fig. 4.10). Note how at 2 °C the same steady state level is achieved more rapidly. After a 30 min period at 660 atm decompression produced a rapid decline in the level of ATP at 35 °C but not at 2 °C (Fig. 4.11). ADP was constant throughout and no AMP was detected.

Primary amnion cells, whose normal culture generation time was 30 hours and which failed to show both blebbing when initially exposed to pressure and contraction when subsequently decompressed, showed no change in the level of ATP at 660 atm (Fig. 4.12). The curious constancy of ADP led Landau and Peabody (1963) to consider the remote possibility that pressure may cause a new source of ATP to be detected.

The question which arises from the ATP results is, how is the assumed liquefaction of the cytoplasm at high pressure related to the rise in ATP in the cell? In FL amnion cells the relationship looks close. Perhaps it is significant that immersing amoebae and sea urchin eggs in solutions of ATP raises their gel strength. We are left with the vague notion that supplementary ATP leads to an

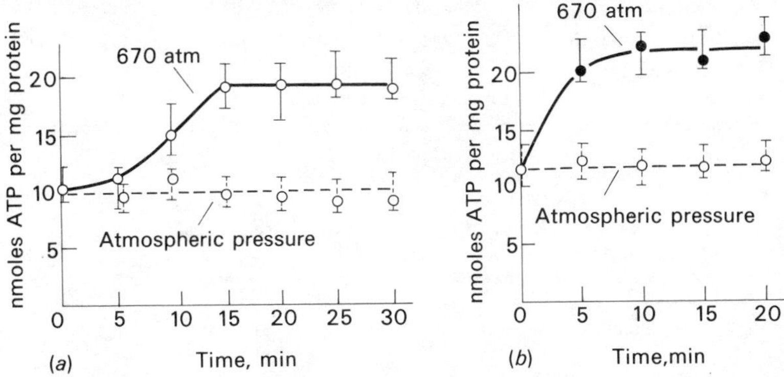

Fig. 4.10 (*a*) The change in ATP level in FL amnion cells under 670 atm hydrostatic pressure at 35 °C. Each point represents the average of 6 experiments with the vertical extensions representing the range. Maximum level is achieved between 10 and 15 min. (*b*) The change in ATP level with time in FL amnion cells under 670 atm hydrostatic pressure at 2 °C. Maximum level is achieved within 5 min. (Landau & Peabody, 1963)

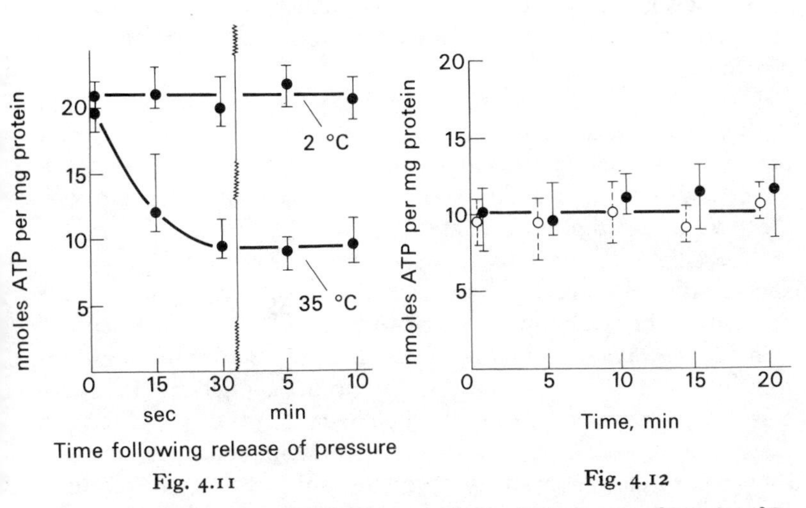

Fig. 4.11

Fig. 4.12

Fig. 4.11. The change in ATP level upon release of pressure at 2 °C and 35 °C. The cells had been previously subjected to 670 atm for 20 min. Note the return to normal level within 30 sec at 35°C. (Landau & Peabody, 1963)

Fig. 4.12. The ATP level in primary amnion cells during pressure treatment at 35 °C. Each point represents the average of 10 experiments. Note that there is little or no change as a result of pressure application. ○, atmospheric pressure; ●, 670 atm. (Landau & Peabody, 1963)

TABLE 4.2. *Rate of oxygen consumption of* Tetrahymena pyriformis *in nutrient broth at selected pressures and* 25 °C. (Macdonald & Gilchrist, 1969)

Means of three experiments

Pressure	Q_{O_2} at pressure measured at temperatures between 25 and 28 °C and calculated for 25 °C, assuming a Q_{10} of 2, independent of pressure (%) precompression rate)	Q_{O_2} after decompression measured at 25 °C ±0.2 °C (% precompression rate)
250	85	94
550	32	44
750	38	11
100	27	11

increase in gel strength, and when pressure reduces gel strength more ATP becomes detectable in FL amnion cells, but not in primary cells.

Nothing is known of the ways in which pressure may directly interfere in the energy releasing reactions at the centre of the streaming and cleaving processes.

Measurements of respiration at high pressure are clearly of interest in this context. They can be carried out on intact and otherwise normal cells by determining the amount of dissolved oxygen which is consumed over a convenient time interval. Techniques for the purpose are dealt with in Chapter 7.

The ciliate *Tetrahymena pyriformis* at a temperature of 28 °C shows a diminished rate of oxygen consumption at pressures above 200 atm. Table 4.2 shows the relative rates of the oxygen consumption of *Tetrahymena* obtained from experiments lasting approximately 20 min in which pressure was rapidly applied and temperature simultaneously measured. The changes in the rate of oxygen consumption were abrupt and showed no oscillations. Pressures up to 550 atm show a partially reversible inhibition of respiration. Pressures of 750 atm or higher show no recovery, with decompression coinciding with the disintegration of the cells.

Although the accuracy of these measurements is not high the results are clear enough. 250 atm, a pressure which immediately inhibits the cleavage of *Tetrahymena* inhibits oxygen consumption by only 15 per cent. Severe pressure treatment, 750 or 1000 atm,

TABLE 4.3. *Some effects of pressure on* Tetrahymena pyriformis *from log phase cultures, in nutrient media at 25–28 °C*

Process	Pressure (atm)	Immediate or short-term effect
Cell division[1]		
(i) preparation for	175	Initial severe inhibition diminishing after several hours
(ii) cleavage process	250	Immediate and total inhibition
Cell growth in mass[1]	175	Unaffected, as determined by the increase in cell volume at pressure
Oxygen consumption[2]	250	15% reduction, reversed on decompression
	1000	73% reduction irreversible – see Table 4.2
Ciliary locomotion[3]	267	Slight reduction in swimming speed, increasingly obvious within 5 min
Reversal reaction[3]	267	Some inhibition of reversal marked
	530	inhibition of reversal

[1] Macdonald (1967) [2] See text [3] Macdonald (1973)

only reduces respiration to about a third but causes lysis. In the short term, aerobic respiration is much more resistant to pressure than the cell as a whole. High pressure lysis may arise from a failure of the cells to osmoregulate and it is significant that at lower pressures the contractile vacuole of *Tetrahymena* fails to discharge, perhaps due to an inhibition of the contractile part of its cycle. The inhibition of aerobic respiration may be a consequence of a reduced rate of ciliary-driven locomotion; thus data in Table 4.2 may reflect the behaviour of the cell rather than any direct effect of pressure on oxidative metabolism, although this seems unlikely. Table 4.3 summarises the differential sensitivity to pressure in some aspects of *Tetrahymena*'s physiology.

Cells from the dog thymus gland and dog lymphocytes show a Q_{O_2} which is little influenced by pressures of less than 300 atm, but pressures above 500 atm cause inhibition of oxygen consumption (Kono, 1958). *Bacillus coli communis* shows a reduction in oxygen consumption which is linearly related to pressure (Fig. 4.13). Bacterial anaerobic respiration at pressure is discussed in Chapter 3.

The first measurement of aerobic respiration at high pressure was made by Fontaine (1929*a*) who used squares of the alga *Ulva*

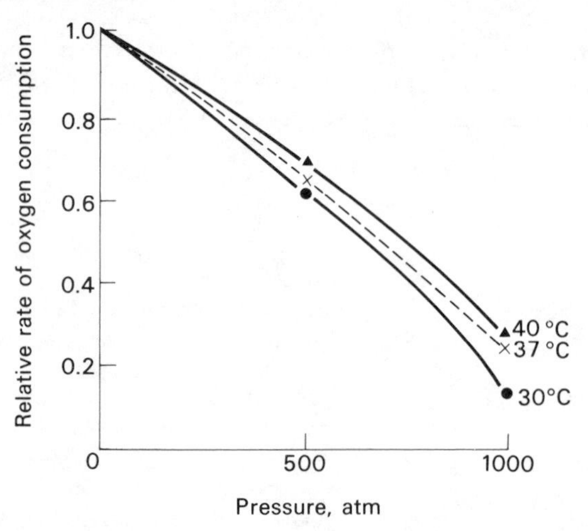

Fig. 4.13. The rate of oxygen consumption of *Bacillus coli communis* at high pressure. (Kono, 1958)

lactuca, a preparation singularly devoid of the behavioural complication possessed by *Tetrahymena*. The quantity of oxygen consumed during a 5 hour incubation at temperatures between 18 and 20 °C was determined by Winkler titration (Table 4.4).

The respiration of some selected tissues of the frog (species not stated) has been measured at high pressure at 24.5 °C by Fenn and Boschen (1969). The experiments involved the incubation of tissue for 3–4 hours at pressure. The P_{O_2} of the test medium was determined by electrode before and after compression, and was normally 1 ATA at the start. Liver showed a reduced oxygen consumption at pressures in the 100–500 atm range, the homogenised liver being approximately twice as sensitive to pressure as intact pieces of liver. The results of other experiments by these authors are given in Table 4.5. Frog skin respiration was unaffected by 100 atm and maximally inhibited by 300 atm (Fig. 4.14(*a*)). It will be recalled that *Tetrahymena* also showed little further decrease in Q_{O_2} with increase in pressure above 550 atm. These results hint at the existence of two components of aerobic respiration, one much more sensitive to pressure than the other.

Intact muscle was the only frog tissue whose oxygen consumption was found by Fenn and Boschen to be stimulated by pressure (Fig. 4.14(*b*)). Kono (1958) has also reported the stimulatory effect

TABLE 4.4. *Oxygen consumption of the alga* Ulva lactuca *at high pressure and* 18–20 °*C*. (Fontaine, 1929*a*)

Pressure, kg cm^{-2} (approx atm)	Inhibition over a 5 hour period (% control rate, expt. series 1)	Inhibition over a 5 hour period (% control rate, expt. series 2)	Rate after decompression (% control, expt. series 2)	Rate next day (% control, expt. series 2)
50	−11	—	—	—
100	−14	—	—	—
200	−13	−13	−6	−2
300	−31	—	—	—
400	−32	−34	−14	+5
500	−44	−46	−10	no change
675	−66	−66	−27	−48
800	−60	−60	−47	−84

Pressures higher than 500 atm caused irreversible damage.

TABLE 4.5. *The inhibition of the oxygen consumption of frog tissues by high pressure*. (After Fenn & Boschen, 1969)

Tissue	*n*	Pressure (atm)	Rate (% control)
Kidney	5	440–570	88 ± 4
Kidney (air)	4	370–460	89 ± 12
Lung	4	370–460	70 ± 5.7
Heart	2	370	39
Muscle – minced	3	440	73
Muscle – ground	1	425	39

n = number of experiments.

of 100–500 atm on the oxygen consumption of, not only excised skeletal muscle, but also heart, cerebrum and kidney from *Rana nigromaculata*. In the experiments of Fenn and Boschen there is every reason to think that the muscles were caused to contract by the applied pressure and that contraction led to an increase in oxygen consumption. The differences between Kono's results and those of Fenn and Boschen may, perhaps, be attributed to differences in the design of the experiments. Kono used an electrode mounted inside the pressure vessel and thereby monitored the rate

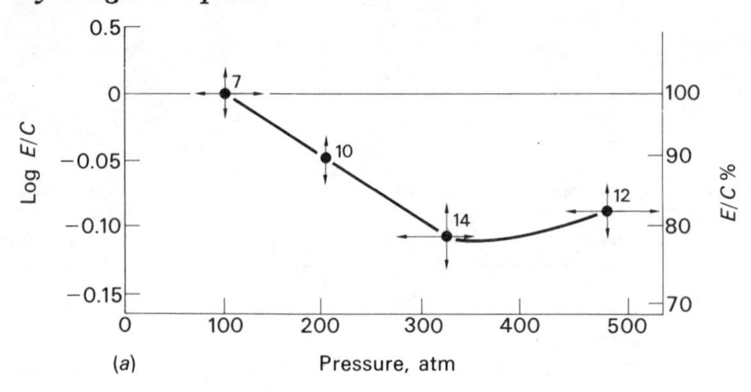

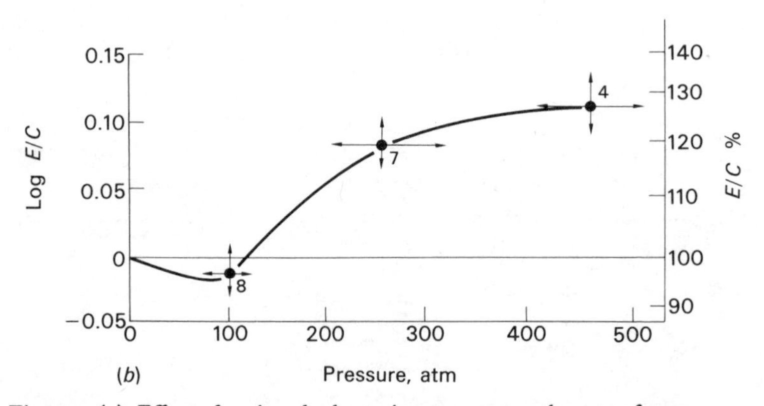

Fig. 4.14 (*a*) Effect of various hydrostatic pressures on the rate of oxygen consumption of frog skin plotted as log E/C where E is the experimental rate under pressure and C is the control rate. The average control rate in 15 experiments was 1.31 ± 0.09 mm^3 per g per min. The experimental rate under pressure is indicated on the right in per cent of the control rate or 100 E/C. Vertical arrows indicate the standard errors and horizontal arrows show the range of pressures averaged together. Figures at each point show the number of experiments averaged for that point.

(*b*) Rate of oxygen consumption of frog muscles under various hydrostatic pressures plotted as in (*a*). (Fenn & Boschen, 1969)

of oxygen consumption continuously. Fenn and Boschen's results are averaged values obtained from P_{O_2} determinations carried out before and after incubation, and may have missed an initial stimulation by pressure. Their experiments raise an interesting problem because the tissues were exposed to a P_{O_2} of approxi-

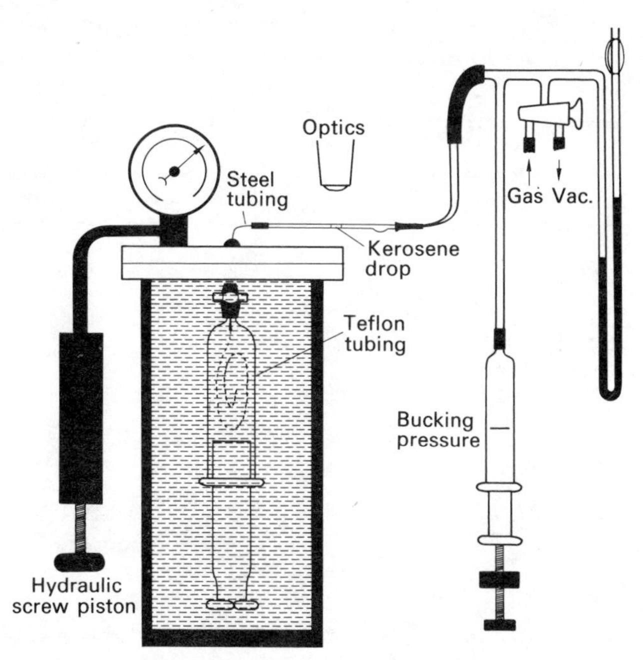

Fig. 4.15. Pressure vessel used to measure the effect of hydrostatic pressure on the equilibrium pressure of dissolved gases. See text. (Scholander & Bradstreet, 1965)

mately 1 ATA before compression. Normally aerobic respiration would be little affected by a P_{O_2} of 1 ATA. Hydrostatic pressure, however, alters the solubility of oxygen and other atmospheric gases, and consequently also alters the equilibrium pressure or activity of the dissolved gas.

This effect has been studied experimentally by Enns, Scholander and Bradstreet (1965) using the apparatus shown in Fig. 4.15. When the water in the pressure vessel is pressurised, previously dissolved gas tends to come out of solution. It diffuses across the wall of the teflon tubing and creates a pressure, the equilibrium pressure, which is then measured by applying a bucking pressure. Water vapour pressure corresponding to the experimental temperature is deducted from the bucking pressure. It will be seen in Table 4.6 that oxygen, nitrogen, argon, helium and carbon dioxide showed similar increases in equilibrium pressure of approximately 14 per cent, per 100 atm.

TABLE 4.6. *The effect of hydrostatic pressure, P, on the equilibrium pressure of dissolved gases, p.* (After Enns, Scholander & Bradstreet, 1965)

Gas	p at P = 102 atm (mm)	p at P = 68 atm (mm)	p at P = 34 atm (mm)	p at P = 0 atm (mm)
O_2	839.5	805	771	734.5
	892	855	819	781
	410.5	390	373	359
	508	484	464	443
O_2 at 0.5 °C	789			682
O_2 in seawater	842	806	775	737
N_2	843	803	773	733
	811	777	744	705
	844	806.5	769	732
N_2 in detergent	811	774	742	712
Ar	849	814	779	741
	839	803	771	734
He	815[a]	779	748	719
	828.5	797.5	765	732
CO_2	817	779	742	705
	849.5	811.5	774	737

[a] P = 102.7 atm.

Two equations are presented by the authors:

(i) $$\ln(p_1/p_2) = \bar{v}(P_1 - P_2)/RT$$

which relates equilibrium pressures (p) to hydrostatic pressures (P) and

(ii) $$\ln(k_1/k_2) = \bar{v}(P_1 - P_2)/RT$$

which relates changes in Henry's Law constant (k) to hydrostatic pressure. Both equations assume $\bar{v}$, the partial molar volume of the dissolved gas, to be little changed by the pressure, which is substantially true.

Extrapolation of the experimental results of Enns *et al.* suggests that the equilibrium pressure of gases dissolved in water which is then compressed to 1000 atm is increased approximately fourfold. The significance of this effect in aerobic respiration is not clear. Water which is equilibrated with air at atmospheric pressure contains oxygen at a partial pressure of approximately 0.2 ATA. At 1000 atm the partial pressure or activity of the gas will be approximately 0.8 ATA. This would seem to be a partial pressure of oxygen which, at atmospheric pressure, exerts little effect. However, 1 ATA of dissolved oxygen becomes effectively 4 ATA at 1000 atm, entering the range of partial pressures of oxygen normally associated with oxygen toxicity at normal atmospheric pressure. Accordingly, in the experiments of Fenn and Boschen, new and unintentional variables may have been introduced. These may still be relevant to deep sea conditions, bearing in mind the gas-secreting tissue in the swimbladders of deep sea fish (Chapter 5).

Oxygen electrodes do not register the increase in oxygen activity arising from hydrostatic pressure as they respond to the diffusional supply of oxygen down a concentration gradient. Biochemical reactions which are similarly sensitive to oxygen supply will presumably be unaffected by the change in oxygen activity. There are only two cases known to the author in which the raised activity of dissolved oxygen has been invoked in a tentative way to account for an effect of hydrostatic pressure. One is on a photosynthetic oxygen-liberating process and the other on seed germination (Vidaver, 1972). We obviously need to know more of this obscure but fundamental phenomenon, and more analytical studies are needed to clarify the pressure sensitivity of oxidative metabolism.

SOME CONCLUSIONS

Deep sea cell biology has a most exciting future. Deep sea Protozoa, and the eggs and tissues of deep sea animals are going to be difficult but rewarding to study, providing their healthy retrieval is possible. It remains to be seen how difficult their successful retrieval is. Conceivably certain tissues may survive for study in animals retrieved in a moribund condition. Of special interest in our present state of knowledge are the properties of cell membranes which, along with intact, healthy animals, may present the most difficult areas of deep sea physiology to investigate in any detail.

Cell membranes

Three areas for study readily identify themselves when cell membranes under high pressure are considered: (i) the structure of aqueous solutions, in bulk and at interfaces, (ii) the structure and passive properties of membranes, and (iii) the metabolic properties of membranes. With such an exciting prospect before the reader it is sad to preface this section with the remark that there is little modern work to report.

SOLUTIONS

Some of the ways in which pressure alters solutions have been described in Chapter 2. An interesting example of how pressure affects the physiology of cell membranes is provided by Podolsky's experiment on cat erythrocytes (Podolsky, 1956). These cells exhibit a Na^+ efflux which declines progressively when Br^-, NO_3^-, and I^- ions are substituted for Cl^- in buffered medium. Podolsky found that 80 atm reduced the rate of Na^+ efflux in KCl media to that found in KNO_3 media at 1 atm. Yet the same pressure exerted no effect in the presence of K^+ acetate, propionate or butyrate, so presumably pressure was not simply affecting the cell membrane. The behaviour of Cl^- ions at 80 atm resembled NO_3^- ions at 1 atm, a phenomenon which Podolsky suggested may be accounted for if pressure changed the effective hydration diameters of the ions, such that NO_3^- at 1 atm matched Cl^- at 80 atm. Podolsky's hypothesis remains to be tested. It is of considerable significance because 80 atm is such a widespread pressure in the oceans. It would probably be better to view the phenomenon as a shift in the selectivity of the membrane for the ions, caused by hydration changes around both the ions in question and the pores through which they move in the membrane. Apart from changes in hydration we can expect pressure to enhance the dissociation of ions and lower the pH of physiological solutions (Distèche, 1972). Further, pressure may alter the structure of water around micro-surfaces, the vicinal water of Drost-Hansen (1972). Such effects might influence the properties of membranes.

EFFECTS OF PRESSURE ON MEMBRANE STRUCTURE

It will be recalled that no gross changes in membrane structure have so far been detected in those cells which have been fixed at high pressure and subsequently examined under the electron microscope (p. 129).

The response of a membrane to hydrostatic pressure is difficult to predict. The generally accepted but generalised lipoprotein sandwich, involving many weak bonds, might be highly susceptible to pressure. The fact is, however, that many cells seem to carry out membrane processes which are little altered at high pressure. From this we may infer that a relatively normal membrane structure is maintained at pressure for a short time at least and membranes are certainly not pressure-labile in the manner of certain microtubules or protein polymers.

Perhaps the first question to ask is, does a membrane behave as a bulk phase, responding to compression by a compaction related to the bulk compressibility of its chemical ingredients? There is indirect evidence that the membranes of both living cells and liposomes respond to compression as a bulk phase (see Chapter 7), but no direct compressibility measurements have been reported. It is possible that, on compression, instead of diminishing in bulk volume, cell membranes may dissociate and 'expand' by hydration. Of course, changes in the molar volumes of the constituent molecules would nevertheless be negative. It will be recalled that micelles at high pressure behave as a bulk phase under some conditions and as a hydrated solute under others (p. 45). It does not seem possible to predict how even a simple membrane will respond to pressure. The teflon membrane used in oxygen electrodes becomes less permeable to oxygen at high hydrostatic pressure, but how is such a permeability change brought about? Should we imagine a sort of compressed sponge with a reduced porosity? Are certain pores occluded by differential compressibilities? Biological membranes at high pressure present a number of basic questions which seem to invite experiments (Macdonald & Miller, in press).

INEXCITABLE MEMBRANES – METABOLIC ACTIVITIES

Okada (1954) reports that pressures of up to 1000 atm increase the electrical conductivity of both red cells and bladder tissue. Unfortunately the brief English summary discloses nothing more of this interesting Japanese work.

Experiments with the frog skin by Brouha, Pequeux, Schoffeniels and Distèche (1970) have brought to light a number of interesting effects of pressure which may have wide significance. The experiments took the form of measuring the potential difference across a piece of frog skin mounted in a pressure vessel. With ordinary

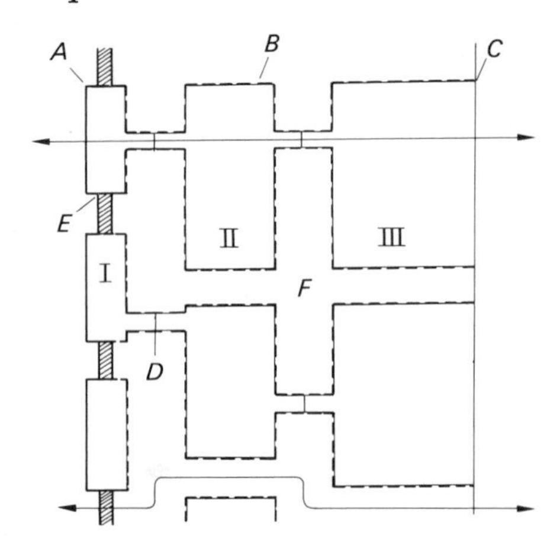

Fig. 4.16. Diagram of the frog skin. I stratum corneum, with *E*, zonulae occludentes; II, stratum granulosum; III, stratum germinivatum. *D*, desmosomes, *F*, extracellular spaces. The external surface of the skin is on the left. (Farquhar & Palade, 1964)

Ringer bathing either side of the skin the observed potential difference, usually at least 50 mV, with the inside surface of the skin positive, approximates to the sum of a sodium diffusion potential and a potassium diffusion potential, both partly shunted by the diffusion of chloride. The skin is layered as shown in Fig. 4.16. An active transport of Na^+ takes place across the cell boundaries, driving sodium into the intercellular spaces and towards the inside surface of the skin. The sodium diffusion potential therefore arises at the outer surface of the skin and chloride ions follow passively; the potassium diffusion potential arises at the inner surface of the skin (Koefoed-Johnson & Ussing, 1958). Fig. 4.17 shows the effect of 100 atm on the skin potential in the presence of ordinary saline. Pressure was applied and removed abruptly at the points *A* and *B*. Compression caused a brief depolarisation followed by a steep and sustained hyperpolarisation. Decompression caused a transient change in potential followed by a slow decline, with an overshoot, to the original potential. Brouha *et al.* interpreted the effect of pressure as an increase in the permeability of the skin to Na^+. When isotonic sucrose bathes the external surface of the skin and ordinary saline the inside surface, pressure causes no hyperpolarisation. The presence of isotonic sucrose on the outside surface reduces the

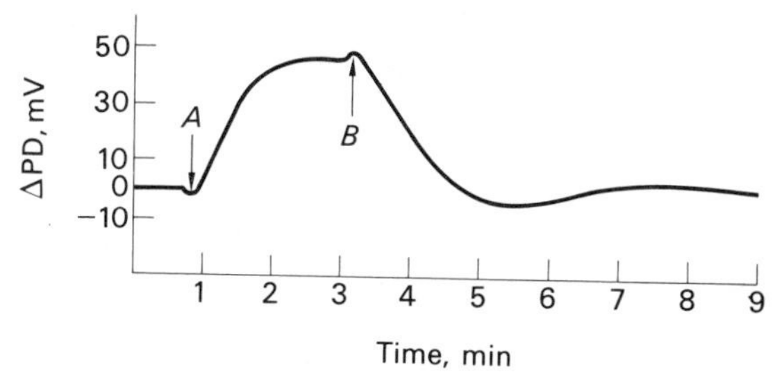

Fig. 4.17. Effect of a pressure step of 100 atm on the potential difference across a skin bathed on both sides by Ringer's solution. (Pressure is applied at *A*, removed at *B*). Horizontal scale, time in minutes; vertical scale, variations of the potential difference in mV. (Brouha *et al.*, 1970)

skin potential by about half and abolishes its sensitivity to pressure, because, according to Brouha *et al.*, the Na^+ gradient at the outer surface of the skin is greatly reduced. A change in the permeability of that surface to Na^+ will not affect the gross potential which is now presumably reduced to the potassium diffusion potential.

Ouabain at a concentration of 5.1×10^{-5} M bathing the inside surface of the skin reduces the skin potential at 1 atm and alters the pressure-induced hyperpolarisation in two ways. The height of the hyperpolarisation is reduced and instead of being sustained it declines quite rapidly. These results are consistent with pressure causing an increase in the permeability of the skin to Na^+.

Oxytocin added on the inside surface of the skin is thought to increase the permeability of the outer layer to Na^+. Oxytocin and pressure act additively, and may therefore work on the same target. Other interesting comments from Brouha *et al.* on their frog skin pressure experiments are (*a*) while the permeability to Na^+ appears to be increased no evidence was obtained to show that Cl^- or K^+ permeabilities were altered, and (*b*) no evidence came to light showing that 100 atm directly influenced the Na^+ transport mechanism. No explanation was put forward to account for the transient effects of compression and decompression. It was also pointed out that a membrane composed of inert pores is not likely to be affected by pressure, and equally unlikely to be able to differentiate between K^+, Na^+ and Cl^- ions.

At higher pressures, Pequeux (1972) found some evidence for interference in the sodium pump and associated enzymes in the isolated gills of the eel, *Anguilla anguilla*. The preparation proved sensitive to pO_2 and to standardise the compression procedure the gills were incubated in pre-oxygenated seawater over a 75 min period. During this period at 1 atm the internal concentration of K^+ ions remained constant, while the internal Na^+ and Cl^- concentrations rose by 45 per cent and 49 per cent respectively. Thus the Na^+ extrusion mechanism fails to hold its own against the diffusion of sodium from the external seawater. The effect of 250 atm over 75 min is to diminish the entry of Na^+ and Cl^- substantially, indicating an increase in the net rate of Na^+ extrusion. 500 atm causes a reverse effect; the entry of Na^+ and Cl^- ions is considerably increased and the tissue loses K^+. This would suggest an inhibition of the Na^+ extrusion which is coupled with K^+ entry.

The distribution of Na^+ and K^+ ions in toad sciatic nerves following pressure treatment has been measured by Gershfield and Shanes (1958) during the course of some pharmacological studies. Table 4.7 shows the extent to which the Na^+ content of nerve increased over 3 hours at 167 atm and 670 atm pressure. The K^+ content of the nerves was reduced over the same period, and to an increased extent at the higher pressure. Note how each of the two pressures which were used produced similar effects, unlike the fish gill experiments.

There seems to be only a very general conclusion to be drawn from the experiments on erythrocytes, frog skin, fish gill and toad nerve. Deep sea pressures exert multifarious effects on the physiology of 'ordinary' membranes and in most cases the work of outlining the effects in a preliminary fashion has only just begun. A similar situation exists in the study of excitable membranes.

EXCITABLE MEMBRANES

The responses of the giant axon of the squid, *Loligo pealii*, have been observed at high pressure by Spyropoulos (1957*a*). Only the short term effects of rapidly applied pressures at temperatures between 10 and 15 °C have been reported.

Some of the properties of the nerve changed in a spectacular fashion while others were hardly affected. Big changes were seen in the duration of the action current, particularly the falling phase, which was prolonged, perhaps by pressure affecting the K^+

TABLE 4.7. *Comparison of the Na^+ and K^+ contents of paired toad sciatic nerves following exposure to 167 atm and 670 atm for 3 hr. X, experimental; C, 1 atm control.* (After Gershfield & Shanes, 1968)

Pressure (atm)	No. of pairs	Na^+ (μmoles (g wet wt)$^{-1}$)			K (μmoles (g wet wt)$^{-1}$)		
		X	*C*	*X*−*C*	*X*	*C*	*X*−*C*
167	6	71.2 ± 2.2	67.7 ± 2.1	$+3.5 \pm 1.0$	38.1 ± 1.3	41.5 ± 1.7	-3.4 ± 0.6
670	6	71.7 ± 2.3	67.9 ± 1.7	$+3.8 \pm 1.2$	29.4 ± 0.8	36.6 ± 0.4	-7.2 ± 0.7

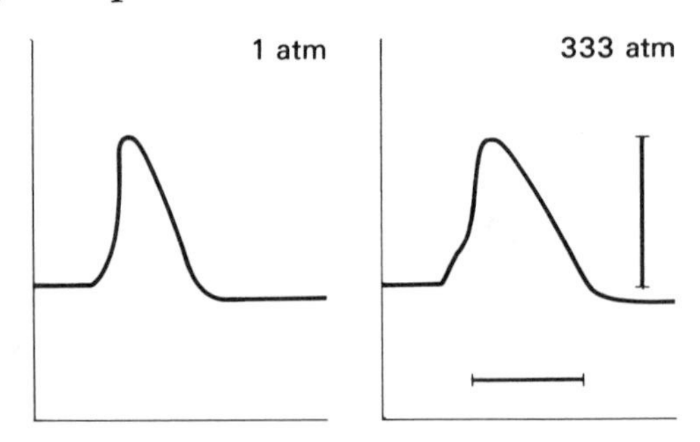

Fig. 4.18. Membrane action potentials of a squid giant axon recorded at atmospheric pressure and 333 atm. Vertical bar at right subtends 100 mV; time marking 5 msec. Temperature 10–15 °C. (After Spyropoulos, 1957*a*)

conductance. Conduction velocity and spike height were little altered (Fig. 4.18). Threshold, which was determined by the application of rectangular current pulses from an internal electrode, declined at high pressure, and reached zero at 200–460 atm depending on individual axons. A reduction in the threshold membrane current was detected by as little as 6 atm pressure. Resting potentials remained unaffected by pressures up to 333 atm beyond which measurements were obliterated by spontaneous spikes. Axons from deep water squid should prove interesting preparations in the light of these findings.

Pressure experiments on single fibres from myelinated motor nerves of *Bufo marnius* show some similarities with the squid giant axon (Spyropoulos, 1957*b*). Pressure increases the duration of the action current and in particular the falling phase (Fig. 4.19). In these measurements, care was taken to exclude temperature changes following compression. Spike amplitude and velocity were little affected. In contrast to the squid giant axon, the myelinated motor fibre showed little change in threshold at pressure.

Grundfest's pioneering study of the amphibian myelinated nerve subjected to high pressure contains much interesting data (Grundfest, 1936). He demonstrated an enhanced excitability at pressure and a gradual diminution in the amplitude of action potentials when a nerve bundle was exposed to 533 atm over a 10 min period (Fig. 4.20). Unfortunately Grundfest immersed the nerves in liquid paraffin and on compression the temperature would have risen by several degrees and dissipated only very slowly. It is therefore difficult to be sure of the conditions of the experiments

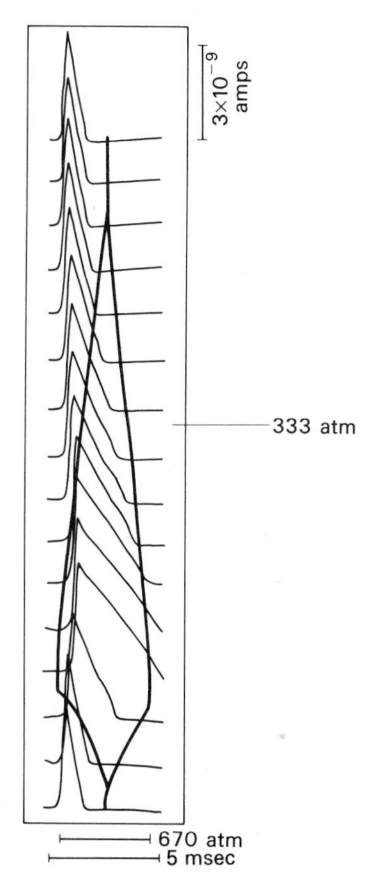

Fig. 4.19. Effects of hydrostatic pressure upon the action current at the node of Ranvier in Bufo motor nerves. The action current was recorded with the 'three-compartment' method. The node was stimulated once every 5 sec and photographs were taken continuously while pressure was varied. One beam (the wedge-shaped traces) of the oscilloscope recorded the pressure; this was the output of a transducer system, the sensitive element of which was placed in the pressure vessel. The other beam gave a record of the response of the node. The pressure was obtained by measuring the separation of the two lines at the point at which they were intersected by the beam recording the response. Temperature: 23–25 °C. (After Spyropoulos, 1957*b*)

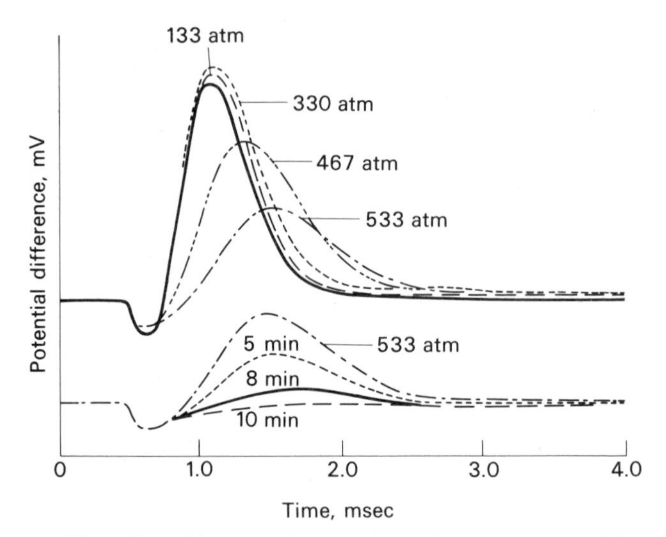

Fig. 4.20. The effect of hydrostatic pressure on frog motor nerve. Two sets of tracing from records obtained on the same nerve.

Lower set. Decrease in the excitability of frog nerve during continued application of high pressure. Strongly supermaximal shocks were used and the shock artifact is the first downward deflection. The uppermost curve shows the spike immediately after applying 533 atm pressure. Then follow responses at 5, 8 and 10 min continued exposure to the pressure.

Upper set. Change in the form of the spike with different pressures. In this nerve, the spike height and duration were increased at 133 atm pressure. In other experiments no change was found until pressures of 267 atm were applied. Lowering of the height and slowing of conduction also occur at different pressures in different nerves. In this case these phenomena become evident at 467 atm, but the variation is from 400 atm to 670 atm for different nerves. The shock artifact appears smaller at 467 atm and 533 atm due to the slowed conduction. Following the main spike at 333 atm and 467 atm pressure there is seen a small elevation caused by repetitive activity. (After Grundfest, 1936)

although the results are in broad agreement with those of Spyropoulos (1957*b*).

While American physiologists were investigating the properties of nerve (Grundfest, 1936) and muscle (p. 156) at high pressure some 40 years ago, so too were the Germans led by Ebbecke. With Schaefer, Ebbecke published in 1935 a number of results relating to the activity of frog nerve and muscle at high pressure (Ebbecke & Schaefer, 1935). They found that pressure reduced the response of ischial nerve bundles to maximal stimulation while sub-maximal stimuli were rendered more effective at pressure, the amplitude of

TABLE 4.8. *The effect of hydrostatic pressure on the frog ischial nerve subjected to sub-maximal stimuli.* (Ebbecke & Schaefer, 1935)

Pressure (atm)	Spike height (mV)	% response
0	2.24	—
100	2.37	+6
200	3.95	+76
400	6.97	+211
500	7.10	+217
0	0.40	—

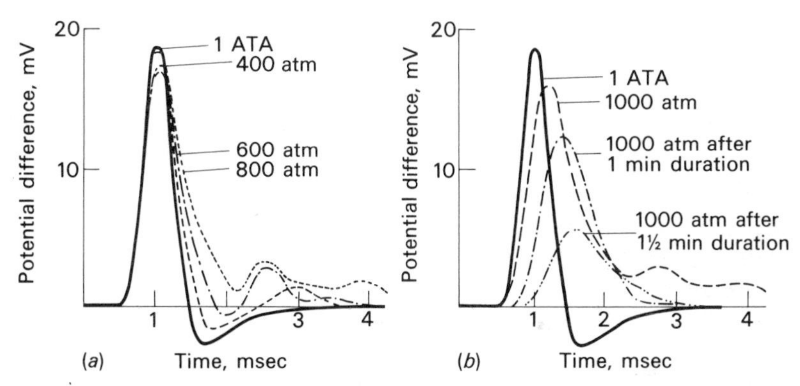

Fig. 4.21. The effect of hydrostatic pressure on the ischial nerve of the frog subjected to maximal stimulation. (*a*) and (*b*) show results from the same nerve; pressures are as indicated. (Ebbecke & Schaefer, 1935)

the action potential elicited being quadrupled by 400 atm. Even 500 atm, a pressure which reduced the response of the nerve to a full stimulus, enhanced the response to sub-maximal stimuli (Table 4.8; Fig. 4.21).

After-potentials were influenced by compression in a complicated way illustrated in Fig. 4.22. In contrast, striated muscle from the same animal showed no double waves in the after-potential, and only a slight decrease in threshold to sub-maximal stimuli at pressure. The amplitude and velocity of action potentials in muscles were reduced by pressure to a greater extent than in nerves (Table 4.9).

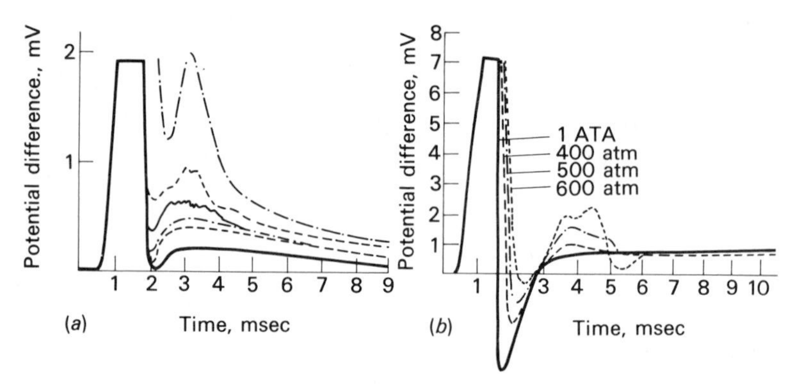

Fig. 4.22. Action potentials and after-potentials from the ischial nerve of the frog subjected to maximal stimulation at high hydrostatic pressures. (*a*) The first spike goes off scale. From bottom to top, 1 ATA, 100, 200, 300, 400 and 700 atm. Note the increase in negativity at 100 and 200 atm, and the series of small waves superimposed on the after-potential wave at 300 atm. (*b*) Pressures as indicated. Note the second wave of activity forming without further stimulation at 600 atm. (Ebbecke & Schaefer, 1935)

TABLE 4.9. *Action potentials from the ischial nerve and sartorius muscle of the frog, subjected to maximal stimulation at high hydrostatic pressures.* (Ebbecke & Schaefer, 1935)

Pressure (atm)	Time of ascending phase (msec)	Total duration (msec)	Latency time (msec)	Maximum spike height (mV)
		A. Nerve		
1	0.51	1.0	0.45	18.6
100	0.51	1.0	0.45	18.0
200	0.53	1.0	0.45	17.4
400	0.51	1.15	0.48	17.1
600	0.51	1.46	0.48	17.4
800	0.51	1.63	0.51	16.6
1000	0.56	1.85	0.53	15.4
1000: 1 min duration	0.56	2.02	0.79	12.1
1000: ½ min duration	0.67	—	1.10	2.3
After decompression	0.53	1.04	0.73	8.4
	B. Sartorius muscle (direct stimulation)			
1	2.75	7.2	6.7	7.8
200	3.0	9.0	7.5	6.3
400	3.25	—	8.7	3.4

Muscle

BULK COMPRESSIBILITY AND VOLUME CHANGES IN MUSCLE

Henderson and Brink (1908) determined the compressibility of rabbit skeletal muscle in saline and compared the results with those obtained for the compressibility of gelatin solution (Fig. 4.23). Unlike cytoplasm and myosin, gelatin solutions undergo an increase in viscosity at high pressure which could mean that intermolecular cross linkages are being formed or, more likely, that an increase in the degree of asymmetry of the molecules is favoured. Despite the complexity of muscle, its compressibility curve closely matches that of gelatin but nothing can be deduced from these data about the molecular changes involved (Fig. 4.24). There is no evidence to show that high pressure disrupts the fine structure of muscle.

The volume changes which occur when frog skeletal muscle contracts have been followed by Abbott and Baskin (1962). A twitch yields the volume change shown in Fig. 4.24. The volume increase may be caused by the excitation-coupling reaction and the volume decrease may reflect changes involving the active state. The relationship between bulk volume changes and the effects of pressure on equilibria and reaction rates has been discussed in the context of glycolysis in Chapter 3, and more generally in Chapter 2.

SKELETAL MUSCLE

Ebbecke was the first physiologist to study muscular contraction at pressure in a twentieth century context although Regnard published the first paper on the subject in 1887. Ebbecke (1935*b*) discovered that a moderate pressure caused an intact muscle to develop contracture. Pressure contracture has since been demonstrated in frog skeletal muscle, the slowly contracting retractor penis muscle of the turtle and in glycinerated preparations of rabbit psoas. The retractor penis muscle, at 4 °C, shows an increase in tension at a pressure of 133 atm and at constant pressure the tension fades to half the initial maximum within 5 min (Brown & Edwards, 1932). Longer compression allows the tension to disappear. In the short term the maximum tension occurs at 533 atm when it is approximately 90 per cent of that elicited by electrical stimulation at the same pressure. No pressure contracture has

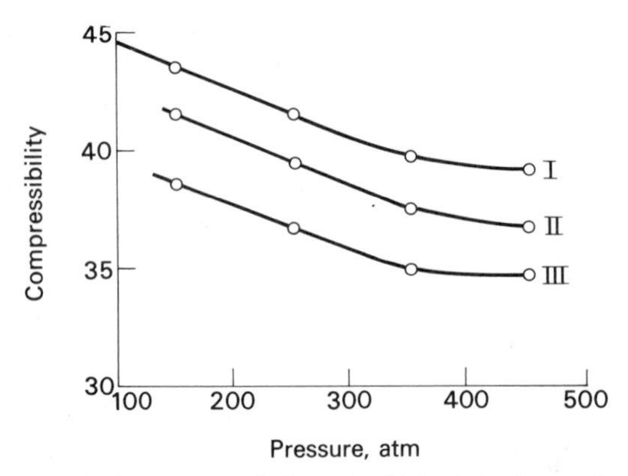

Fig. 4.23. The bulk compressibility of rabbit skeletal muscle. Vertical axis, compressibility in arbitrary units. The compressibility of water averages 42 units between 100 and 500 atm on this scale; horizontal axis, pressure in atm. I, 0.2% gelatin solution; II, 10% gelatin solutions; III, rabbit's muscle. (Henderson & Brink, 1908)

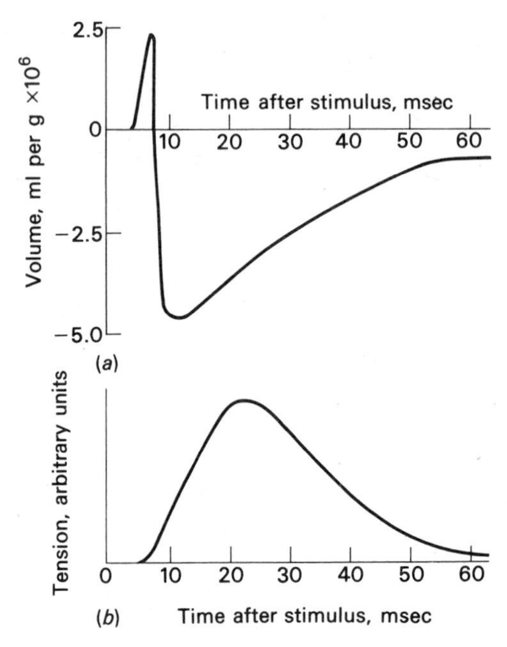

Fig. 4.24. (*a*) Volume changes as a result of a single twitch of a frog sartorious muscle. The muscle was mounted isometrically at reference length. (*b*) Tension change recorded from the same muscle as in (*a*). Temperature 20 °C. (Abbott & Baskin, 1962)

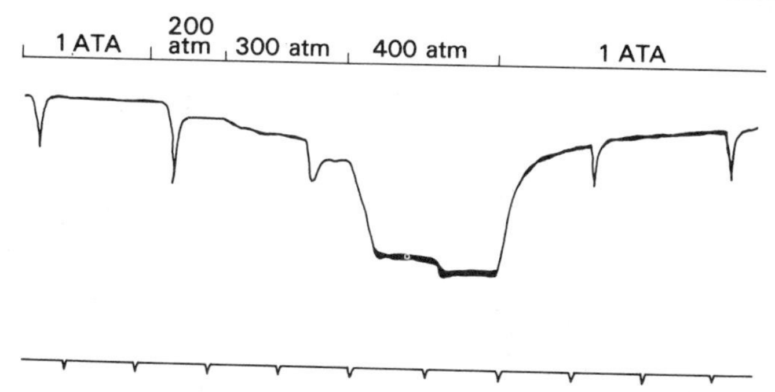

Fig. 4.25. Effect of hydrostatic pressure on the contraction of the frog gastrocnemius muscle. Vertical axis, twitch amplitude at 1 ATA, 200 atm, 300 atm, 400 atm and 1 ATA. Shortening of the muscle is shown by a downward deflection. Horizontal axis, time marks in seconds. Note the increase in the twitch response to stimulation at 200 atm and the decrease at 300 and 400 atm. Note also the incomplete relaxation. (Ebbecke, 1935*b*)

been reported so far in crustacean muscle although when the claw muscle of *Callinectes sapidus* was stimulated at 200 atm it failed to relax to less than 25 per cent of the tension developed (Brown, 1957). Ebbecke (1935*b*) described a similar effect in the frog gastrocnemius muscle stimulated at pressures below the level required to induce contracture (Ebbecke, 1935*b*; also Fig. 4.29). The isotonic twitch shown in Fig. 4.25 is enhanced by 200 atm and fails to relax completely. In contrast, tetanic tension is relatively little affected by pressure.

When a pressure contracture was sustained for some minutes it required a similar recovery period at 1 atm before electrical stimulation had any effect. Conversely electrically stimulated and fatigued muscle responded only weakly or not at all to compression. Not surprisingly perhaps, both types of contraction have precursors in common. Electrical stimulation of the frog gastrocnemius muscle via its motor nerve provided a few cases of enhanced excitability to sub-effective stimuli at 100 atm which were probably due to the nerve and not the muscle (Ebbecke & Schaefer, 1935). At higher pressures the property of the muscle determined the nature of the response to electrical stimulation of its motor nerve (Tables 4.8 and 4.9). At pressures above 300 atm the strength of the direct electrical stimulation of the muscle could be increased to offset diminished excitability. At pressures sufficiently high to diminish muscular activity it was futile to increase the stimulus to

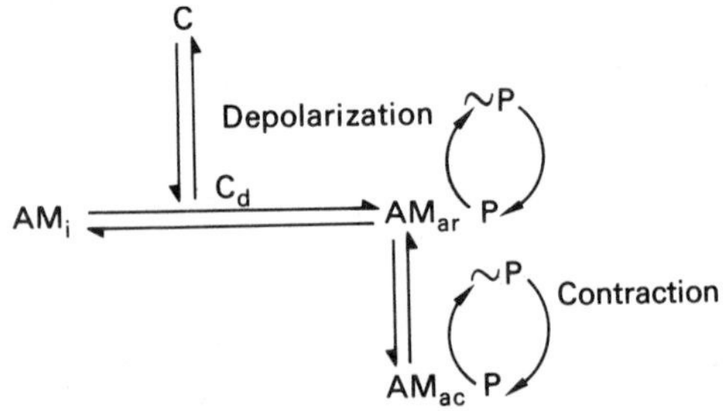

C
Depolarization
$\sim$P
C_d
AM_i
AM_{ar}
P
$\sim$P
Contraction
AM_{ac}
P

Fig. 4.26. Scheme of muscular contraction, showing effects of pressure. AM_i, inactive actomyosin; AM_{ar}, activated contractile unit, relaxed; AM_{ac}, activated contractile unit, contracted; C_d, activator; P, phosphate. (Brown, 1957)

a nerve to get a response. All levels of pressure seemed to slow relaxation.

Brown has more recently proposed a scheme of muscular contraction in which the effects of pressure may be set out and which is shown in Fig. 4.26 (Brown, 1957). Depolarisation causes inactive actomyosin (AM_i) to be converted to the contractile form $AM_{ar}P$ which then shortens to $AM_{ac}P$. It is not the purpose here to discuss the validity of the model but merely to use it as a practical guide. According to Brown (1957), pressure contracture is initiated by a shift in the equilibrium $AM_i \rightleftharpoons AM_{ar}P$. Data in Fig. 4.27 yield a calculated $-\Delta V$ of 350 ml mole^{-1} for the hypothetical reaction. Comparisons with the muscle from deep sea animals would be interesting.

Pressure, as we have seen, can augment the tension developed by electrically stimulated muscle. Ebbecke used an isotonic preparation (Fig. 4.25*a*) while others have used isometric muscles. Using a pressure pulse technique, Brown was able to demonstrate that pressure need only act during the initial tenth of an isometric twitch to yield the extra tension obtained during continuous exposure to pressure. The associated temperature change is probably, but not certainly, unimportant in these rapid compression experiments. These results and other observations led Brown to argue that pressure can affect the active state (alpha state) by favouring the activator C_d (Fig. 4.26). Podolsky (1956) has suggested that as the pressure effect on isometric twitch tension can be mimicked by substituting Cl^- with NO_3^- ions, pressure may act by way of the muscle membrane to prolong the active state.

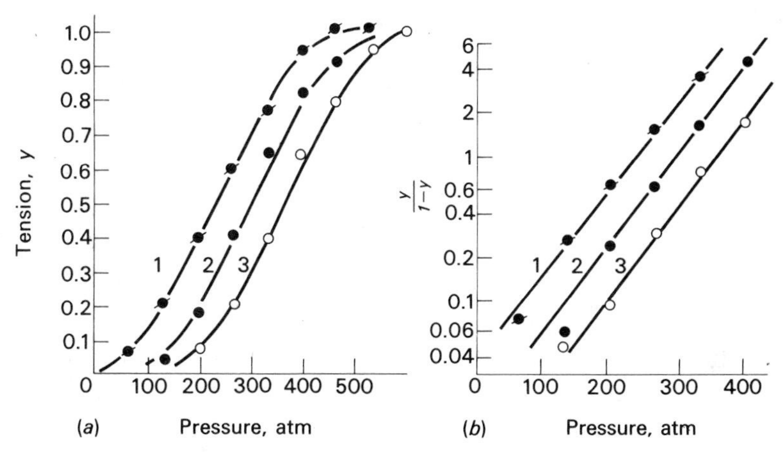

Fig. 4.27. The tension induced by high pressure in the glycerated psoas muscle fibre of the rabbit and in the retractor penis muscle of the turtle. Vertical scale, tension (y); horizontal scale, pressure. Note that in (a) tension (y) is plotted on a linear scale and in (b) $\log_{10}(y/1-y)$ is plotted. Curves 1. 10 mM ATP, pH 5.6, rabbit psoas. Curves 2. Contracture tension of the retractor penis muscle of the turtle. Curves 3. 10 mM ATP, 0.16 mM, pH 6.4, creatinine phosphate, rabbit psoas. (Brown, 1957)

Temperature is an important variable in these experiments. Pressures which augment the contraction of the frog gastrocnemius muscle at room temperatures (130 atm) diminish the tension obtained from a single twitch, and to a lesser extent from a tetanus, at temperatures below 2 °C (Cattell & Edwards, 1932).

Some of the earliest experiments on muscular contraction at high pressure were measurements of heat production and tension during an isometric twitch of the gastrocnemius of the frog *Rana pipiens* (Cattell & Edwards, 1928). Fig. 4.28 shows how heat and tension rise together as pressure increases to 67 atm. Higher pressures cause relatively more heat to be liberated during the development of the twitch tension at temperatures of 6–9 °C. Up to a pressure of 130 atm the tension:heat ratio is constant, both tension and heat increasing some 30 per cent. Between 130 and 330 atm the ratio declines, that is, relatively more heat is liberated for the tension developed (Fig. 4.29). Indeed Cattell (1935) reported that certain individual preparations showed an increased liberation of heat while simultaneously developing less tension. Above 330 atm both the amount of tension developed and heat evolved declined in unison.

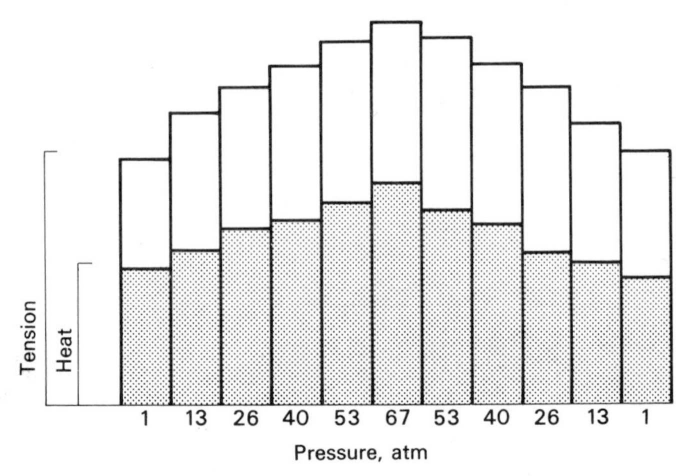

Fig. 4.28. Effect of pressure on the tension and heat produced in a twitch of the frog gastrocnemius muscle. Observations were made on one muscle, first increasing then decreasing the pressure. (After Cattell & Edwards, 1928)

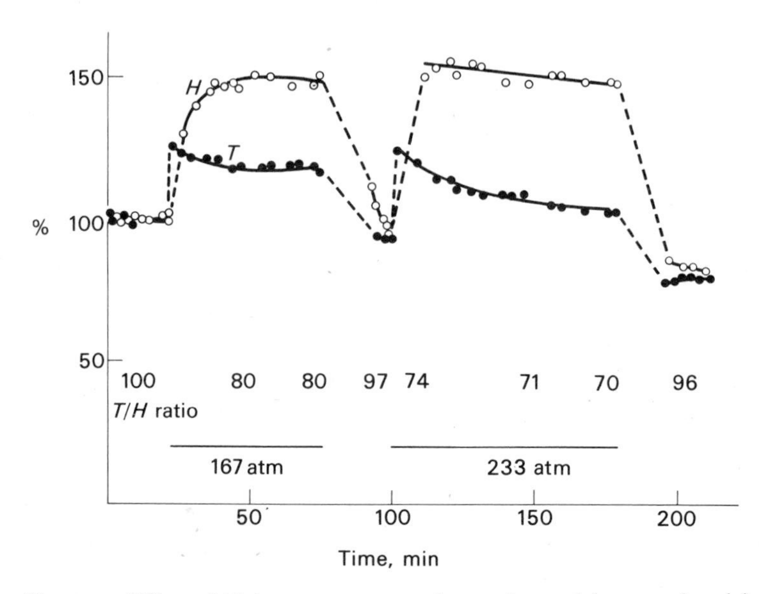

Fig. 4.29. Effect of higher pressures on the tension and heat produced in a twitch of the frog gastrocnemius muscle. (After Cattell, 1935)

TABLE 4.10. *Percentage change in wet weight of* Anguilla *muscles incubated in saline at high hydrostatic pressure*

	Percentage change	
	1 hr	6 hr
600 atm	+5.2	+12.3
Tetanised	+5.0	+8.1
Relaxed, 1 atm	+2.0	+2.5

In summary, in the short term a moderate pressure at low temperature decreases tension. A moderate pressure at intermediate temperatures or above increases the twitch tension of frog muscle without altering the efficiency of the process, while higher pressures bring about a decrease in the efficiency with which the augmented contraction is carried out.

Only one experiment on fish muscle is known to the author. Fontaine (1928) incubated skeletal muscles from *Anguilla* at 600 atm and compared them with electrically tetanised muscles. Both gained weight, presumably by osmosis, and both acidified their medium (Table 4.10). Fontaine concluded that the metabolic state of the pressurised and stimulated muscles was similar. Fenn and Boschen (1969), who found the oxygen consumption of pressurised muscles to be increased, came to the same conclusion (p. 142).

HEART MUSCLE

As the early physiologists recognised, cardiac muscle is a convenient one to deal with because it is self stimulating.

Isolated heart preparations

Using the ingenious apparatus described in Chapter 7, Edwards and Cattell (1928) showed that the application of a pressure in the region of 60 atm enhanced both the rate and amplitude of contraction of the frog heart at room temperature. The effect on the amplitude of contraction became maximally developed within several seconds of applying pressure and tended to diminish over a five minute period. Heart rate was accelerated by only 10 per

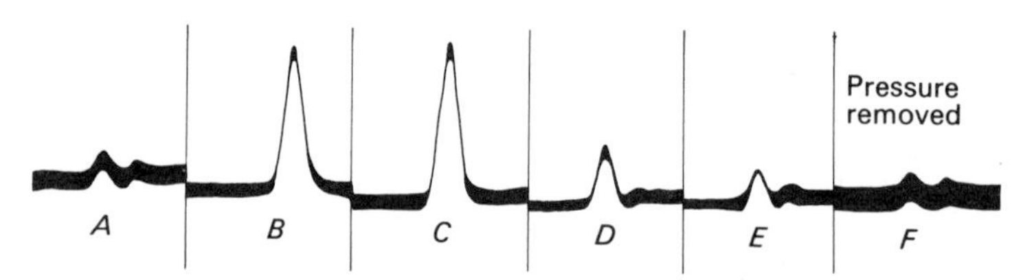

Fig. 4.30. Photokymograph recording of the contraction of the isolated ventricle of the dog fish heart using the apparatus shown in Chapter 7. Each frame was obtained at 1 min intervals. Upstroke denotes contraction. *A*, atmospheric pressure, showing a two-fold wave of contraction which is fused into one intensified wave after 1 minute at 70 atm, (*B*). The effect fades over the next two minutes and in *F* the normal contraction is resumed at atmospheric pressure. (Edwards & Cattell, 1928)

TABLE 4.11. *Tension and time course of contraction of the isolated terrapin heart at high hydrostatic pressure; heart stimulated at 1 min intervals at room temperature.* (Edwards & Cattell, 1930)

	Pressure (atm)		Increase in isometric tension (% pre-compression)	Time course at pressure	
	Mean	Range		Contraction (% pre-compression)	Relaxation
Ventricle	66	64–67	+41.8	+1.5	+9.4
	103	106–100	+68.2	+8.1	+15.6
Auricle	100	96–103	+42	+9.5	+35

cent by the same compression, an observation which has been confirmed by Yasuda (1959). Edwards and Cattell also observed how, in individual hearts which were fortuitously showing two auricular contractions to a single ventricular contraction, the application of 60 atm caused the heart to resume normal symmetrical beating.

Dogfish and turtle hearts showed similar responses to pressure; Fig. 4.30 (Edwards & Cattell, 1928). Further studies by these workers employing the ventricle of the terrapin heart connected to an isometric lever showed how the tension increased at pressure (Table 4.11). The increased isometric tension caused by 67 atm did not involve any significant change in the time required to generate it, but 100 atm, which caused an increase in tension of 68 per cent over the control, involved a longer period of contrac-

TABLE 4.12. *Combined effects of temperature and hydrostatic pressure on the isometric tension of the terrapin ventricle.* (Summary of data from Cattell & Edwards, 1930)

Pressure	Temperature (°C)		
	20	13–12	5
1 ATA	100	125	116
64 atm	135	153	
80 atm	135		116
99 atm	147	147	

Tension expressed relative to that obtained at 1 ATA and room temperature (20 °C).

tion. The time course of relaxation is increased at both pressures. There is something very interesting in the way 67 atm increases the rate at which the muscle develops tension. Analysis of time/tension curves from pressurised ventricles showed that pressure did not favour one particular part of the contraction cycle.

Temperature exerts as an important effect in heart muscle as in skeletal muscle. Reducing the temperature of a strip of terrapin ventricle enhances the tension obtained at atmospheric pressure (Table 4.12). A moderate pressure at room temperature (above 20 °C) also increases tension. Combining the two, 64 atm at 13 °C causes an increase in tension of 53 per cent, which is close to the algebraic sum of the two separate effects. Such an additive effect is not seen at higher pressures. At 5 °C pressure fails to enhance tension at all. Brown (1934*b*) subsequently showed that at temperatures below 5 °C an increase in pressure diminished the isometric twitch tension. Thus temperature alters the sign of the volume change in the rate determining step; at low temperature the limiting reaction determining tension has a $+\Delta V^{\ddagger}$.

Under conditions when pressure enhances the tension in strips of terrapin auricle there occurs an increase in the amplitude and plateau voltage of the action potentials (Edwards & Brown, 1934). At pressures higher than 330 atm (far higher than were used in the original experiments with terrapin heart) the spike height declines and the action potential changes from monophasic to biphasic; tension similarly declines (Fig. 4.31). At pressures in the region of 670 atm electrical activity disappears in $1\frac{1}{2}$ min. Originally,

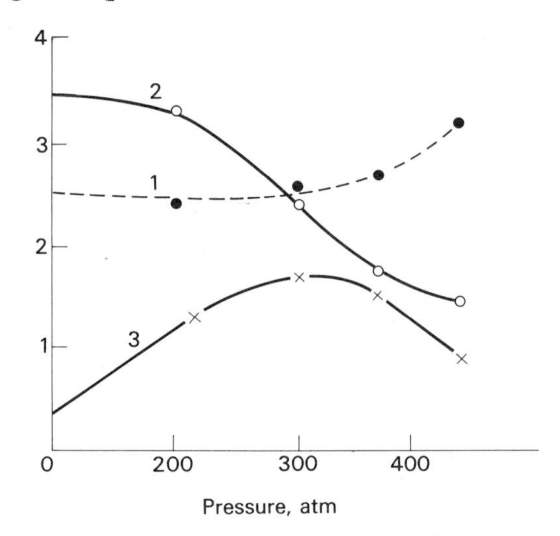

Fig. 4.31. The changes in the rate of contraction (curve 1) and relaxation (curve 2) and the amount of total tension (curve 3) of the isometric contraction of the isolated auricle of the terrapin heart. Values on the ordinate are arbitrary units. (After Edwards & Brown, 1934)

Edwards and Cattell (1928) suggested that a moderate pressure enhanced conduction velocity in the terrapin heart but higher pressures which induce a biphasic action potential may act, according to Edwards and Brown (1934), by slowing conduction. The effect is reversible within limits, as heart preparations which have been inactivated by high pressure resume spontaneous activity after decompression. Yasuda (1959) working with the frog heart (*Rana micromaculata*) found the ventricular action potential to be increased in amplitude at pressure but saw no evidence for a change in conduction velocity.

Cultured heart cells

The effect of pressure on the rhythmic beating of cultured heart tissue from *Rana pipiens* is shown in Fig. 4.32*A*. Pressure was raised in 133 atm steps at intervals of 20–30 min. Note the similarity to the changes in frequency of *Mytilus* cilia subjected to abrupt change in pressure (Fig. 4.7). The effect of temperature on the explant heart tissue is shown in Fig. 4.32*B*; at low temperatures pressure depresses heart rate but at high temperatures it

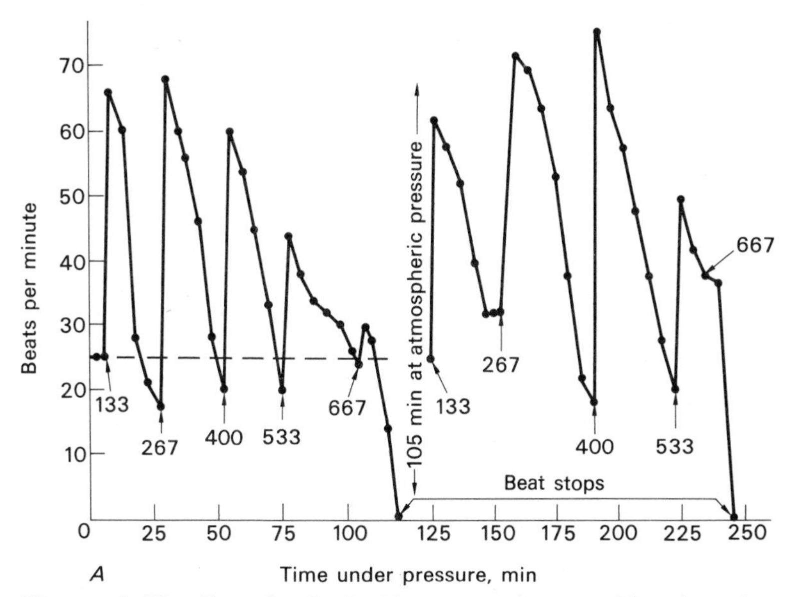

Fig. 4.32*A*. The effects of maintained increments of pressure (given in atm) on the rate of beating of heart cells cultured from *Rana pipiens*. Typical experiment at 25 °C.

accelerates it. Landau and Marsland (1952) account for the transient pressure effects and for the temperature effect in the following manner. Reactants A are converted to B, an excitory substance which induces contraction at a critical concentration.

$$A \xrightarrow{+\Delta V^{\ddagger}} B \qquad E_D \underset{-\Delta V}{\overset{+\Delta V}{\rightleftharpoons}} E_N \xrightarrow{-\Delta V^{\ddagger}} E_{D'}$$

After contraction, B disappears. This main driving reaction involves a $+\Delta V^{\ddagger}$ and is therefore inhibited by pressure when the essential enzyme is saturating as at low temperatures. E_N however may limit the reaction $A \rightarrow B$ at high temperature because the enzyme exists in equilibrium with its thermal denaturation product. Pressure therefore accelerates heart rate at high temperature because it shifts the equilibrium $E_D \rightleftharpoons E_N$ to the right and more

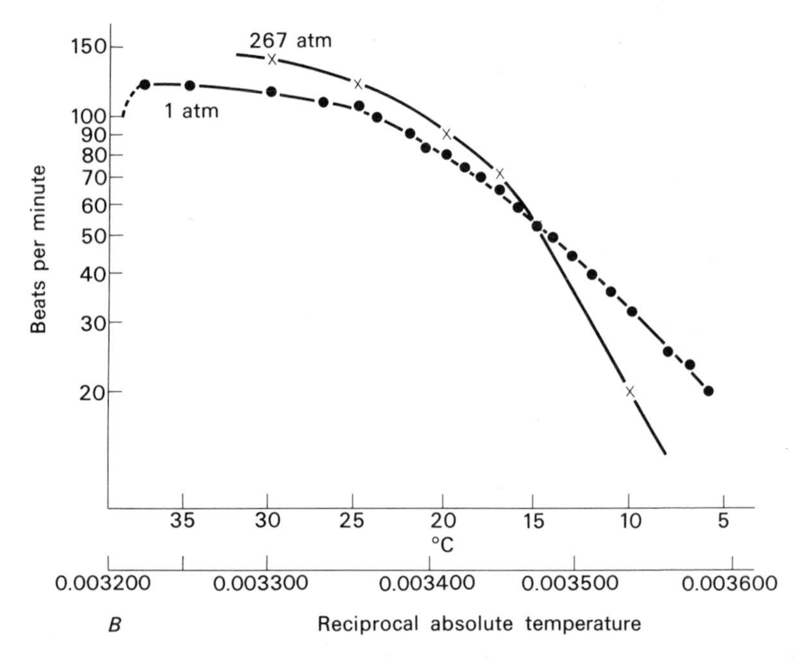

Fig. 43.2*B*. Temperature–pressure effects on the rate of beating of heart cells cultured from *Rana pipiens*. Each point is based on values derived from 6 different heart preparations. At atmospheric pressure it was necessary to select preparations displaying high initial rates of beat in order to cover the whole range of temperature with the same culture. The initial rates of pressurised preparations were multiplied by factors which adjusted them to the level of the atmospheric preparations. For convenience, comparative centigrade values are given in the abscissa. Note that the sign of the pressure effect changes with temperature, below 15 °C the rate is decreased, above 15 °C it is increased, and at about 15 °C pressure has little, if any, effect. (*A* and *B*, after Landau & Marsland, 1952)

than offsets the effect of $+\Delta V^{\ddagger}$ on the main reaction path. At 15 °C pressure exerts no net effect as these two hypothetical reactions cancel out. The irreversible change in the enzyme $E_{D'}$ is required to account for the gradual decline in heart rate seen after compression at high temperatures.

In a later section the behaviour of hearts in intact animals which have been subjected to pressure will be described. Some show transient rate changes following a pressure change while others, including the heart of deep sea Crustacea, do not. The reaction scheme suggests that if a pacemaker mechanism is to acquire insensitivity to pressure as might be expected in a free swimming deep sea animal, reactions A $\rightarrow$ B and to a lesser extent $E_N \rightarrow E_{D'}$ may probably have to change and involve much smaller activation volumes.

SMOOTH MUSCLE

Two brief but conflicting reports have been discovered in the literature. The terrapin stomach preparations used by Edwards (1935) showed a reduced amplitude and rate of contraction when electrically stimulated at pressures of from 12 to 100 atm at 22 °C. More recently Miki (1960) observed that 100 atm increased the frequency and tension of the spontaneous contractions in the isolated frog intestine, but 300 atm or more diminished activity.

DEEP SEA MUSCLES

No physiological studies on deep sea muscles have been reported.

Biochemical investigations on the myosin extracted from deep sea fish were, however, carried out during the *Alpha Helix* cruise to the Galapagos islands in 1970. The effect of pressure on the chemical reactions involved in the contraction of 'normal' muscles has been investigated in several laboratories, principally from the standpoint of using pressure as an analytical tool. The pressure-depolymerisation of myosin (Josephs & Harrington, 1968), acto-myosin, and actin (Ikkai & Ooi, 1969, Ikkai, Ooi & Noguchi, 1966) are clearly important observations and have been mentioned in Chapter 2. The use of pressure in the study of the kinetics of myosin ATP-ase activity (Laidler & Beardell, 1955, Brown *et al.*, 1958) is another basic study. The link between the information obtained from these studies and the identification of the pressure-sensitive component of intact muscle is, however, obscure. It is therefore not surprising that the preliminary biochemical studies of deep sea muscle have not demonstrated any clear cut adaptive features. We should remember that correlations between chemical performance *in vitro* and muscular activity in nature are by no means plentiful. Perhaps the best correlation known is that between contraction velocity (expressed as muscle length per unit time) and myosin ATP-ase activity which Barany (1967) describes in different muscles in the same animal and between muscles from

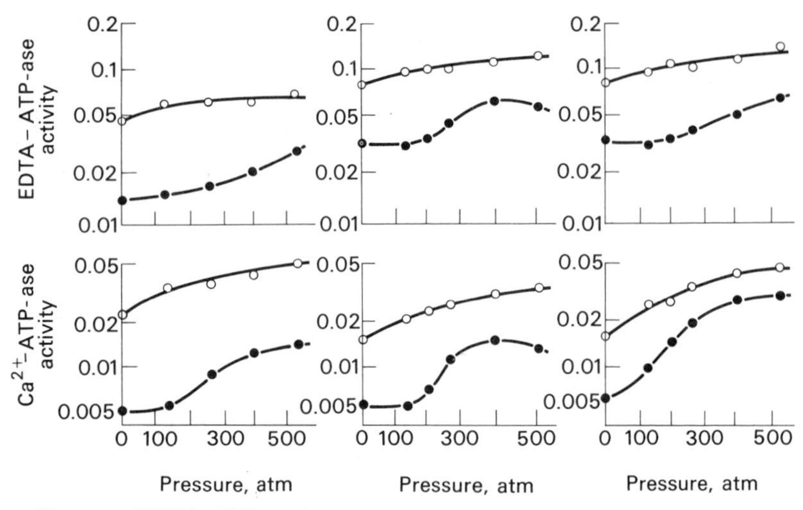

Fig. 4.33. EDTA–ATP-ase (top) and Ca^{2+}–ATP-ase (bottom) of *Coryphaenoides* myosin plotted on logarithmic scale against pressure in atm, for experiments at pH 6.8 (left), pH 7.6 (middle), and pH 8.5 (right). Temperatures: 25 °C (○) and 2 °C (●). (Dreizen & Kim, 1971)

animals as diverse as squid, dogfish and tortoise. Guthe (1969) has extended Barany's findings by showing that the myosin ATP-ase obtained from the muscle of the sea cucumber, *Stichopus moebii*, a slow moving creature like all sea cucumbers, hydrolyses ATP at 6 per cent of the rate observed in rabbit skeletal muscle myosin in comparable conditions. Deep sea cucumbers would be worthy of attention.

Dreizen and Kim (1971) and Kim and Dreizen (1971) extracted myosin and actomyosin from the white skeletal muscle of the deep sea fish *Coryphaenoides* and measured its ATP-ase activity at pressure. Myosin–ATP-ase was activated by calcium or by EDTA. Mg^{2+}–ATP-ase activity was very low and not studied. High substrate concentrations were used which, according to the kinetic arguments previously set out in Chapter 2, suggests that the activation of the enzyme–substrate complex is the rate-determining step. Pressure increased the ATP-ase activity under most conditions (Fig. 4.33). The physiological significance of high substrate concentrations is not clear but the activation volumes ($\Delta V^{\ddagger}$) for Ca^{2+}–ATP-ase and EDTA–ATP-ase are listed in Table 4.13 from Dreizen and Kim (1971). Note the values are calculated

TABLE 4.13. *Activation volume, $\Delta V^{\ddagger}$, in cm^3 $mole^{-1}$, for* Coryphaenoides *myosin ATP-ase at* 0 *to* 134 *atm pressure.* (After Dreizen & Kim, 1971)

pH	Ca^{2+}–ATP-ase		EDTA–ATP-ase	
	2 °C	25 °C	2 °C	25 °C
6.8	−14	−66	−16	−44
7.6	0	−57	0	−36
8.5	−20	−74	0	−35

from two points over the pressure range 0–134 atm whereas *Coryphaenoides* was caught at 220 atm pressure and probably frequents somewhat greater pressures. A $\Delta V^{\ddagger}$ of zero is probably of some adaptive significance but the curves rising at pressures above 200 atm at pH 7.6 (Fig. 4.33) are puzzling. *Coryphaenoides* appears to live at a pressure in which its myosin ATP-ase is highly susceptible to pressure changes, at least under test conditions.

Actomyosin–ATP-ase is probably of greater physiological significance. Fig. 4.34 shows the activity of the Ca^{2+} and Mg^{2+}-activated enzymes at high pressure. The plateau at pressures below 130 atm in the curve for Ca^{2+}-activated ATP-ase at 2 °C is striking and it may be of adaptive value but *Coryphaenoides* lives at pressures beyond the plateau.

Both myosin– and actomyosin–ATP-ase activity is low under all conditions investigated. *Coryphaenoides* has been filmed at depth where it seems to move sluggishly. The white muscle in *Coryphaenoides* from which the enzymes were extracted is a muscle which, in shallow water fish, undertakes short bursts of rapid contraction. The sluggish activity of *Coryphaenoides* muscle is also reflected in the low activity of its cytochrome oxidase (Whitt & Prosser, 1971). Its ionic content resembles that found in other cold water fish, in being rich in Na^+ and poor in K^+ ions and differs from that of surface, warm water species (Table 4.14). The authors of Table 4.14 suggest that the cessation of the blood circulation during the recovery of the deep sea fish would not have seriously affected their data.

A number of deep sea fish have been reported to have a flabby or feeble musculature (Blaxter, Wardle & Roberts, 1971). Factors in the muscular development of such fish, for example the life-

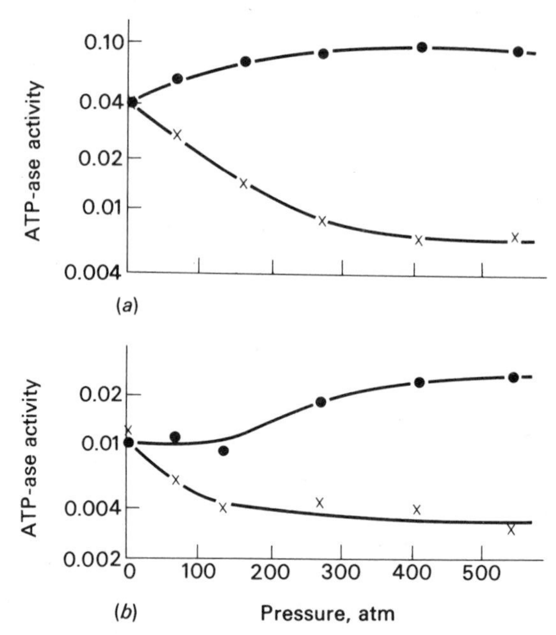

Fig. 4.34. Pressure dependence of *Coryphaenoides* actomyosin ATP-ase in μM P_i per min per mg. Mg^{2+}–ATP-ase (×) and Ca^{2+}–ATP-ase (●) in 0.5 M KCl, 0.025 M Tris, pH 7.6, plotted on logarithmic scale against pressure (atm) for experiments at (*a*) 25 °C and (*b*) 2 °C. (Kim & Dreizen, 1971)

TABLE 4.14. *Sodium and potassium concentrations in the muscles of oceanic fish.* (Simplified from Whitt & Prosser, 1971)

	% dry weight	Concentrations of Na^+ (mmole per kg dry wt muscle)	Concentrations of K^+ (mmole per kg dry wt muscle)
Mugils, genus not stated (shallow water)	19.8 ± 1.7	143.6 ± 9.1	636.6 ± 41.3
Rat tails, *Coryphaenoides* sp. (2000 m)	12.8 ± 2.1	316.0 ± 48.2	479.0 ± 68

style, nutrition, age, or breeding cycle, may exert as much influence on muscle biochemistry as high pressure. Electron micrographs of the white skeletal muscle of the *Coryphaenoides* have been prepared by Herman and Dreizen (1971). The familiar organisation of striated muscle is apparent but the muscle possesses rather short sarcomeres with thick filaments averaging 1.4 μm and thin filaments 0.7 μm in length. The suggestion that short filaments are a result of the high ambient pressure present during their assembly is probably an over simplification.

Pressure, in shifting a monomer ⇌ polymer equilibrium to the left, might conceivably reduce the mean length of a population of filaments seeded *in vitro* but it is unlikely that filament morphogenesis is such a simple process. If so, it appears a curiously labile, maladapted feature which would be suceptible to damage if the animal transgressed a certain rather limited depth range. The suggestion of Herman and Dreizen provides a good example of the problems of interpretation in molecular studies of deep sea organisms. Kinetic analysis of deep sea muscle ATP-ases and other deep sea enzymes are beginning to bring to light features in which they differ from other enzymes. Setting aside the technical difficulties of making valid comparisons between enzymes, what criteria should be adopted in deciding whether or not a difference is adaptive?

Different attitudes to this problem can usually be distinguished among biologists. One view is that all differences are adaptive since all features of an organism are, by definition, adaptive. Alternatively, one may argue that a particular feature is only adaptive if good evidence exists to show it favours that organism's performance in nature. Some features are more obviously related to the organism's environment or its way of life, than others. These we call adaptive. Other features are not regarded as adaptive (or anything else) until their special relationship to the organism as a whole is understood. The latter view is also subscribed to by the author and is probably typical of most experimentalists' outlook. The term 'adaptation' assumes more importance when the question of 'how adaptive?' is tackled; this lies a long way in the future in deep sea biology.

Intact animals

INTRODUCTION

One approach to the problem of identifying adaptive features in physiology is to study the intact organism. In situations in which the nervous system is particularly important, as for example in the short term response of an animal to pressure, the intact polysynaptic system may show different responses from those seen in isolated preparations. In particular, the intact organism may possess compensatory mechanisms which are absent in an isolated organ. Experiments with intact animals will help to identify some of their pressure-sensitive components and define sensible experimental conditions for subsequent analytical studies.

As the pioneer pressure physiologists observed, subjecting an animal to hydrostatic pressure elicits changes in behaviour and locomotor activity. Some marine organisms respond to such small pressures that it is reasonable to postulate the presence of a special pressure-sensing system. The distinction between an animal's sensory response to pressure and a response brought about by pressure affecting the normal functioning of the animal is by no means clear. Accordingly the responses of animals to small applied pressures is considered as a preamble to the effects of high pressures.

RESPONSES TO VERY SMALL INCREASES IN HYDROSTATIC PRESSURE

A wide variety of marine animals are sensitive to an increase in hydrostatic pressure of less than 1 atm. Those animals which possess a bulk gas phase, such as a swimbladder, would appear to possess an organ well suited to providing sensory information about pressure. Animals which are sensitive to small pressure changes and lacking an obvious gas phase are therefore of particular interest. How, for example, do Crustacea sense the very small pressure changes observed by Hardy and Bainbridge (1951) and, most recently by Digby (1972) and Lincoln (1971)? Digby has proposed that the cuticle in Crustacea and perhaps other tissues in other animals, behave as a semiconductor, shunting the electrical potential which exists between the external seawater and the animal's body fluids. By analogy with model electrodes, Digby argues that a flow of current through the semiconductor will promote, by electrolysis, the formation of a steady state hydrogen gas layer on the external surface of the cuticle. Such a hypothetical layer would be too thin to see but, it is postulated, would serve as a pressure sensitive element. Exactly how changes in the thickness or extent of the hypothetical layer of hydrogen relays information about pressure to the animal's nervous system is a question which will have to be taken up seriously once convincing evidence for the existence of the layer of hydrogen is established. It appears an exceedingly difficult entity to isolate, being volatile, externally placed and less than about 0.3 μm thick.

The more general question arises, how might a high hydrostatic pressure affect the electrolysis phenomena postulated by Digby? Is it a transducing system limited by electrochemical laws to work at a low ambient pressure, or can current flow be adjusted to sustain an electrolytically generated layer of gas at high ambient pressure? We simply do not know if deep sea animals can detect small pressure changes in an ambient pressure of several hundred atmospheres. One might imagine that the ability to do so would be important to free swimming mid-water animals and, if such a sensory capacity can be demonstrated, the whole question of low pressure transduction should be revitalised. In the course of an investigation into the effect of moderate pressures on the laboratory scale vertical migration of *Calanus heligolandicus*, Lincoln (1971) discovered that a pressure change of 0.6 ATA or 1.6 ATA at an ambient pressure of 21 ATA caused the animals to undertake a brief upward swimming excursion. In experiments of this type it is always difficult to eliminate all extraneous sensory cues to which the animals may respond, but in this example it is reasonable to invoke the change in pressure as the most likely cause of the animals' swimming response. The observation implies that the hypothetical electrolysis mechanism of pressure-sensitivity works at pressures in which the tendency for gas to dissolve is increased by a factor of 20, although the rate of solution of gas is probably determined by factors other than partial pressure, such as stirring and the area of the gas–water interface.

We should not overlook the observation of Spyropoulos (p. 154) who found 6 atm was sufficient to cause detectable changes in the threshold membrane current of giant axons of the squid.

SURVIVAL OF SHALLOW WATER ANIMALS AT HIGH PRESSURE

P. Regnard was the first man to observe animals contained within a high pressure vessel (Chapter 1) but Ebbecke's observations (1935*a*) established the stimulatory effect of a moderate pressure (< 200 atm) and the inhibitory effects of higher pressures. Ebbecke observed the animals listed below.

Physiological aspects

Actinia equina (Anthozoa)
Sarsia sp. (Hydrozoa)
Cyanea capillata (Scyphozoa) excised sub-umbrella tissue
Beröe (Ctenophora)
Tomopteris helgolandica (Polychaeta)
Ophiuris fragilis (Echinodermata)
Echinus esculentus (Echinodermata)
Littorina obstusata (Gasteropoda)
Shrimps and mysids (Crustacea)
Branchiostoma lanceolatum (Hemichordata)
Fish – various (Vertebrata)

Examples of Ebbecke's observations include the tentacles of *Actinia* moving about at 100 atm and increasingly so at 300 atm, but retracting at 400 atm. The rhythmic beating of the excised subumbrella tissue from *Cyanea* was accelerated at pressures up to 200 atm, thereafter beats were inhibited. It is difficult to imagine an agitated *Echinus esculentus* but Ebbecke reports that 500 atm was insufficient to diminish the pressure-induced activity. *Beröe* also responded in an interesting way. Pressures up to 150 atm inhibited comb plate (compound cilia) activity and at 300 atm the body became wrinkled. These observations prompted Ebbecke's physiological experiments which have already been mentioned.

Workers at the Marine Institute in Kiel, Germany, have measured the survival of a number of North Sea and Baltic Sea animals at high pressure (Naroska, 1968). After 1 hour at pressure the animals were allowed 24 hours to recover and their heart beat was then used as a criterion of recovery. The pressure which killed 50 per cent of the animals (LD_{50}) is tabulated in Fig. 4.35. The low osmotic pressure of the dilute (Baltic) seawater complicates the interpretation. One would like to know the extent to which the data in Fig. 4.35 are a reflection of osmoregulation at pressure particularly as *Gammarus oceanicus* tolerates increased pressures for longer periods in progressively higher osmotic pressures (Fig. 4.36). *Mytilus* gill tissue equilibriated in Ca^{2+}-enriched dilute seawater shows an increased pressure tolerance which may be due to the Ca^{2+} affecting the permeability of tissue to water (Ponat, 1967).

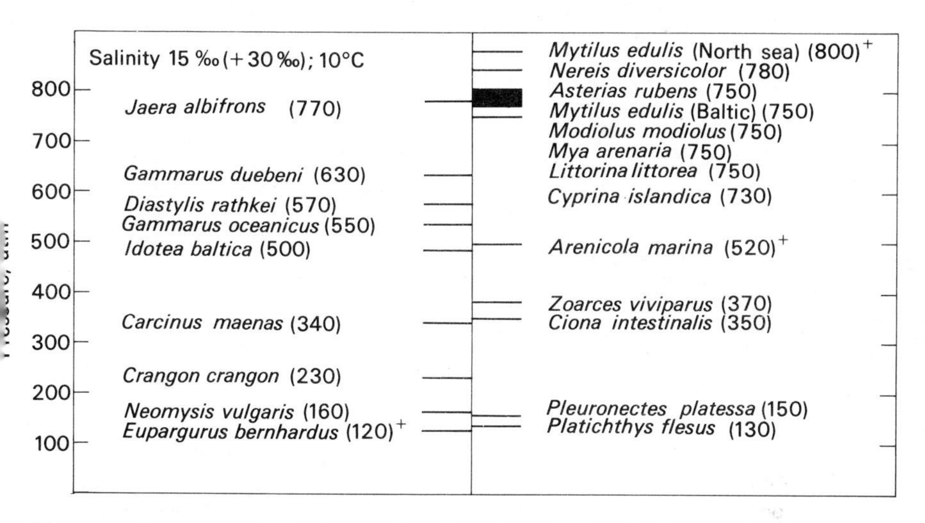

Fig. 4.35. Pressures, which when applied for 1 hour, kill 50% of a sample of the animals. Note seawater of salinity 15‰ was used in most cases. Temperature: 10 °C. (Naroska, 1968)

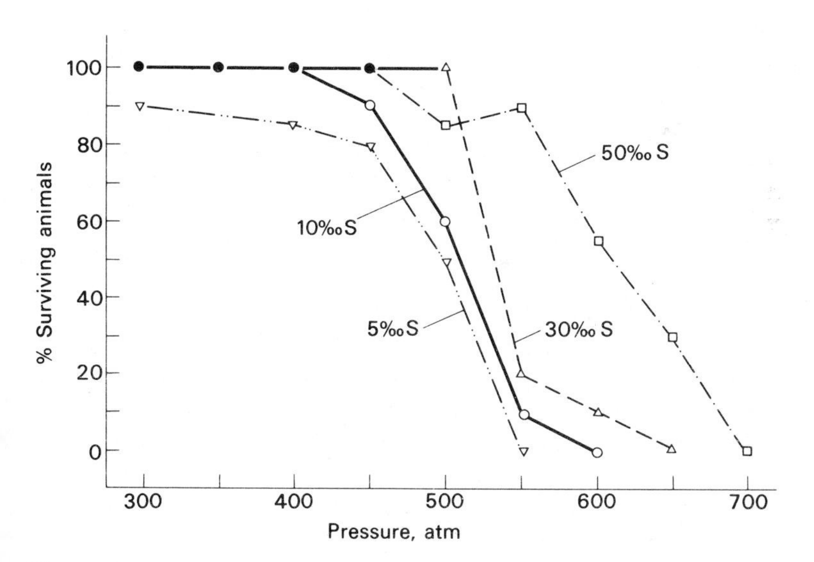

Fig. 4.36. Survival of *Gammarus oceanicus* at high pressure and at different salinities. Groups of animals, adapted to the appropriate salinity, were exposed to selected pressures for 1 hour. Temperature: 10 °C. (Naroska, 1968)

ACTIVITY AND RESPIRATION OF SHALLOW WATER ANIMALS AT HIGH PRESSURE

Fontaine (1929*b*) carried out some of the earliest quantitative experiments on the effect of pressure on intact animals. By measuring the concentration of oxygen in seawater by the Winkler technique, the oxygen consumption of small *Pleuronectes platessa* was determined over 20 min and 60 min periods of incubation at high pressure (Table 4.15). Fontaine (1930) also described how *Pleuronectes* exposed to 125 atm for 20 min showed an increased ventilatory rhythm; the experimental temperature was not recorded. Experiments at 100 atm on the following species also showed an increased rate of respiration.

Observations on the locomotor activity of a variety of animals at pressure (see below) suggest that respiration is probably stimulated through exercise and not by a direct effect of pressure on aerobic metabolism. Although Fontaine did not actually observe his experimental animals at pressure some of his results are consistent with pressure exerting a locomotor effect. He found, for example, that when three small *Pleuronectes* or *Palaemon* were incubated together at pressure, they showed a greater increase in respiration at 100 atm than those animals which were incubated individually and which may have lacked the stimulation of other occupants.

Some animals reveal a considerable degree of normality when exposed to quite high pressures for long periods (Naroska, 1968). Animals such as *Psammechinus miliaris* (Fig. 4.37) which normally lead an inactive existence sometimes show little change in Q_{O_2} at pressures which either stimulate (100 atm) or inhibit (300 atm) the respiration of others (Fig. 4.37, *Nereis, Carcinus*). No control measurements showing the contribution of microbial oxygen consumption were described for these experiments. Dead animals or crustacean exoskeletons can yield a low Q_{O_2}. It is also unfortunate that a steady rate of respiration at atmospheric pressure is not reported from these experiments; diurnal rhythms of activity may be affecting the results. For example is the decline in the Q_{O_2} of *Carcinus* or *Nereis* at 200 atm due to pressure or some diurnal factor (Fig. 4.37(*b*))? Note also the use of dilute seawater in some of the experiments.

Naroska's extensive investigations include some measurements

TABLE 4.15. Pleuronectes platessa *oxygen consumption at high pressures.* (Fontaine, 1929*b*)

Pressure (atm)	% precompression rate over 20 min exposure	% precompression rate over 60 min exposure
25	—	+28.0
50	+20	+38.8
100	+27	+58.0
125	+30	+54.0
150	−50 (dead)	−39.0 (dead)

TABLE 4.16. *Effect of* 100 *atm pressure on the respiration of selected marine animals.* (Fontaine, 1928*b*)

	% increase in oxygen consumption at 100 atm	Duration of pressure (atm)
Ammodytes lanceolatus	+27	(25)
Gobius minutus	+53	30
Crangon vulgaris	+114	60
Palaemon serratus	+35	30

of the effects of pressure on the heart rate of *Gammarus* and *Ciona* (Fig. 4.38(*a*), (*b*) (*c*)). Note how an increase in pressure causes a transient change in heart rate in both animals with *Gammarus* showing acceleration at 5 °C (also at 21 °C, not shown) and *Ciona* a slowing down at 21 °C. Animal 1 in Fig. 4.38(*b*) shows a spectacular change in heart rate immediately following the application of 200 atm but a near normal rate is resumed within five minutes. The similarity to the changes in ciliary frequency following compression is striking.

The activity of fish at high pressure has been observed by few workers. Nishiyama (1965) has described how pressures of 30 atm or less caused *Salmo irideus* (rainbow trout) and *Carassius auratus* (goldfish) to swim in an upright posture and on occasions to drop to the bottom of the experimental tank. *Misgurnus anguillicaudatus* (loach) underwent seizures. In all three species the opercular rhythm increased and became irregular. However, these activity changes may have been enhanced by changes in the blood–gas

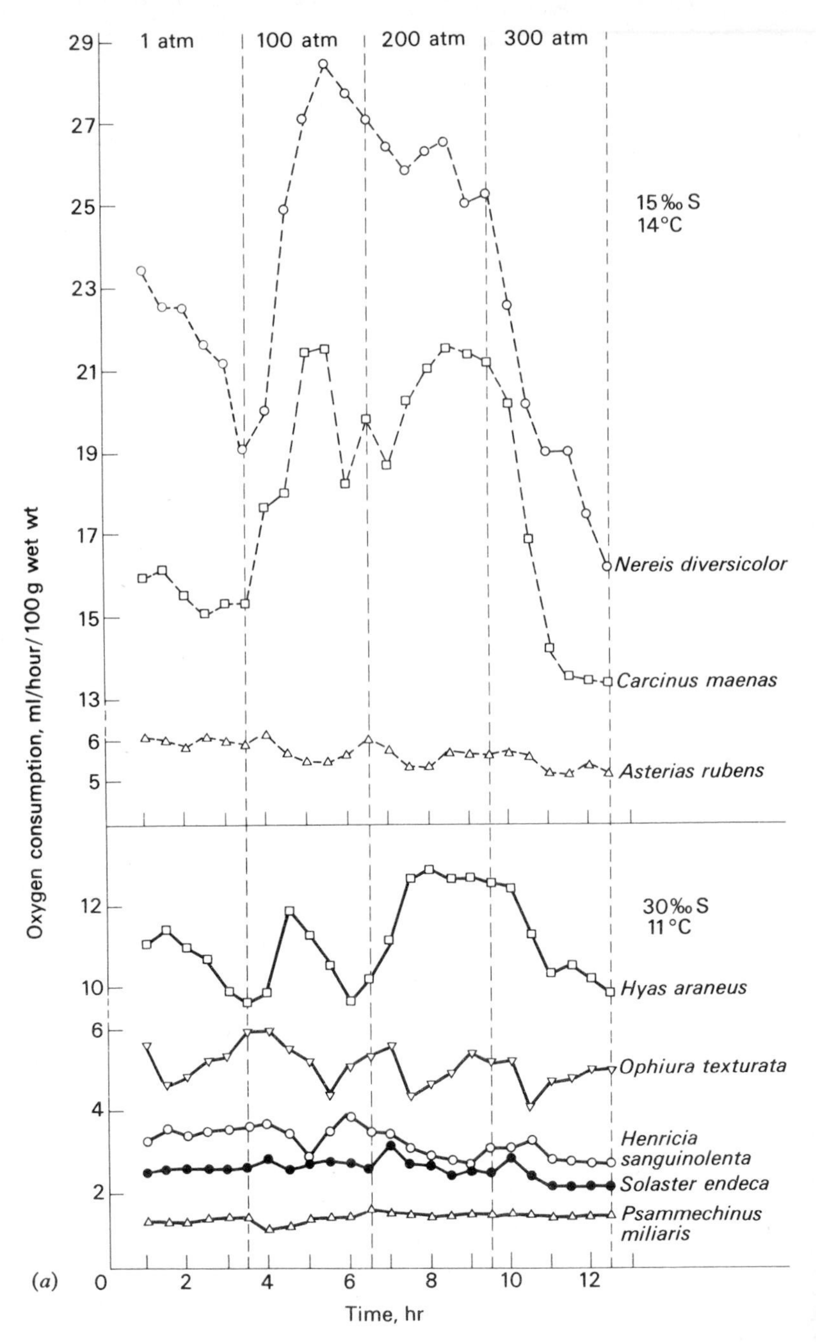

Fig. 4.37. The oxygen consumption of marine animals at high hydrostatic pressure. (Naroska, 1968)

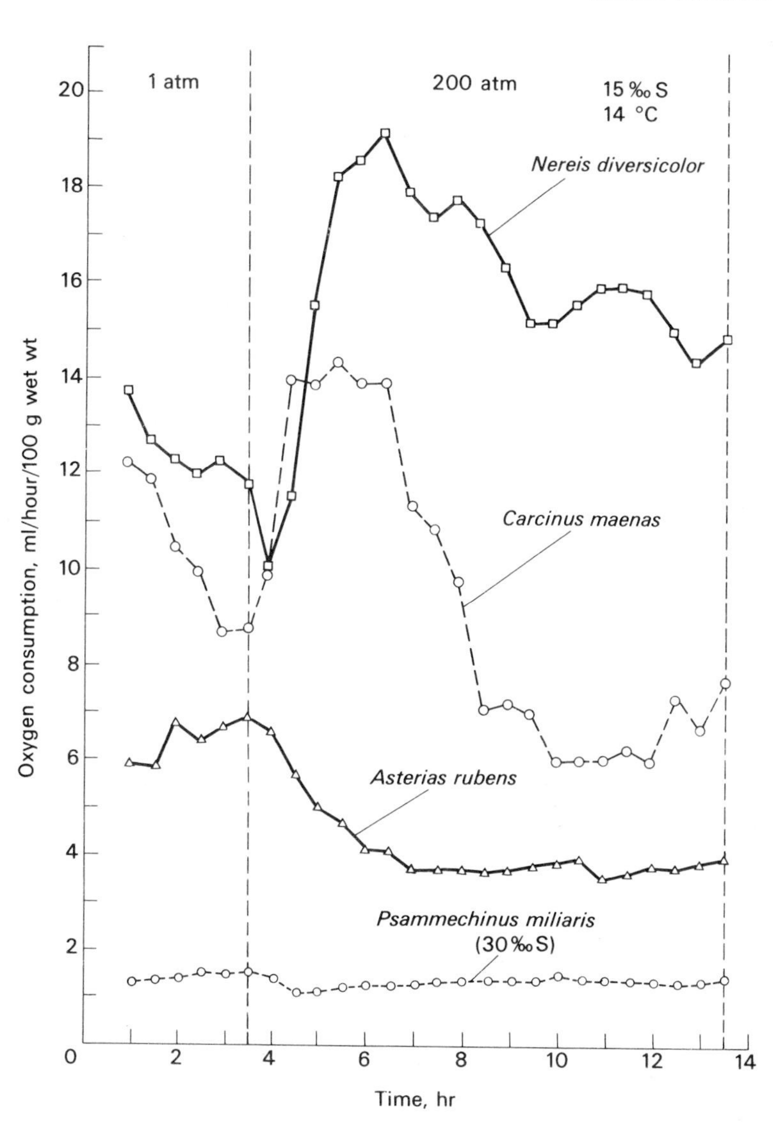
1 atm
200 atm
15 ‰ S
14 °C
Nereis diversicolor
Carcinus maenas
Asterias rubens
Psammechinus miliaris
(30‰ S)
Oxygen consumption, ml/hour/100 g wet wt
Time, hr
0
2
4
6
8
10
12
14
16
18
20

(*b*)

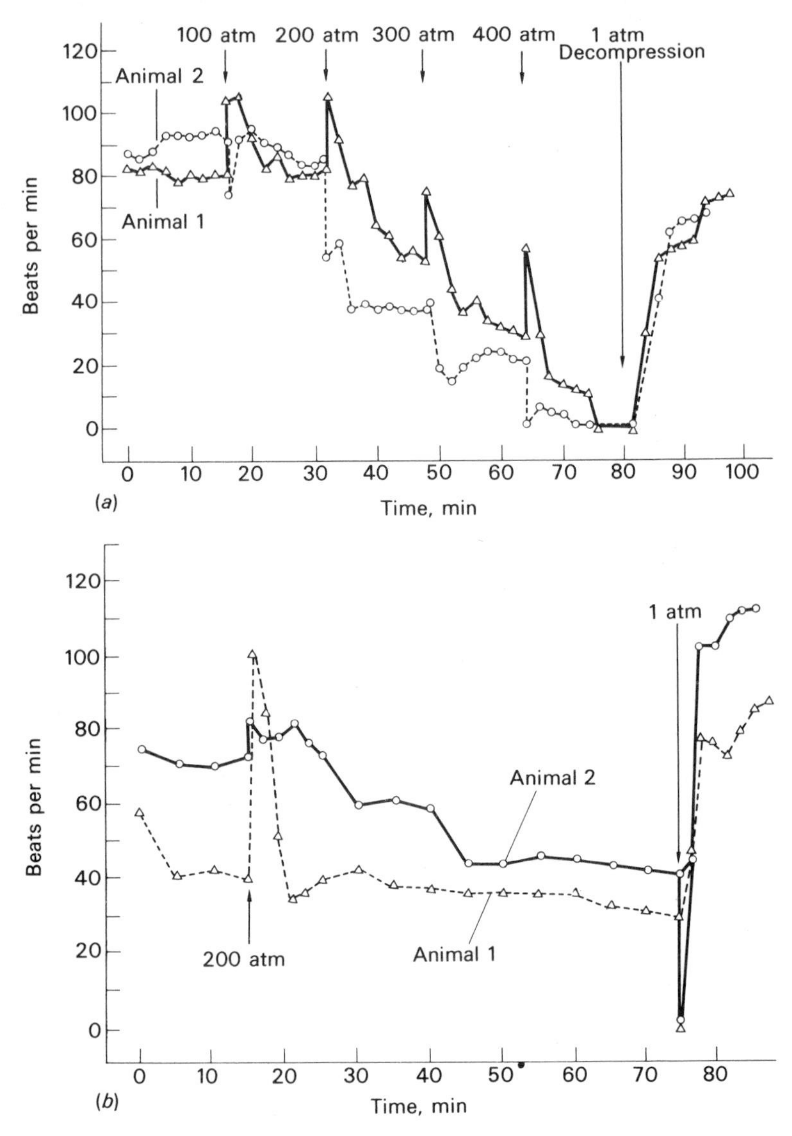

Fig. 4.38 (part). For legend see page opposite.

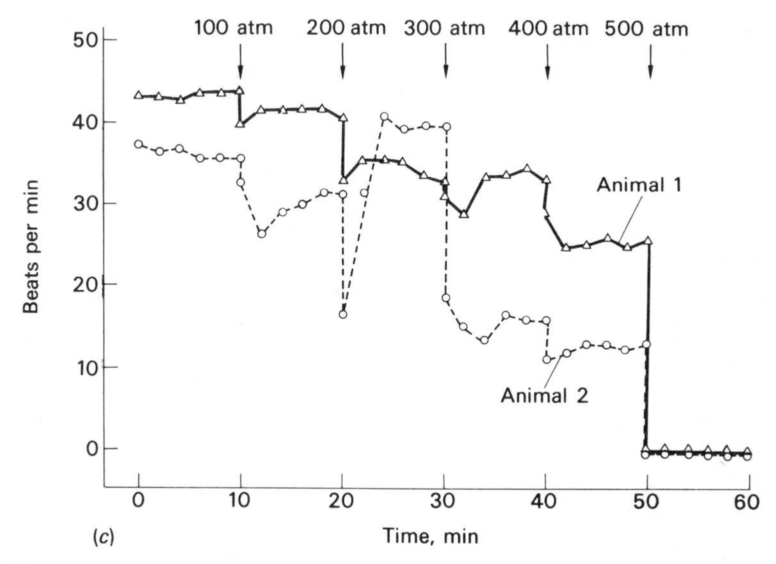

Fig. 4.38(*a*). Heart rate of *Gammarus oceanicus* subjected to high hydrostatic pressure. Temperature: 5 °C. Salinity: 15‰. (*b*). Heart rate of *Gammarus oceanicus* subjected to high hydrostatic pressure. Temperature: 5 °C. Salinity: 15‰. (*c*). Heart rate of *Ciona intestinalis* subjected to high hydrostatic pressure. Temperature: 21 °C. Salinity: 15‰. ((*a*), (*b*) and (*c*), Naroska, 1968)

tension. Air in the swimbladder might have entered the bloodstream in solution and the resultant elevated blood P_{O_2} could have induced the kind of behavioural activity described. It is significant that on decompression Nishiyama reported gas bubbles emerging from both gill openings and anus. Brauer (1972) avoided this complication and reports the following convulsion pressures for four species of shallowwater fish: *Achinis fasciatus* 130 atm, *Paralychtis dentalus* 105 atm, *Anguilla rostrata* 102 atm and *Symphurus palguisa* 86 atm.

MARINOGAMMARUS MARINUS

Marinogammarus is an amphipod which is available on British coasts all the year round. Its tolerance to pressure has been described in some detail in the hope that it will provide a basis for comparisons with deep sea animals. Amphipods inhabit the greatest depths of the oceans.

Casual observation of *Marinogammarus* subjected to high hydrostatic pressure soon reveals three types of changes in its activity. These changes or responses have been designated Types I, II and III and their detailed and quantitative description is given below. Like many Crustacea *Marinogammarus* is sensitive to small hydrostatic pressures. 3 atm at 3 °C normally causes the animal to uncurl its body, extend its antennae and then move around. This activity does not seem to increase as the pressure is raised to 10 atm, and is regarded as a response separate from that seen at higher pressures.

Compression to 100 atm causes an increase in normal and abnormal locomotion (the Type I response). As rate of compression influences the extent of an animal's response to high pressure it will be assumed in the following account that a step-wise compression of 50 atm applied in a few seconds at 5 min intervals was used unless otherwise stated. Fig. 4.39 shows the change in activity (defined in the legend) at 100 atm at 3 °C. Coordination is typically poor and the animal's swimming and crawling movements are jerky. Contractions (spasms) of the whole body occur in a dorsal–ventral plane. During slow compression less activity is elicited but a significant amount is apparent at 25 atm. Above 50 atm spasms become increasingly common. Hyperexcitability diminishes abruptly with sudden decompression. Prolonged exposure to 100 atm shows a decline in the spasms and locomotor activity in general. To some extent this hyperexcitability is reflected in an elevated Q_{O_2} (Table 4.17).

Legend to Fig. 4.39

Fig. 4.39. Changes in the locomotor activity of *Marinogammarus marinus* at 3 °C during sudden and gradual compression to 100 atm. Activity, plotted on the vertical scale, is the percentage of a 3 min observation period during which the animals moved about. Hydrostatic pressure in atm is shown separately above it.

(*a*) Upper curve, pressure increases in two steps and is held constant for two hours. Lower curve, activity declines after setting up the experiment but rises with the application of pressure. Solid points refer to animals held at pressure for two hours. Circles refer to animals which were decompressed after twenty minutes, as shown by the vertical dashed line. Each point is the mean of not less than six experiments each involving one animal, ± the standard error. The decline in activity following decompression at 20 min is highly significant.

(*b*) Upper curve, pressure increases in approximately 6 atm steps to 100 atm and is held constant for 1 hour. Lower curve, as in (*a*).

(*c*) Upper curve, pressure increases at the rate shown over a period of 5½ hours. Lower curve, as in (*a*), but without a preliminary settling-down period. The mean level of activity at 100 atm (5½ hr) is not significantly different from zero, but the activity at 3 and 3½ hours after the start of compression is significantly above zero. ((*a*), (*b*) and (*c*), Macdonald, 1975)

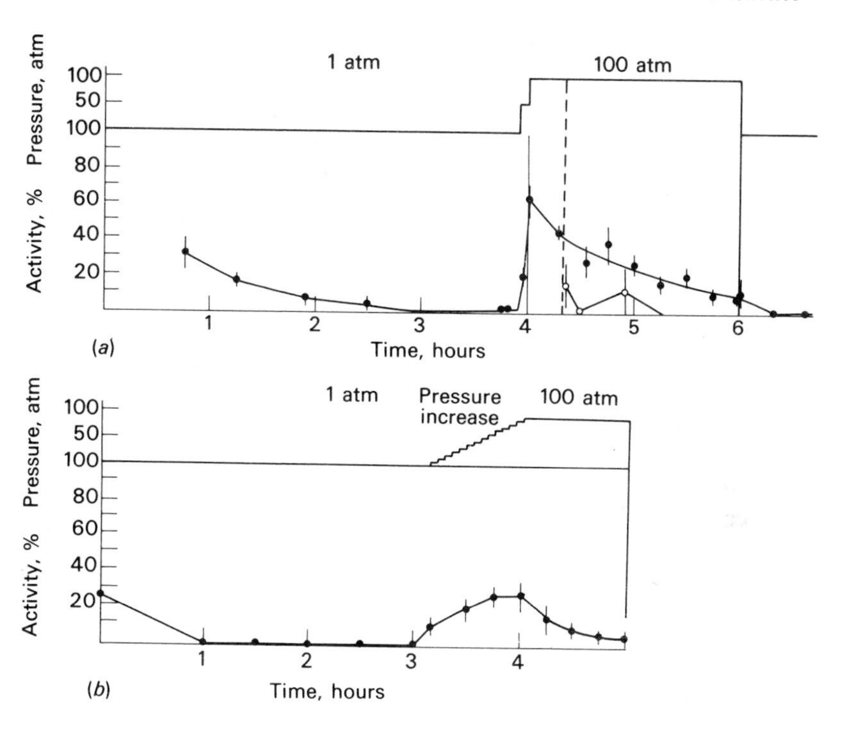

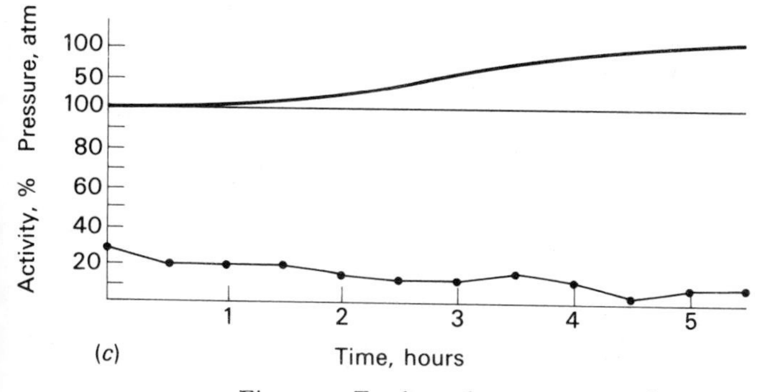

Fig. 4.39. For legend see page opposite.

TABLE 4.17. Marinogammarus marinus: Q_{O_2} *at 3 °C at high hydrostatic pressure.* (Macdonald, 1975)

The figures are ml O_2 at NTP per g wet weight per hour. Relative rates of oxygen consumption, uncorrected for extraneous oxygen consumption, ignore errors due to changes in Q_{O_2} with declining P_{O_2}.

	Pre-compression (2 hr period)	At pressure (1st hr)	At pressure (2nd hr)	After decompression
Rapid compression to 100 atm (50 atm steps at 5 min interval): 5 animals per expt	0.059	0.077	0.042	—
	0.072	0.081	—	—
	0.050	0.105	0.045	—
M.R.R.	100%	145%	71·6%	—
		During increase in pressure	First hr at 100 atm	
Slow compression to 100 atm (6 atm steps at 4 min intervals taking 1 hr overall): 5 animals per expt	0.093	0.093	0.050	—
	0.088	0.075	0.044	—
	0.045	0.043	0.035	—
M.R.R.	100%	93·4%	57%	—
		Over a 4 hr period at pressure		
Rapid compression to 200 atm (50 atm steps at 5 min intervals): 9 or 10 animals per expt	0.053	0.005		0.0246
	0.040	0.013		—
	0.090	0.013		0.04
	0.070	0.002		0.015
M.R.R.	100%	12%		41%
Control expts	0.018	—		—
Heat-killed animals	0.014	0.010		After 14 hr at 200 atm 0.008
Exoskeleton	0.016	—		
Seawater only	No consumption of oxygen detectable			
Animals at 1 atm: 5 per expt	0.064 constant over first 3 hr then declining to 56%			
	0.053 constant over first 3 hr then declining to 41%			

M.R.R. = mean relative rate.

The Type II response is a severe inhibition of locomotion at 200 atm. The characteristics of the Type II response at 3 °C are first, the degree of inhibition progresses little with time, at least over 2 days of exposure, and second, recovery after decompression is rapid and not related to the duration of the exposure to 200 atm. Figs. 4.40 show these features. The rate of compression influences the degree of inhibition of locomotor activity, slow compression causing fewer animals to be immobilised (Macdonald, 1972). The criterion used in the experiments illustrated in Figs. 4.40 is the presence or absence of moving pleopods. In animals confined in glass tubes the pleopods generate a continuous respiratory current and under such conditions their rate of continuous beating may be measured at high pressure (Fig. 4.41).

In *Marinogammarus* which are exposed to 200 atm at 3 °C only the feeblest of limb movements is to be seen. Not surprisingly a low Q_{O_2} is detected under these conditions (Table 4.17). Note how oxygen consumption increases after decompression.

The Type III response in *Marinogammarus* is seen when the animal is totally immobilised by rapid compression to 500 atm. The recovery process at atmospheric pressure is much slower after 15 min treatment than it is after many hours exposure to 200 atm (Fig. 4.42). The same figure also shows that the rate of recovery has a normal, positive temperature coefficient but the severity of the effects of 500 atm are increased by lowering the temperature, that is, the action of pressure has a negative temperature coefficient. This description of *Marinogammarus* at pressure should enable comparisons to be made with deep sea Crustacea. Two salient points emerge from the *Marinogammarus* studies. One is the difficulty of using Q_{O_2} as an index of activity. It will be clear from an inspection of Table 4.17 and Fig. 4.39 that the Q_{O_2} is only a rough guide to locomotor activity. Secondly the hyperexcitability phenomena are influenced by the rate of compression and the duration of pressure treatment. The question now arises, how do animals which normally live at several hundred atmospheres pressure respond to changes in pressure?

OCEANIC ANIMALS, INCLUDING DEEP SEA ANIMALS

Euphausia krohni (Macdonald, Gilchrist & Teal, 1972) which lives no deeper than 400 m (Fig. 4.43) shows a Type I response including

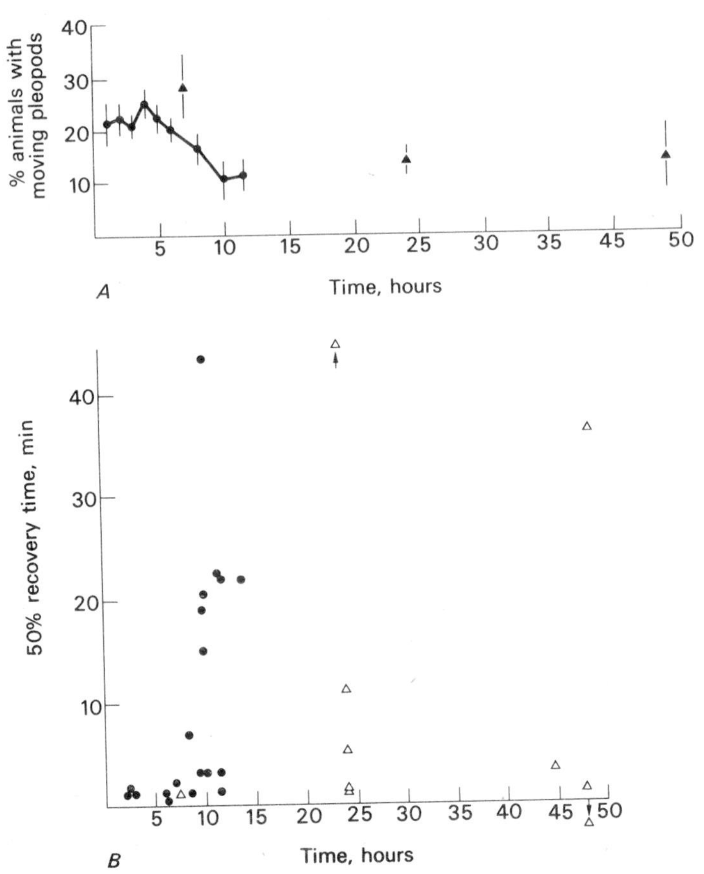

Fig. 4.40*A*. The effect of prolonged exposure at 200 atm on movement in *Marinogammarus marinus* at 3 °C. Groups of ten animals were confined in a pressure vessel and the number with moving pleopods is plotted against the time of exposure to 200 atm. Solid points refer to animals confined in 100 ml seawater, and triangles to animals kept in flowing seawater. In the latter case each point refers to not less than 3 experiments ± the standard error.

B. Recovery of *Marinogammarus* from the inhibitory effects of 200 atm at 3 °C. The time for half a group of ten animals to recover some movement of the pleopods at atmospheric pressure following decompression from 200 atm is plotted on the vertical scale. The duration of exposure to 200 atm is shown on the horizontal scale. Solid points refer to animals which were confined in 100 ml seawater, triangles to animals which were maintained in flowing seawater. The arrow at the top of the figure points to a recovery time of 160 minutes. The arrow at the bottom of the figure refers to the failure of the animals to recover in one experiment. (*A* and *B*, Macdonald, 1975)

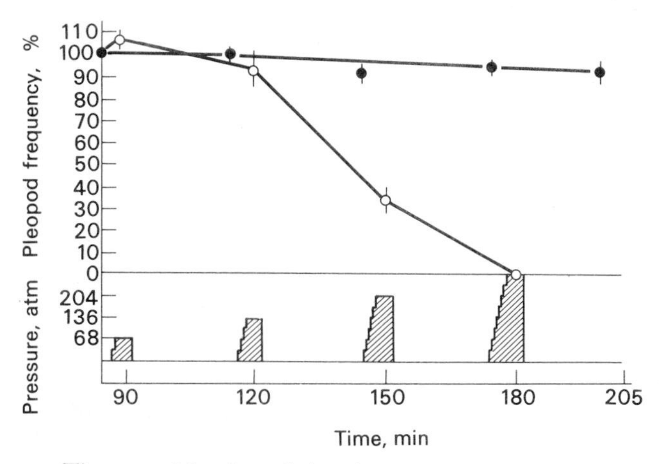

Fig. 4.41. The rate of beating of the pleopods of *Marinogammarus marinus* confined in glass tubes and subjected to various pressures at 10 °C. The time scale starts on mounting the animals in the pressure vessel. The pleopod frequency obtained from individuals at time 85 min is expressed as 100 and subsequent frequencies expressed as a percentage. Each point represents the mean pleopod frequency from 9 individuals, ± the standard error. ●, values obtained at the pressure indicated at the base of the graph and by the right-hand ordinate; ○, values obtained at 1 atm and 5 min before compression. (Data of A. F. Youngson; from Macdonald, 1972)

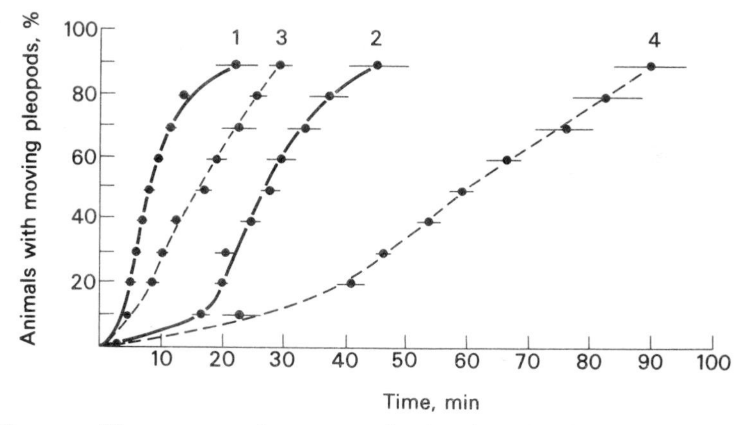

Fig. 4.42. The recovery of movement in the pleopods of *Marinogammarus marinus* after exposure to 500 atm for 15 min. Each point refers to the mean of 6 experiments, ± the standard error. Ten individuals were used in each experiment. Curve 1 relates to animals exposed to 500 atm at 13 °C and allowed to recover at 13 °C, whilst curve 2 refers to animals similarly compressed but which recovered at 3 °C. Curve 3 relates to animals compressed at 3 °C and which recovered at 13 °C, and curve 4 relates to animals which were both compressed and allowed to recover at 3 °C. (Macdonald, 1972)

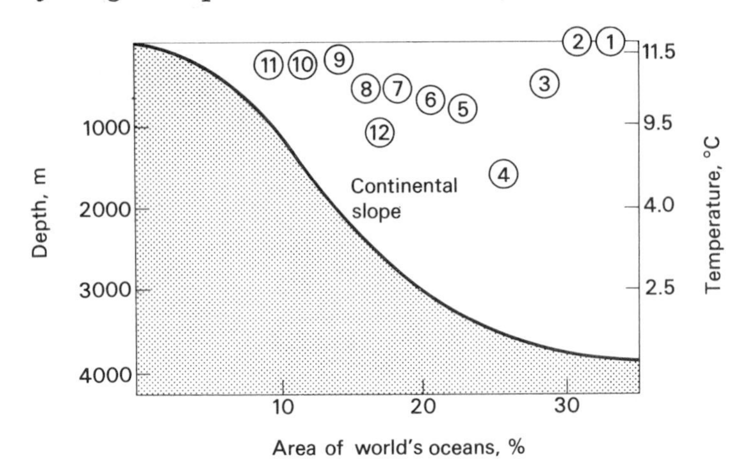

Fig. 4.43. The vertical distribution of various species of oceanic plankton observed at pressure. A portion of the hypsographic curve is shown, in which the depth of the sea, in metres, is plotted against percent surface area of the world's oceans. The figure depicts the continental slope descending to the ocean floor. Temperatures plotted on the right-hand ordinate correspond to those prevailing at depth in the area where the animals were obtained. Numbers refer to species listed and indicate typical depths of occurrence. 1. *Anomalocera patersoni* (Copepoda), 2. *Parathemisto* sp. (Amphipoda), 3. *Conchoecia* sp. (3 species) (Ostracoda), 4. *Gigantocypris mülleri* (Ostracoda), 5. *Pareuchaeta gracilis* (Copepoda), 6. *Pleuromamma robusta* (Copepoda), 7. *Euaugaptilus magna* (Copepoda), 8. *Megacalanus longicornis* (Copepoda), 9. *Euphausia krohni* (Euphausiacea), 10. *Amalopeus elegans* (Decapoda), 11. *Acanthephyra pelagica* (Decapoda), 12. *Cyphocaris anonyx* (Amphipoda). (Macdonald, Gilchrist & Teal, 1971; Macdonald, 1972. See also illustrations, Chapter 1)

violent body flexing. At 200 atm pleopod activity declines over a period of an hour. The decapods *Acanthephyra pelagica* and *Amalopeus elegans* which live at comparable depths also showed a Type I response and a decline in activity at 200 atm. Results obtained from *Systellaspis debilis*, a decapod which also lives no deeper than 1000 m, emphasises the importance of the locomotor-effect of pressure on respiration. Fig. 4.44 shows how respiration may be varied by pressure change rather than by pressure itself. Data on the respiration of oceanic decapods and euphausiids at pressures to 100 atm are reported by Teal and Carey, 1967, and Teal, 1971.

Amphipods live at the greatest depths but so far only *Parathemisto*, a surface dwelling oceanic species, and *Cyphocaris anonyx* a red form which presumably lives at a depth of at least 1000 m have been observed at pressure (Macdonald *et al.*, 1972, Macdonald,

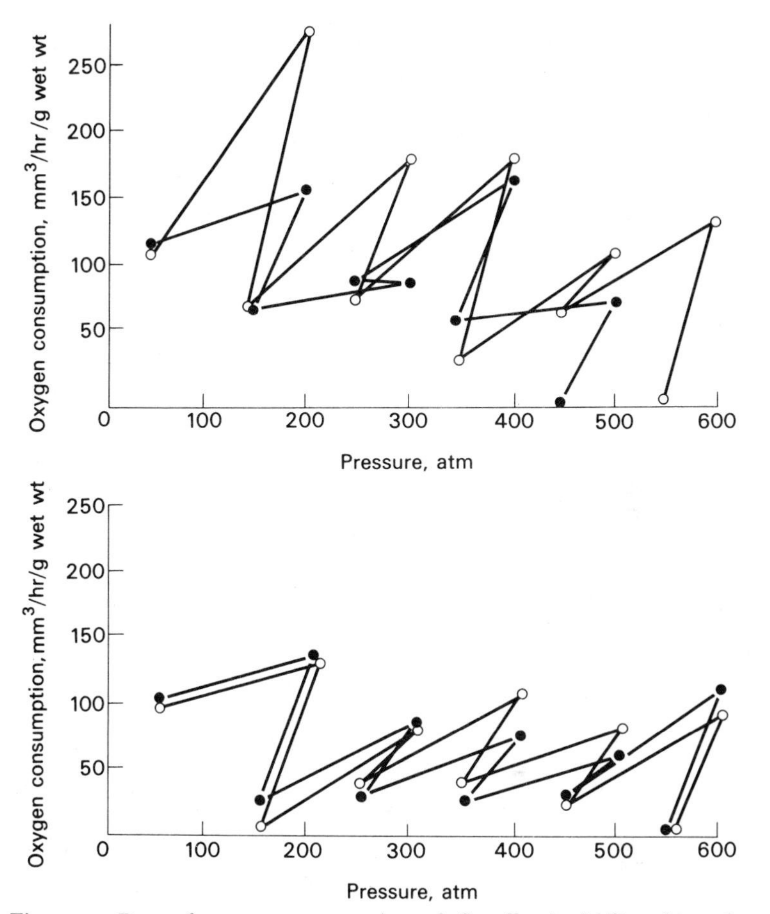

Fig. 4.44. Rate of oxygen consumption of *Systellaspis debilis* subjected to changes in hydrostatic pressure at 10 °C. The results of four experiments each involving one individual are plotted. Vertical scale: rate of oxygen consumption in mm³ O_2 per g fresh wt per hour. Horizontal scale, pressure in atmospheres. (Macdonald, Gilchrist & Teal, 1972)

1972),* *Parathemisto* proves even more sensitive to pressure than *Marinogammarus*. Six individuals of *Cyphocaris* were observed in one experiment and continued to beat their pleopods rapidly over a 1 hour period at 200 atm at 3 °C, thereby demonstrating a greater tolerance to pressure than *Marinogammarus*.

* A deep sea amphipod, more tolerant than *Cyphocaris*, has recently been studied at sea (Macdonald & Teal, 1975).

TABLE 4.18. *The pressure tolerance of* Gigantocypris mülleri *at* 3 °*C*. (Macdonald, 1972)

Pressure (atm)	No. of expts	Total no. of animals	Change in locomotor activity during compression	Change in locomotor activity at pressure
200 (50 atm steps at 5 min intervals)	4	17	None	None. Longest experiment 8¼ hours, shortest experiment 1 hour
300 (as above)	4	16	None below 200 atm. 3 of 4 expts showed decrease in activity at pressures higher than 200 atm	Reduced level of activity sustained for 4 hours. In one experiment activity declined thereafter to feeble paddling motions which were observed for a total of 7¼ hours
400 (as above)	3	11	None below 200 atm, activity reduced at pressures higher than 200 atm	Severe reduction in activity after 30 min
500 (68 atm steps at 1 min intervals)	5	12	None below 200 atm, activity reduced at pressures higher than 200 atm	11 out of 12 individuals stationary at 15 min

A few observations of copepods at pressure suggests that future work with these rather difficult animals might be rewarding. For example, *Pleuromamma robusta* which lives at depths of between 200 and 1200 m showed no increase in activity at moderate pressure. *Megacalanus longicornis*, another large red bathypelagic copepod, has been seen to recover simultaneously with decompression after 15 min exposure to 500 atm at 3 °C (Macdonald *et al.*, 1972).

The giant deep sea ostracod *Gigantocypris mülleri* is both a relatively convenient animal with which to work and one which lives at depths normally considered to be bathypelagic. Its vertical distribution in the North Atlantic is indicated in Fig. 4.43 and its morphology is described in Chapter I. Table 4.18 summarises the information available on this animal's locomotor performance at pressure. A typical result of measurements of the heart rate at pressure is shown in Fig. 4.45. Of the total five individuals whose hearts have been observed at high pressure, two showed arrhythmia at 1 atm which was 'cured' by the application of 100 atm. *Gigantocypris* thus shows interesting responses to pressure. Although capable of swimming vigorously in the warm at 1 atm it shows no inclination to do so when exposed to a moderate pressure. Its activity continues undiminished at 200 atm and 3 °C for several hours. Respiration in adults is unaffected by pressure, but is decreased by pressure in the young (Fig. 4.46). Exposure of adults to 500 atm for 15 min at 3 °C shows they are only slightly more resistant to such treatment than *Marinogammarus* (Table 4.18).

Thus far we have a rough picture of the short term effects of pressure on a variety of marine animals. The observations on oceanic animals were all made at sea with material which was fresh but which had been exposed to both pressure and temperature changes during recovery. The locomotor performance of *Gigantocypris* suggests it is adapted to life at about 200 atm (or a depth of 2000 m) but no greater. Although the observations are somewhat limited it is interesting that its heart rate declines markedly at pressures higher than 200 atm. The two cases of cardiac arrhythmia being cured by pressure may be even more significant than the slope of the curve in Fig. 4.45*B*. Further, pressure fails to cause sudden large changes in heart rate in *Gigantocypris*.

The Type I and II responses to pressure seen in *Marinogammarus* were both influenced by the rate of compression. It is therefore necessary to regard the apparent depth limitation of 2000 m for *Gigantocypris* with some caution. Slower compression may produce rather different results and the effect of decompression which *Gigantocypris* experiences during recovery has yet to be assessed. The problem of retrieving deep sea animals without loss of pressure is discussed in Chapter 7.

In summary it appears that shallow water but *oceanic* Crustacea have no greater tolerance to pressure than some littoral species, while mesopelagic and bathypelagic animals show an increased degree of pressure tolerance. Pressure clearly exerts major effects on the excitable tissue of intact animals and, equally clearly, evolution from shallow water into the deep sea has required special

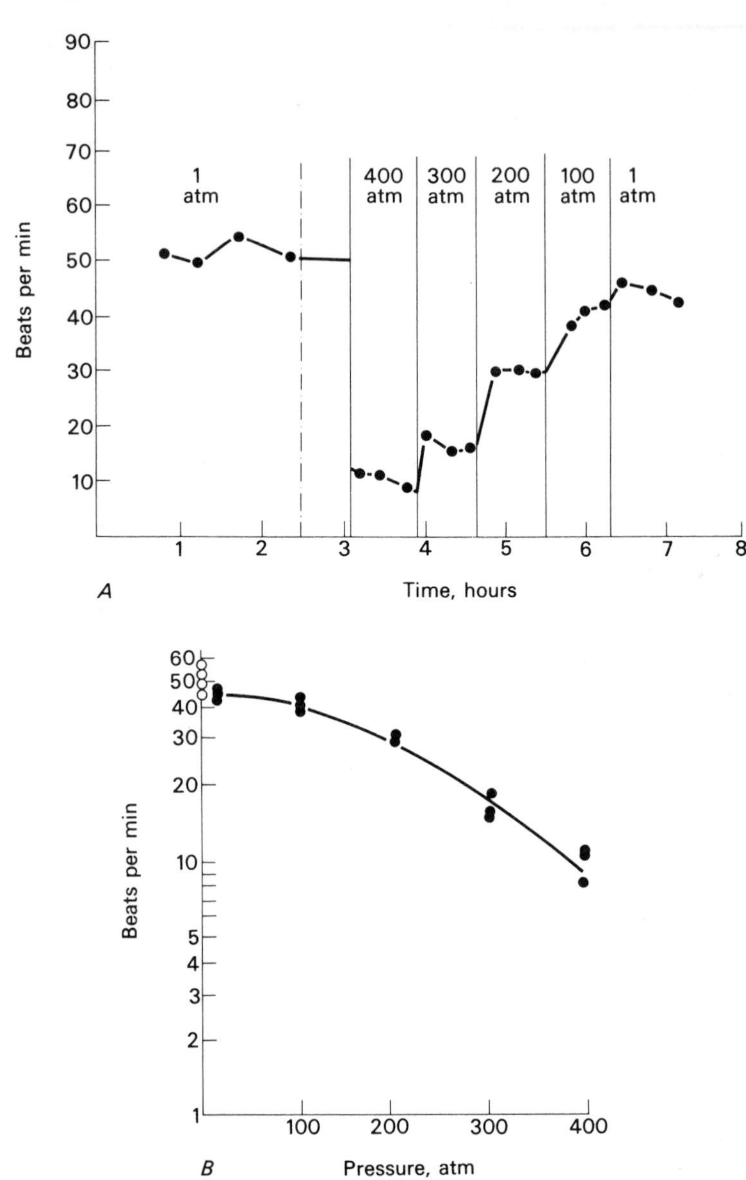

Fig. 4.45*A*. The effect of various pressures on the heart rate of *Gigantocypris mülleri* at 3 °C. Pressure was increased by steps of 50 atm at 5 min intervals and decreased rapidly. The start of compression is indicated by a dotted line. The result of a typical experiment is shown. Each point refers to the mean of 3 or 4 determinations on one animal.

B. Results in *A* are plotted on a log scale against pressure. (*A* and *B*, Macdonald, 1972)

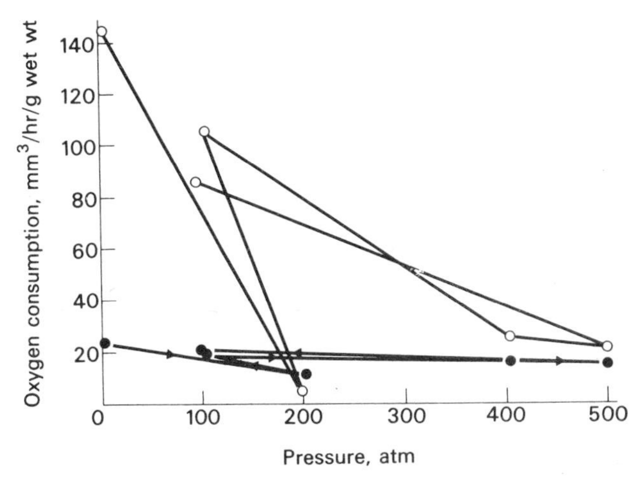

Fig. 4.46. Rate of oxygen consumption of *Gigantocypris mülleri* subjected to changes in hydrostatic pressure at 10 °C. Upper curve, average of two separate experiments; one small individual was used in each experiment, giving results which differed by less than 20%. Lower curve, average of two separate experiments in which a large individual was used in each case with results differing by less than 20%. Vertical scale, rate of oxygen consumption in mm^3 O_2 per hour per g fresh weight; horizontal scale, pressure in atmospheres. (Macdonald, Gilchrist & Teal, 1972)

adaptations of which we know nothing. Whilst the short term effects of pressure are interesting, long term effects have yet to be investigated. We have little idea of how much pressure animals can adapt to without genetic changes.

MARINE MAMMALS AND NON-MARINE ANIMALS AT PRESSURE

A number of non-marine animals have been observed in pressure vessels and in general their responses are broadly similar to those seen in invertebrates. Deep diving marine mammals present interesting problems. Weddell seals for example, naturally expose themselves to a pressure increase of up to 50 atm over a period of a few minutes, and sperm whales probably experience 100 atm pressure. The responses of non-marine animals, including man, to high pressure are considered in Chapter 7.

Some conclusions

Individual cells and the integrated activity of intact animals are both susceptible to deep sea pressures. Cytoplasmic structures, cell transport phenomena and excitability are all highly pressure-sensitive.

The hyperexcitability (Type I) behaviour seen in marine animals appears a widespread phenomena and its absence in deep sea forms such as *Gigantocypris* may be highly significant. The extent and complexity of the adaptations to high pressure required of deep sea organisms should be apparent from this and the preceding chapters. In the next two chapters we consider factors other than pressure which are also particularly important in the deep sea.

5 SENSORY PHYSIOLOGY AND BUOYANCY IN THE DEEP SEA

Introduction

Marine animals are equipped with a full complement of sensory systems. The ability to perceive light, sound vibrations, and dissolved substances is widespread and in the deep sea, where special conditions prevail, sensory systems have special interest.

It will be argued in this chapter that the deep sea provides distinctive conditions for the functioning of sensory systems in at least three ways. First, the stimulus or signal may have special properties, for example, sunlight attenuates rapidly with depth. Secondly, the transduction stage in the sensory process is likely to be influenced by high pressure, quite apart from the functioning of the sensory nerves. Thirdly, the stimulus may have a distinctive role in the organism's relationship to its environment. The immense volume of the deep sea surely poses severe communications problems to some animals and the question of the range and sensitivity of sensory systems is clearly of basic interest. The main sensory stimuli, light, sound and chemical agents are both sensed and emitted by marine animals.

Movement is often the response of an organism to sensory information and accordingly we consider how the deep sea affects locomotion and buoyancy. It is the purpose of this chapter to survey knowledge of deep sea sensory physiology and to attempt some assessment of the magnitude of the adaptive problems which deep sea animals have solved.

On seeing and being seen in the depths

SUNLIGHT – THE STIMULUS

Light in the deep ocean derives from the sun and from bioluminescent reactions. The transmission of sunlight through seawater is greatly influenced by suspended particles which scatter

TABLE 5.1. *Attenuation of sunlight in the sea*

	Path length in which light reaches 1% surface value (m)	Diffuse attenuation coefficient, k
Turbid water, i.e. harbours, shoreline	3	1.5
Coastal water	10–30	0.46–0.15
Clearest deep water, 700 m depth – Mediterranean	100	0.03

it and by water itself which absorbs it; dissolved salts play little part in light absorption (Jerlov, 1968). The diffuse attenuation coefficient k is defined by equation (1) and provides a single number which expresses the fall off in light intensity with distance (Lythgoe, 1972).

$$I_z = I_0 \, e^{-kz} \tag{1}$$

I is the spectral downward irradiance as measured by a suitable photocell, the details of which are given by Lythgoe (1972), z is the distance in metres between I_0 the source or upper reference level and I_z the target or lower level.

Some examples of k are listed in Table 5.1, compiled from the data of Clarke and Denton (1962).

Fig. 5.1 illustrates the fall off in light intensity with depth. At depths greater than 1200 m sunlight is apparently useless for vision. Note also the constant ratio of 0.01 for light shining upwards: light shining downwards. The very dim light in the twilight zone of the sea thus retains its directionality.

The two causes of the attenuation of light in seawater, absorption and scattering, act in a way which is related to wavelength. Very small particles scatter light preferentially in the shorter wavelengths; hence turbid water transmits preferentially in the longer, red wavelengths. Conversely, clear seawater transmits preferentially at shorter wavelengths, 460–480 nm, absorbing the longer and redder wavelengths more heavily. The sunlight which penetrates to the deep sea is consequently blue. Large particles in suspension scatter light independent of its wavelength and contribute to a general blurring of outlines. In short, the sunlight with which deep sea animals have to contend in the upper twilight zone is exceedingly dim but directional, and blue in colour.

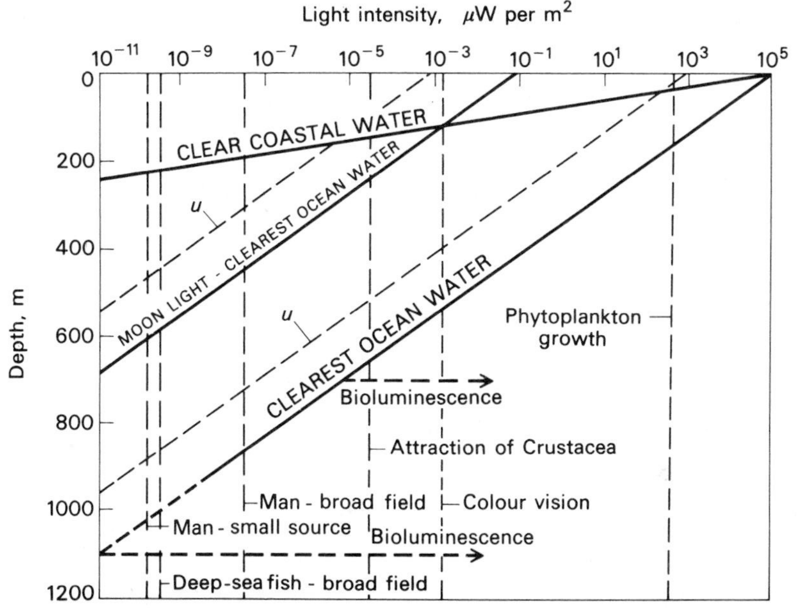

Fig. 5.1. Schematic diagram to show the penetration of sunlight into the clearest ocean water ($k = 0.033$) and into clear coastal water ($k = 0.15$) in relation to minimum intensity values for the vision of man and certain deep-sea fishes. The approximate minimum values for the attraction of Crustacea, colour vision in man, and for phytoplankton growth are indicated, as well as the range of intensity of bioluminescence in the sea. The penetration of moonlight and the approximate intensities of upward scattered (u) sunlight and moonlight in the clearest ocean water are also shown. Since light penetrating deep into the sea is confined to a narrow band of wavelengths, the values given for colour vision represent the approximate maximum depths at which a blue hue would be observed. (Clarke & Denton, 1962)

Animals which live in the twilight zone also experience diurnal and seasonal light rhythms. However, the bulk of the deep sea whose average depth is 3800 m is beyond the range of sunlight and its daily rhythmic influence.

Bioluminescence is an important source of light at depth and is considered on p. 208.

DEEP SEA VISION

There are probably two reasons why animals of the twilight zone have been extensively studied. One is their general accessibility, a deep water haul generally brings up a few viable specimens and interesting measurements can be carried out on their extracted visual pigments. The other reason is surely the fact that the twilight environment is outstandingly interesting. We can readily appreciate some of the evolutionary problems of survival in a dimly lit world. We cannot so easily imagine an environment in which touch, sound or chemical sense conveys important information to animals. Kennedy (1964) comments on the visual world of some selected terrestrial animals as follows: 'To a bee it is a carelessly formed, perhaps Impressionistic world in which movement carries great meaning and in which the richness of colour is extended to endow two flowers, each an insipid white to us, with radically different shades of "bee violet". To a fly, these qualities of colour are apparent in his downward directed field of view; but upward, he sees in black and white. To the octopus, the world may look very much like our own, though somewhat wetter, coarser grained and more horizontally and vertically ordered.' The human observer in a deep submersible obtains a strong impression of the visual aspect of deep sea biology. Beebè has written some excellent accounts of dives in his bathysphere off Bermuda in the early 1930s and his descriptions are well worth detailed study. Of a descent to 560 m in 1934, Beebe reported 'A little after three o'clock when we reached 1700 ft I hung there for a time and made as thorough a survey as possible. The most concentrated gazing showed no hint of blue left. All outside was black black black and none of my instruments revealed the faintest glimmer to my eye' (Beebe, 1935).

Note how Beebe reached the threshold of horizontal vision at a depth of 560 m. Anticipating the first American astronauts, Beebe conducted at least one of his dives in public. In 1932 he relayed a live radio broadcast from a depth of 650 m to the United States of America and Great Britain.

Deep sea eyes

The eyes of deep sea animals, and in particular of fish, exhibit many features which are associated with their functioning in dim blue light. The eye pigments of deep sea representatives of three major groups of animals, Acanthopterygii (bony fish), Elasmobranchii (sharks) and the Euphausiacea (shrimp-like animals) have been examined. The techniques employed in determining the spectral absorption of the pigments differ but the results are clear enough. Most deep sea visual pigments have maximum light absorption at wavelengths in the region of 480 nm (i.e. $\lambda_{max} = 480$ nm). This corresponds to the wavelengths of maximal transmission of deep sea water. The obvious conclusion is that these eyes are adapted to make efficient use of the available light. In contrast, the λ_{max} of the eye pigments of shallow water marine animals centres on 500 nm, corresponding to their ambient spectrum, although as Wald, Brown and Brown (1957) point out, a smooth transition is seen between the two, and blue-absorbing visual pigments are common at relatively shallow depths. Examples of the λ_{max} of the retinal pigments of selected marine animals are shown in Fig. 5.2 and Fig. 5.13. Other data are presented in Table 5.2. See also Lythgoe (1972) and Goldsmith (1972).

There is discussion in the literature about the nomenclature of visual pigments. Denton and Warren (1956) called the retinal pigments they discovered in deep sea fish chrysopsins or visual golds. The pigments of coastal fish are referred to as rhodopsins (rose pigments) and those of fresh water fish porphyropsins (visual purples). Rhodopsins are composed of retinene 1 and porphyropsins of retinene 2. Chrysopsins are based mainly on retinene 1 and according to Munz (1965) should be written VP (visual pigment), followed by the λ_{max} wavelength with a subscript denoting the retinene, which may be a mixture or, as in the case of the bathypelagic hatchet fish *Argyropelecus affinis*, pure retinene 1. This animal's visual pigment is $VP478_1$. Differences in λ_{max} are due to differences in the protein moiety, the opsin of the visual pigment, and the nature of the chromophore (Wald, Brown & Brown, 1957).

The eyes of deep sea fish are often big with a relatively large and highly transparent lens well placed to transmit a lot of light on to the retina which is typically composed of rods, sometimes in two layers, and capable of considerable summation (Munk, 1966). The rods are usually long and narrow and occur in abundance. Light may be reflected internally along their length and thereby concentrated at the base where it is absorbed by the visual pigment. Some deep sea fish possess tubular eyes (Fig. 5.3) which may provide stereoscopic vision, a wide search area and/or an increase in light sensitivity when working in binocular fashion (Weale, 1955). Deep sea elasmobranchs, in common with some other animals which use their eyes in dim light, have a reflecting tapetum

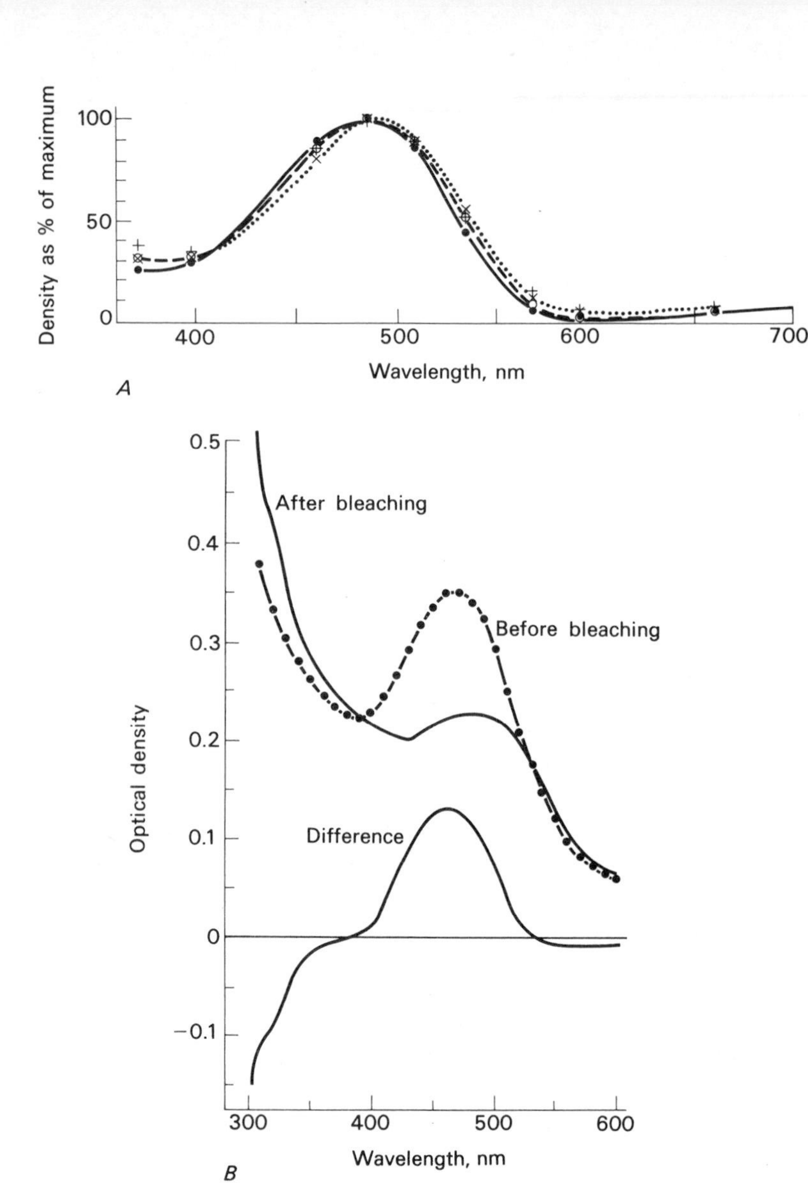

Fig. 5.2*A*. Spectral curves of unbleached retinae from deep sea fish: ●, *Xenodermichthys copei* Gill; ○, *Gonostoma elongatum* Günther; ×, *Chauliodus sloanei* Schneider; the +s represent corresponding results obtained on *Conger conger*. (Denton, 1959)

B. Upper curves. Absorption spectrum of an extract of dark-adapted *Euphausia pacifica* eyes, and of the same solution after 20 min in darkness at 24 °C are seen to coincide (●, experimental points after 20 min in darkness). The absorption spectrum of the same solution after a 20 min exposure to sunlight is also shown. Lower curve. Difference spectrum obtained by subtracting optical densities after exposure to light from those before exposure to light at each wavelength. (Kampa, 1955)

TABLE 5.2. *Light absorption of the visual pigments of some oceanic animals*

Animal	Depth	VP λ_{max} (nm)	Method	Reference
Euphausia pacifica	Epipelagic	462	Aqueous extract difference spectra	Kampa (1955)
Stomias boa	500 m approx	475	Difference spectra	Denton & Warren (1956)
Myctophum punctatum	Similar	475	Fresh retinae bleached with white light	Denton & Warren (1956)
Flagellostomias sp.	Similar	475	Fresh retinae bleached with white light	Denton & Warren (1956)
Argyropelecus olfersii	Relatively shallow	475	Fresh retinae bleached with white light	Denton & Warren (1956)
Conger vulgaris	Relatively shallow	475	Fresh retinae bleached with white light	Denton & Warren (1956)
Bathylagus wesethi	Caught between 560 m and 760 m depth	484–486	Digitonin extracts	Munz (1958)
Argyropelecus affinis	Caught between 560 m and 760 m depth	478	Digitonin extracts	Munz (1958)
Sternoptyx obscura	Caught between 560 m and 760 m depth	485–487	Digitonin extracts	Munz (1958)
Stomias atriventer	Caught between 560 m and 760 m depth	490	Digitonin extracts	Munz (1958)
Lampanyctus mexicanus	Caught between 560 m and 760 m depth	490	Digitonin extracts	Munz (1958)
Melamphaes bispinosus	Caught between 560 m and 760 m depth	488	Digitonin extracts	Munz (1958)
18 species bony fish	Most caught between 200 and 1500 m	Typically 485	—	Denton & Warren (1957), Denton (1959)
Centrophorus squamosus	Caught at 50 m	472–484	Digitonin extracts	Denton & Shaw (1963)
Centroscymnus coelolepsis	Caught at 50 m	472–484	Digitonin extracts	Denton & Shaw (1963)
Deania calcea	Caught at 50 m	472–484	Digitonin extracts	Denton & Shaw (1963)
Alepisarus ferox (lancet fish)	400	480	Digitonin extracts	Wald, Brown & Brown (1957)

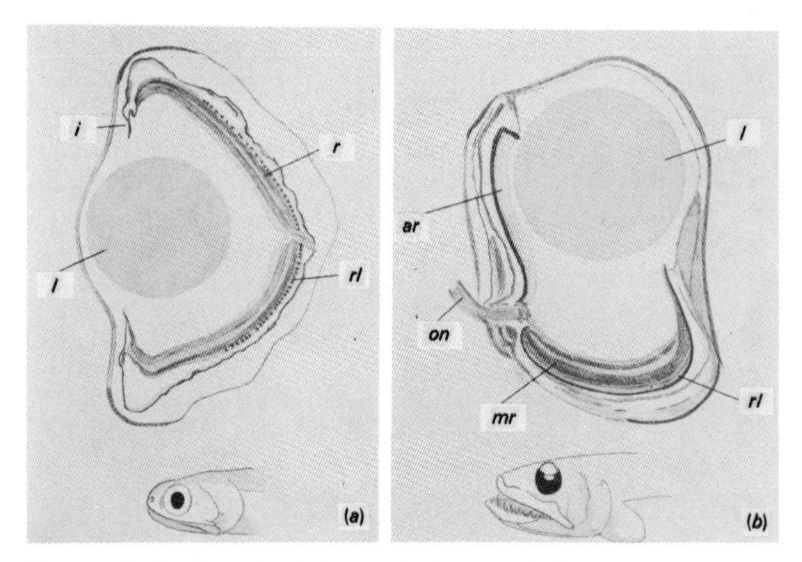

Fig. 5.3. Section through (*a*) the eye of a lantern-fish (*Lampanyctus warmingi*), and (*b*) that of a tubular-eyed fish (*Scopelarchus*). *ar*, accessory retina; *mr*, main retina; *r*, retina; *rl*, rod layer; *l*, lens; *on*, optic nerve; *i*, iris. (Marshall, 1954)

behind the retina (Denton & Nicol, 1964). The tapetal plates are composed of reflective guanine crystals and are so aligned to reflect light back through the retina and either out through the pupil or into the non-reflecting iris. The retina is thus given a second chance to absorb light which is not allowed to scatter within the eye. In fishes living in the upper, brighter water, eye-shine occurs when light is reflected out through the pupil and this is normally prevented by the migration of pigmented cells over the tapetal plates. Deep water elasmobranchs, for example *Deania calcea*, a shark, do not have occlusible tapeta, presumably because they live in light so dim that eye shine would fail to yield significant light since the visual pigments would absorb most of the light which has passed through the unbleached retina before leaving the eye. These same fishes possess retinal pigments in optical densities much lower than that found in deep water teleosts which lack a tapetum. Denton and Nicol (1964), who have studied these eyes in detail, concluded that the presence of a tapetum resulted in an absorption of light by the retina equal to that achieved by the deep sea teleost with extra retinal pigment.

In dim light the presence of excessive retinal pigment might be counter productive. A small flux of light produces a photochemical effect which must exceed any spontaneously occurring effects, or 'noise'. Reducing the total amount of visual pigment might reduce the 'noise' level. However 'noise', if it is to interfere with vision, has to involve the isomerisation of the visual pigment and it is possible that the spontaneous thermal breakdown of the visual pigments is a denaturation of the protein component which fails to intrude on the visual reactions (Lythgoe, 1972). Until more is known of the chemistry of the bleaching reaction, and in particular of the effects of high pressure and low temperature on the process, the relative quantities of visual pigments in deep sea eyes will probably lack a satisfactory explanation.

As Denton and Nicol (1964) suggest, the presence of a tapetum and a reduction in the amount of visual pigment may confer unknown advantages to the animal. A visual pigment with a λ_{max} of 480 nm only absorbs 3 per cent more light at a wavelength of 480 nm than a pigment with a λ_{max} of 502 nm. This is a small difference but evolutionists believe small differences exert decisive selective pressure in the long term. Denton and Warren (1957) have compared the light sensitivity of the human eye with that estimated for deep sea fish whose large pupils improve illumination of the retina, increasing the sensitivity of the eye by a factor of no more than 5 and probably by about 2.5. Low absorption of light in the eye, especially the lens, confers a further gain in sensitivity over the human eye of about $\times 2$ and the high density of visual pigments probably accounts for a further increase in sensitivity of $\times 3$. These three contributing factors are relatively well understood and suggest that the retina of deep sea fish absorbs between 15 and 30 times as much light as the human retina, under conditions of dim blue illumination.

In both the human and deep sea fish retinae the limitation to light sensitivity is thought to be set by the signal:noise ratio. Denton and Warren argue that the noise level will be lower in the deep sea fish than in the human eye conferring a further increase in sensitivity to light of $\times 4$ that of the human retina. This means the greatest depth at which fish may perceive daylight is 1150 m, and their eyes would be between 60 and 120 times more sensitive than those of humans. If low temperature and high pressure and perhaps other unknown effects favour the signal:noise ratio further, increasing sensitivity by a factor of 100, then the depth to which optically useful sunlight penetrates is not more than 1300 m. In other words, eyes have little to gain by becoming more sensitive, especially, as Denton and Warren point out, at such depths bioluminescence becomes relatively much brighter than sunlight.

Electrophysiological and behavioural studies are needed to substantiate the anatomical and spectrophotometric work. It is one thing to predict a fish is capable of detecting certain light–dark differences at depth; it is another problem to demonstrate that this property is used in normal life. Prediction, however, has an esteemed place in this area of physiology for it was in 1936 that Clarke and, separately, Bayliss, Lythgoe and Tansley (1936) predicted the existence of the characteristic absorption maxima of the visual pigments in deep sea animals. As we have seen their predictions were confirmed by Kampa (1955), by Denton and Warren (1956), and by Munz (1958).

BIOLUMINESCENCE IN THE DEEP SEA

Many deep sea organisms emit light. The evidence for this comes from several sources. Beebe (1935) and subsequent observers in deep submersible vehicles have seen animals giving off light at depth. A special sensitive instrument, a bathyphotometer, has been lowered from surface ships to record the frequency and intensity of the flashes of light occurring at depth. Many deep sea animals possess characteristic organs of light emission or photophores, and in a few cases light emission has been recorded from animals brought up from depth.

Beebe was the first to see deep sea bioluminescence and the spectacle moved him to write 'There seems no doubt but that light organs function as light organs to the highest degree, some steady, others fading and increasing in intensity, and still others eclipsed by occasional winking of dermal blinders.' Also, 'Here I began to be inarticulate for the amount of life evident from the dancing lights and its activity, the knowledge of the short time at my disposal and the realisation that most of the creatures at which I was looking were unnamed and had never been seen by any man were almost too much for any connected report or continued concentration.' So reported Beebe from a depth of 560 m in the Atlantic.

Other observers in modern vehicles have confirmed the widespread occurrence of deep sea bioluminescence, for example Piccard and Dietz (1957). The bathyphotometer developed by Clarke and colleagues (see Clarke & Wertheim, 1956) has provided objective recordings of the frequency and intensity of deep sea bioluminescence. The instrument comprises a fast response photomultiplier tube housed with batteries in a pressure casing which is lowered from a surface ship by a multiconductor cable. A pressure transducer in the instrument provides a record of depth, along with the light detected by the photometer. The instrument is sensitive to a light intensity of 10^{-7} μW cm^{-2} which approximates 10^{-12} full sunlight, in the 400–500 nm waveband. It is less sensitive than the eyes of deep sea fish, but flashes of light separated by 0.2 sec may be resolved. Some typical recordings obtained by Clarke and Hubbard (1959) in the Western Atlantic are shown in Fig. 5.4. A great number of light flashes are detected; some seem to be elicited by the instrument disturbing the organisms (bathyscaph observers have noticed this) but Clarke and Hubbard also conclude that much was spontaneous bioluminescence. In daytime conditions, flashes of light reaching a maximum intensity of 10^{-3} μW per cm^2 receptor surface were detected against a daylight background of approximately 10^{-3} that intensity.

The frequency of flashing attained a maximum of about 100 min^{-1} and in very deep water it dropped to a few flashes per minute. Generally, luminescence was most intense at 700–900 m depth, and declined markedly at depths below 2000 m. The interpretation of the data is rather limited. Clarke and Hubbard (1959) measured the light output of convenient shallow water animals and concluded that an animal which was capable of flashing light as bright as the brightest flashes from surface animals could be detected 8 m away from the instrument. Conversely, microscopic organisms may produce a detectable flash when in close proximity to the bathyphotometer window. Thus the records shown in Fig. 5.4 are made up of bright and dim lights shining over a variety of distances but probably not much more than 8 m. To assist in interpreting the light flashes a camera has been coupled to the bathyphotometer in such a way that the luminescent organisms cause their own photograph to be taken (Breslau & Edgerton, 1959). This is surely one of the most elegant of deep sea instruments. It has been used in the Mediterranean to show that in some

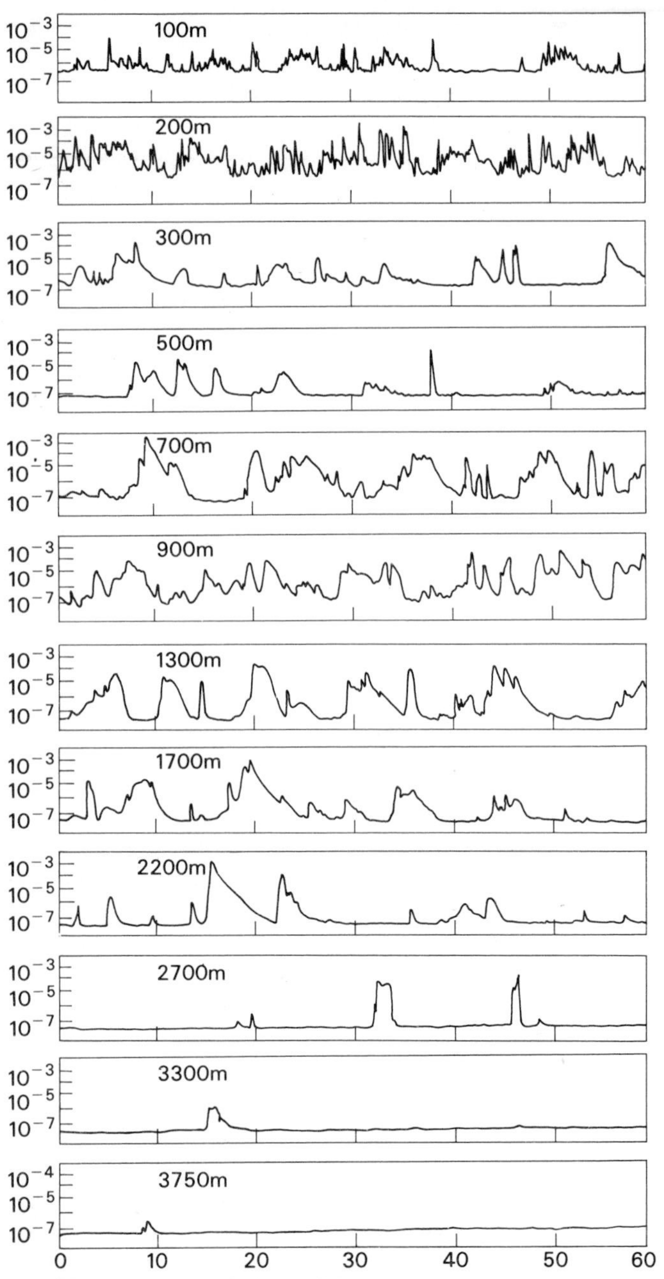

Fig. 5.4. The luminescent flashing of deep sea organisms. Sections of a record obtained on September 15–16, 1957, 2300 to 0100 hours EST. Bottom depth, 3750 m. Intensity of light in μW cm^{-2} is plotted on the vertical scale and time in seconds, on the horizontal scale. (Clarke & Hubbard, 1959)

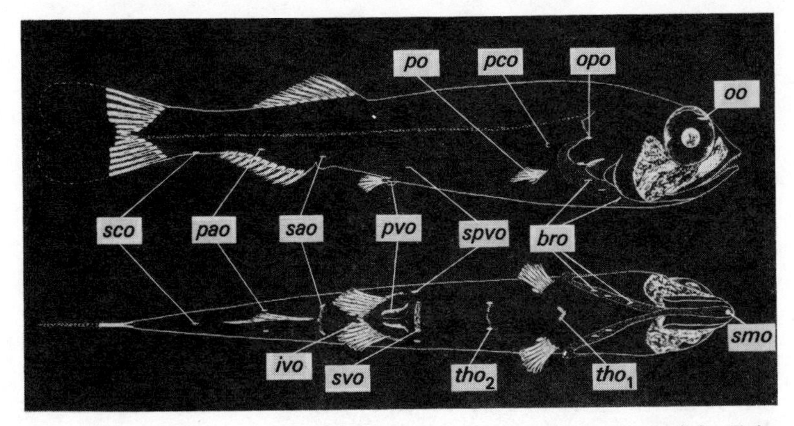

Fig. 5.5. Lateral and ventral view of *Searsia schnakenbecki*. Legend for light organs: *smo*, submental; *bro*, branchiostegal; *oo*, eye organ; *opo*, opercular organ; *po*, pectoral organ; tho_1, transverse thoracic organ 1; tho_2, transverse thoracic organ 22; *spvo*, supraventrals; *svo*, transverse subventral; *pvo*, posterioventral; *ivo*, inferior ventral; *sao*, supra-anal organ; *pao*, postanal organs; *sco*, subcaudal organ; *pco*, postclavicular (shoulder) organ. ½ natural size. (Nicol, 1958)

tests at any rate, most of the recorded light flashes come from organisms considerably smaller than 1 cm (see Chapter 7).

Morphological studies of deep sea fish and comparison with shallow water fish leave little doubt of the selective advantage of bioluminescence at depth. Haneda and Johnson (1962) assert that up to 98 per cent of the population of deep sea fish (i.e. 98 per cent of individuals, not species) possess photophores or photogenic organs. As an example, consider the luminescent secretion of the bathypelagic fish *Searsia schnakenbecki*, shown in Fig. 5.5, which has been studied by Nicol (1958). A living specimen was observed to give off 'myriads of blue–green sparks into the water' when touched. The light was very bright and at 11 °C was calculated to emit a radiant flux of 430×10^{-9} μW per cm^2 receptor surface at 1 m distance in air. The gland which gave off the luminescent secretion (postclavicular gland, Fig. 5.5) contained large (up to 40 μm diameter), red-coloured, acidophilic granules which Nicol regards as the photogenic granules which are released into the water. This form of deep sea luminescence may prove particularly convenient to study experimentally.

PRESSURE–TEMPERATURE STUDIES ON BIOLUMINESCENCE

A number of studies have been carried out on the effects of temperature and pressure on bioluminescent reactions. The first point to note is that the molar volume change in the light-releasing step is probably small. Schematically, the light producing reaction may be written $R \rightarrow P^* \rightarrow P + \text{light}$. P^*, the product of reactants R, exists in the excited state and spontaneously returns to a lower energy state P with the emission of light. The general term luciferin is used for the compound which emits light in such a process, although chemically it varies from organism to organism. Similarly, luciferase is the general term used for the enzyme which catalyses the oxidative reaction $R \rightarrow P^*$ and luciferases differ according to their substrate.

Bacterial luminescence is not controlled by nervous or hormonal mechanisms as may be the case in the complex photophores of animals, and it is therefore a simple case to study.

In the marine bacterium such as *Achromobacter fisheri* the luminescent reaction is essentially as follows. An aldehyde–dihydroflavin mononucleotide ($FMNH_2.RCHO$) is oxidised to yield flavin mononucleotide which emits light. $FMNH_2.RCHO$ is regenerated by DPN oxidase according to the simplified scheme shown in Fig. 5.6.

In bacterial cultures, nutritional factors influence the light intensity, and cell density affects the emission spectra, decreasing the wavelength with increase in cell numbers. A species emits light at a characteristic temperature which is related to the normal environment of the cell. An increase in temperature increases the light intensity up to a certain point, after which the intensity falls off. If the cells are rapidly cooled after heating beyond the optimum the light intensity rapidly returns to the characteristic level for that temperature.

Johnson and his colleagues have explained this effect in terms of a reversible thermal denaturation of a limiting enzyme, luciferase in Fig. 5.6. Thus the overall rate of FMN* synthesis is set by the concentration of effective enzyme catalysing its formation. At low temperatures the luciferase is likely to be present in excess, and under these conditions it is found that pressure diminishes light output by some 70 per cent at 5 °C in *A. fisherii*. If we assume that deep sea bioluminescent bacteria exist and are intent on emitting

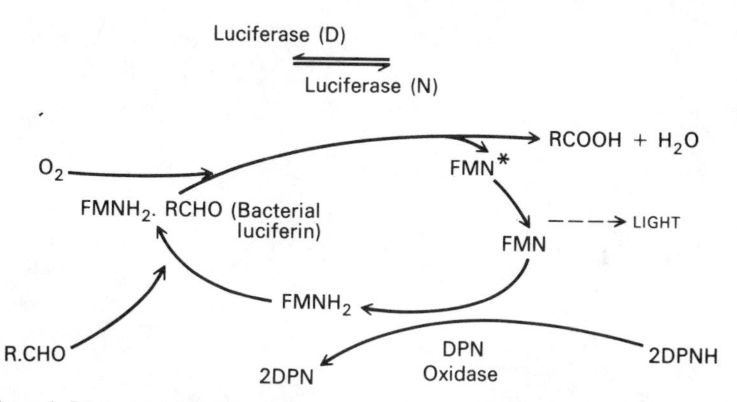

Fig. 5.6. Bacterial bioluminescence: schematic representation. (Adapted from Nicol, 1962)

light efficiently, then some adjustment in the volume changes involved in the rate-determining step in the luciferase reaction is to be anticipated. Exactly why bacteria emit light is not clear, so it is difficult to anticipate the direction in which selective pressures will work.

Bioluminescence in animals seems more purposeful and control of its emission involves an effector system (Nicol, 1969). Three different animals have been studied at high pressure and selected temperatures; they include a worm, a crustacean and a ctenophore, all shallow water marine animals.

In order to consider the effect of deep sea pressures on the bioluminescence of *Chaetopterus variopedatus* a preliminary description of this animal is required. It is an elaborately constructed tube-dwelling worm which lives in shallow water and secretes a luminous substance (Nicol, 1952*a* and *b*). The main area of secretion is segment XII, which consists of large flaps of tissue, and the aliform notopodia whose surfaces are rich in photocytes (Fig. 5.7). The luminous secretion is seen under a microscope as discreet, brightly shining particles. They are mixed with mucus which streams into the ambient seawater, giving a glowing light which lasts for up to 15 min. Other areas of the animal do not secrete any luminescent material in any quantity but give off flashes of light lasting a minute or two. These areas are the peristomial tentacles, the three fans, and the posterior notopodia. Both kinds of luminescent response are under nervous control and

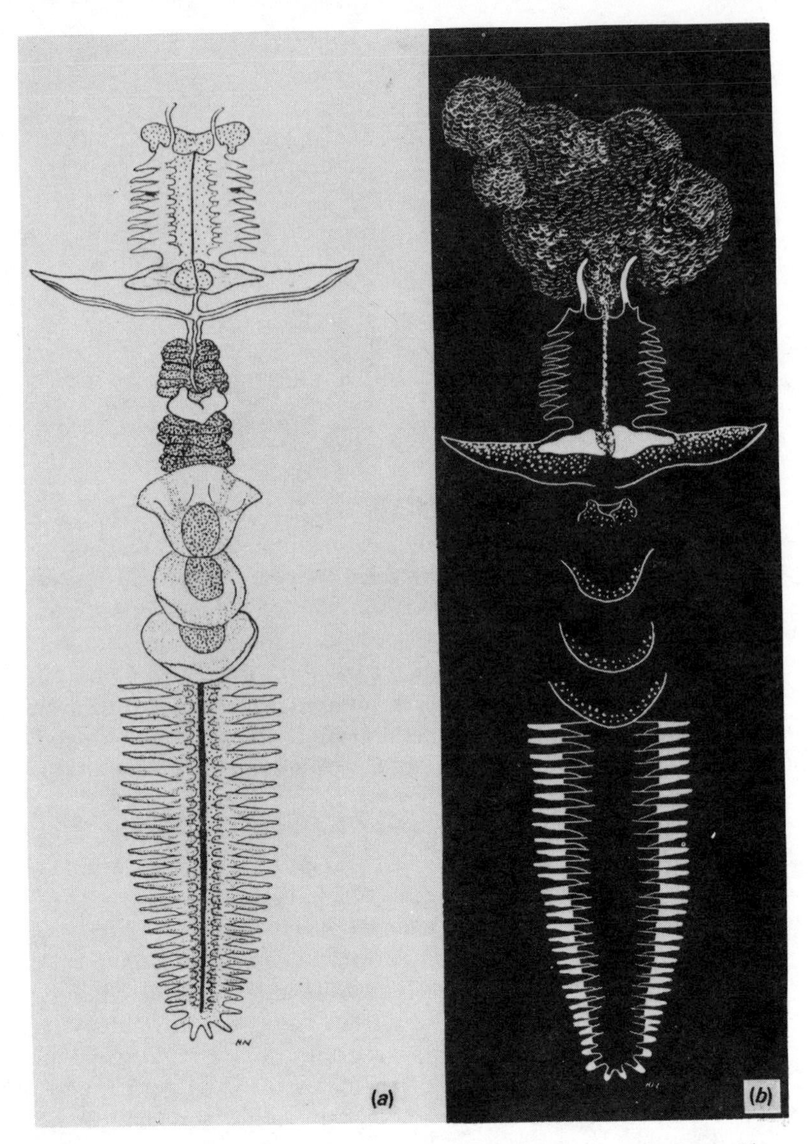

Fig. 5.7. (*a*) Dorsal view of a specimen of *Chaetopterus variopedatus*. About life-size. (*b*) Dorsal view of a luminescing specimen of *Chaetopterus*, as seen in the dark. About life-size. (Nicol, 1952*a*)

can be elicited by mechanical and electrical stimulation (Nicol, 1952*b* and *c*).

Electrical stimulation of segment XII causes a light response in the notopodia with a latent period of some 4 sec; the response is enhanced by increasing the voltage or the frequency of stimulation, which causes summation. Fig. 5.8 shows how increasing the frequency of stimulation causes a spread of excitation. The ventral nerve cord conveys the impulses to the photocytes as shown by the fact that its section prevents excitation spreading. Acetylcholine may act peripherally as well as centrally in the conduction process. Nicol, whose work has provided us with a great deal of knowledge about luminescent effectors, believes the release of the luminescent material may involve a cytoplasmic contractile process in individual cells, and his papers should be consulted for further details.

Sie, Chang and Johnson (1958) have investigated the effect of pressure and temperature on the luminescence of *Chaetopterus*, dividing their study into two parts, (i) the luminous secretion and (ii) the flashes of the posterior notopodia. The material from segment XII has been purified and it was found that luminescence may be generated by the following mixture: a photoprotein, a peroxide, ferrous ions and molecular oxygen. This luminescence system is clearly different from the bacterial luminescence outlined previously. Luminous slime obtained from live worms provides a declining source of light for several minutes. At 10 °C, several hundred atmospheres pressure causes an abrupt inhibition in the emission of light, an inhibition which is rapidly reversed on decompression. At higher temperatures (up to 30 °C), pressure causes an abrupt but lesser inhibition and decompression is accompanied by a small overshoot. Unfortunately, deep sea temperatures were not used but the results suggest that the extracellular luminescent reactions of *Chaetopterus* are markedly pressure-sensitive and without appropriate adjustment would only produce a small amount of light in abyssal conditions (Fig. 5.9).

The slow flashes of light produced by the posterior notopodia lasted several seconds and were greatly diminished by lowering the temperature to below 14 °C. Flashes were elicited by the electrical stimulation of excised notopodia and pressure was applied at various times in relation to the flash response. Pressure which was applied at the peak of the flash reduced the amount of

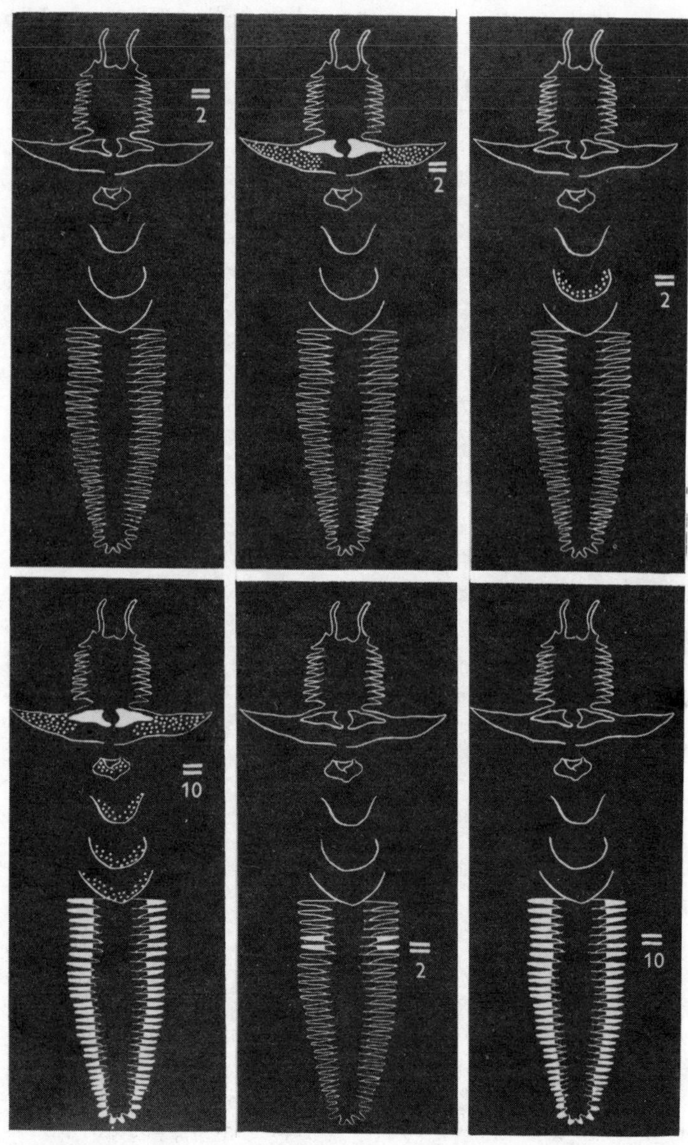

Fig. 5.8. Luminescence in *Chaetopterus*. The effects of stimulating the nerve cord at various frequencies. The double bar indicates the position of the stimulating electrodes. The numbers refer to the frequency of stimulation. Note the tendency for the excitation to spread as the frequency is raised. (Nicol, 1952*b*)

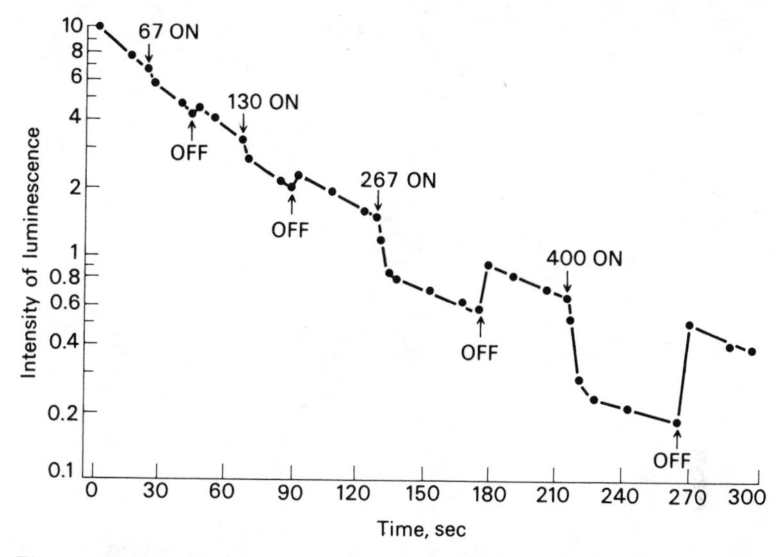

Fig. 5.9. Effect of hydrostatic pressure on the luminescence from *Chaetopterus* slime. Pressure (given in atm) is applied and released as indicated on the curve of light intensity against time. (Sie, Chang & Johnson, 1958)

light emitted (Fig. 5.10*A*). Decompression a few seconds later resulted in a delayed flash, suggesting that pressure inhibits the light emitting reaction, as it does in the bulk secretion (Sie, Chang & Johnson, 1958). A similar inhibitory effect is seen when pressure is applied whilst repetitive electrical stimulation of the notopodia produces flashes; the intensity of the flashes immediately decreases but may subsequently recover. Decompression during electrical stimulation produces a flash of light, greater in intensity than the previous flashes at pressure (Fig. 5.10*B*). This could be due to the effect of pressure on the secretory mechanism rather than the luminescence reaction itself. Notopodial luminescence is more sensitive to temperature than extracellular luminescence and would certainly be seriously impaired at abyssal temperatures. The effect of pressure on notopodial luminescence is broadly similar to its effect on extracellular luminescence; both are inhibited by pressure and both can show overshoot on decompression at temperatures above 10 °C. The situation is obviously complex because the light emitting reaction and the effector mechanism are pressure-sensitive.

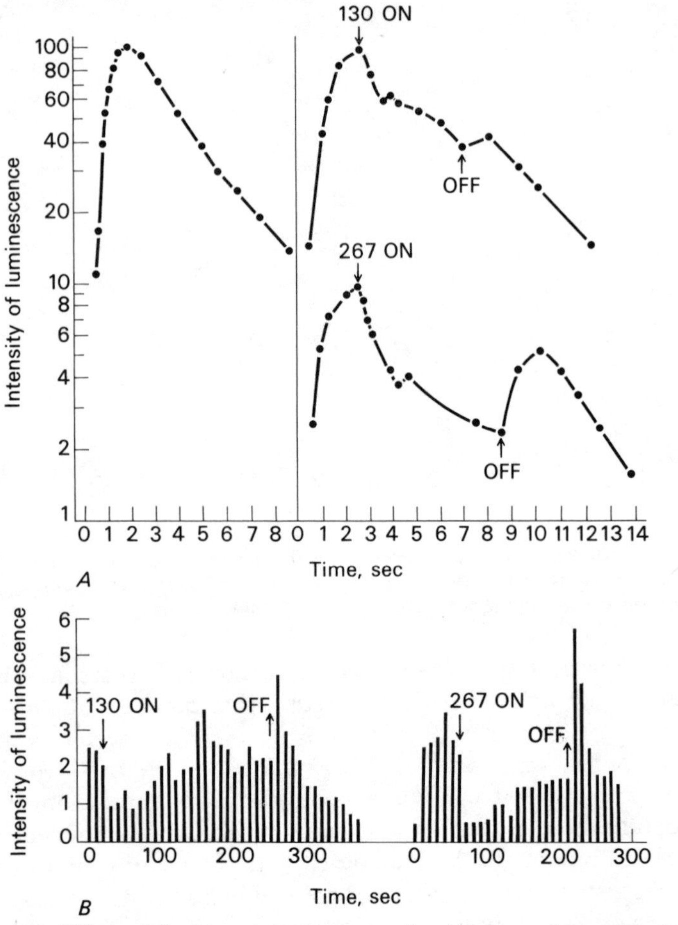

Fig. 5.10. Effect of hydrostatic pressure on luminescence from electrically stimulated excised notopodia of *Chaetopterus*.

A. Single flashes at 14.7 °C. Left-hand curve, control at atmospheric pressure. Right-hand curves, pressure applied as indicated; pressures are given in atm.

B. Consecutive flashes at 25.8 °C; effect of 130 atm and 267 atm pressure. (*A* and *B*, Sie, Chang & Johnson, 1958)

In the study of the mechanism of bioluminescence, the shallow water ostracod *Cypridina hilgendorfii* has figured prominently. Like *Chaetopterus*, it secretes a luminescent material. The light producing reactions involve luciferin and oxygen. If luciferin oxidises spontaneously it yields no light, but when oxidised by luciferase it emits light. Studies of the effects of pressure and temperature on this system are complicated by details of the purification procedures .but Bronk, Harvey and Johnson (1952) concluded the following: (*a*) pressure appears to oppose the spontaneous oxidation of luciferin and may, according to conditions, cause an increase in the light intensity of a particular preparation; (*b*) under other conditions high pressure at low temperatures inhibits luminescence. No investigation has been carried out on the secretion of the luminescent material.

The ctenophore *Mnemiopsis leidyi* is the third animal whose bioluminescence has been subjected to a pressure/temperature study. This animal's luminescent reaction proceeds intracellularly in the meridional and other canals, and bright flashes of light lasting about a third of a second at room temperature may be elicited by mechanical or electrical stimulation. Light inhibits the luminescence. Chang (1954) has shown that summation of electrical stimuli occurs at frequencies of 50 sec^{-1}, also facilitation at 1–3 sec^{-1} and eventually fatigue. In order to investigate these responses at high pressure, Chang and Johnson (1959) used small squares of meridional canal tissue mounted in a pressure vessel fitted with a window. The preparation proved exceedingly sensitive to low temperature and moderate pressures. At atmospheric pressure and 5 °C only a very weak flash could be elicited by strong electrical stimuli and this was abolished by 60 atm. At higher temperatures, 20 to 40 atm pressure diminished the light output perceptibly (Fig. 5.11). Interestingly, stepwise compression caused less inhibition than abrupt compression to the same pressure, and successive stimuli raised the intensity of flashes previously diminished by pressure. In some cases pressure increased the excitability of the system, causing a previously exhausted preparation to yield some light. This investigation demonstrated that the intracellular luminescence of *Mnemiopsis* has some similarity with the extracellular luminescence of *Chaetopterus*; neither system would perform adequately in abyssal conditions. In both animals pressure seems to diminish light output by impairing effector control and some short term adaptation takes place at high pressure. The similarity with locomotor activity in intact animals at high pressure is striking.

No work on the emission of light by fish in simulated deep sea conditions has been reported. In view of the well established nerve

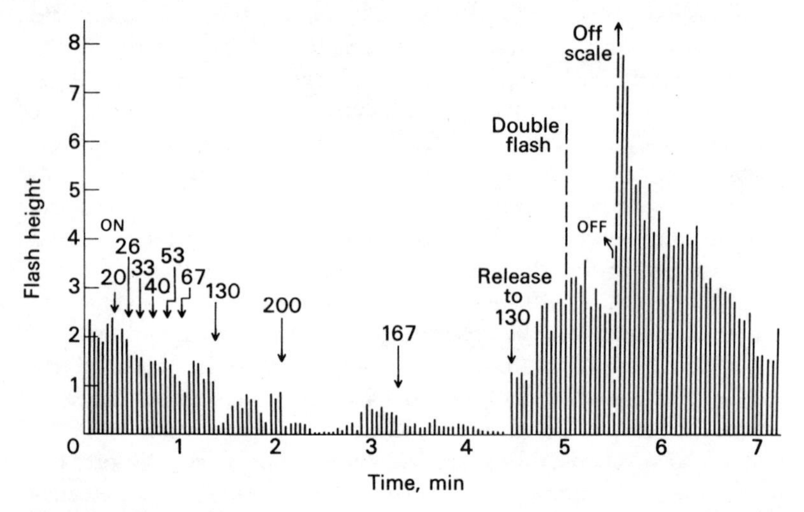

Fig. 5.11. Effect of hydrostatic pressure on the bioluminescence of electrically stimulated meridional canal tissue of *Mnemiopsis* (Ctenophora). Pressure is applied in steps during repetitive stimulation as indicated. Light intensity is plotted on a relative scale on the vertical axis. Temperature: 21.6 °C. (Chang & Johnson, 1959)

connections to photophores and the less well established role of hormones in fish bioluminescence, this area promises to be a complicated and exciting one to investigate.

BIOLOGICAL ROLES OF DEEP SEA BIOLUMINESCENCE

These are probably little different from those seen in shallow water animals. A simple scheme is set out in Table 5.3; for a fuller discussion papers by Nicol (1969) and McAllister (1967) should be consulted.

Camouflage

Clarke (1963) pointed out the extraordinary prevalence of ventral photophores in fish, squid and crustaceans which live in the twilight zone of the sea. Such animals live at a depth where there is sufficient sunlight to render them conspicuous when seen from below. Clarke put forward the idea of luminescent countershading, whereby an animal generates light of the appropriate spectrum and

TABLE 5.3. *Some functions of bioluminescence in multicellular animals.* (Nicol, 1969)

A. Camouflage
(i) Blinding potential predators
(ii) Countershading
B. Communication
(i) Attraction; lures
(ii) Warning
(iii) Shoaling
(iv) Searching

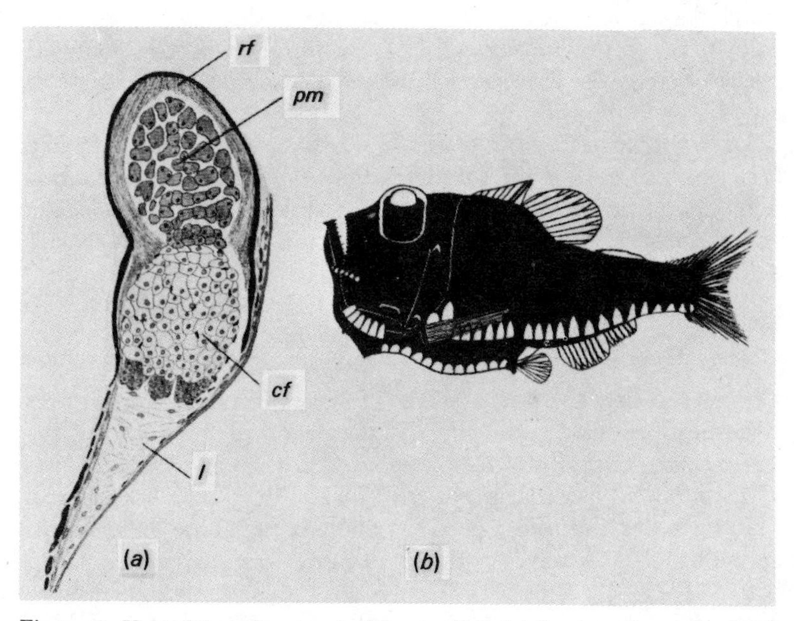

Fig. 5.12. Ventral luminescence in deep sea fish. (*a*) Sections through ventral photophores of the hatchet fish, *Argyropelecus affinis*. *pm*, photogenic mass of glandular cells; *rf*, reflecting layer; *l*, lens; *cf*, colour filter. (*b*) Diagram illustrating luminescent countershading (see text). (Marshall, 1954)

intensity to 'fill in' the shadow it casts downward. *Argyropelecus*, for example, appears well equipped to do this. Its photophores comprise a series of ventrally directed tubes which cause light, generated in canals connecting dorsally to the tubes, to shine downwards (Fig. 5.12). A piece of tissue located in the tubes serves as a colour filter and the peak transmission wavelength is appropriate to the surroundings at 480 nm (Denton, Gilpin-Brown & Wright, 1970). The lens and reflecting surfaces spread the light in a fan-shaped distribution providing countershading when viewed by a potential predator at various angles from below. Whether the animal regulates its light output to match the overhead intensity remains an intriguing problem. Luminescent countershading is also a plausible device in Crustacea which possess ventral photophores, for example *Sergestes richardi*, *S. robustus*, and *S. grandis* (Dennell, 1955). *In situ* observations from deep submersibles as well as laboratory studies are obviously required to investigate this camouflage mechanism more closely. Countershading has probably little need of precise control because the eyes of the potential predators (e.g. fish) are not capable of focussing a sharp image at any distance.

Use of light to blind potential predators is described by Clarke (1963) in the case of the squid *Heteroteuthis* which ejects a luminous cloud from what in other squid is an ink sac. The bathypelagic fish *Searsia*, already mentioned, may blind predators (Nicol, 1958).

Communication

Communication between animals by means of photophores raises two questions. One is the effective range of a photophore and the other concerns the information conveyed by the light signal.

Nicol (1958) has considered the range of deep sea photophores. He concludes that most deep sea animals emit blue light, corresponding to the wavelength of maximal transmission in deep water. Table 5.4 lists data from Nicol (1958) showing the light emitted by animals, many of which live in the twilight zone. The shallow water ctenophore *Mnemiopsis* emits the brightest light. Nicol's calculations of the effective range of photophores involved certain assumptions, namely that the receptor can detect light of 1.6×10^{-10} μW cm^{-2} with a maximal sensitivity at a wavelength of 470 nm. Further, it was assumed that the seawater transmitted either 99% per m or 90% per m and that the light was not focussed to assist its transmission. Interesting notional figures are thus obtained. The data refer to light traversing a 1 m light path in air and impinging on a 1 cm^2 receptor surface. Deep water Hydromedusae give a flux of between 7×10^{-9} and 9×10^{-9} μW per cm^2 receptor surface at 8 °C. *Searsia*, a bathypelagic fish, produces a secretion which emits 430×10^{-9} μW cm^{-2} at 11 °C, at normal atmospheric pressure. The photophores of *Myctophum punctatum*, a mid-water fish, should be detectable over a 10–16 m distance. The emission spectra and the retinal absorption spectra for this animal are given in Fig. 5.13. *Acanthephyra purpurea* may be detected over 5–7 m distance. Astonishingly the Radiolaria *Cytoclades* and *Aulosphaera* would be visible at a distance of 4–6 m. Luminous lures, seen in angler fishes, probably have a lesser range. The lure of *Dolopichthys* has been seen to glow but that of *Ceratias holböelli* has not (see Chapter 1).

Information appears to be transmitted by the nature of light flashes or their spatial distribution. Angler fish lures presumably mimick some kind of prey. Nicol has suggested that euphausiids employ light signals to keep in shoals. Several other cases are described in which photophores emit flashes of red light. The deep sea squid *Thaumatolampas diadema* (Nicol, 1962) and the mesopelagic fishes *Pachystomias* and *Malacosteus* are reported to do this and in the case of *Pachystomias* maximal retinal absorption is at a wavelength of 575 nm (red) although absorption of blue light is still quite significant (Denton, Gilpin-Brown & Wright, 1970). Why should red light be emitted in an environment which readily absorbs that colour? Denton *et al.* suggest the red light will illuminate red animals, particularly Crustacea, which, in normal deep sea light appear black and inconspicuous. In red light their bodies will reflect very strongly and show up clearly. The reflectance of black deep sea fish is less than 5 per cent and the spectral composition of the reflected light is little different from the incident light. The red decapod *Acanthephyra* similarly reflects very little in the blue region of the spectrum but a great deal at wavelengths above 600 nm. *Acanthephyra* and other animals of similar reflectance would obviously be at risk in the neighbourhood of *Pachystomias* and similarly equipped predators.

Boden, Kampa and Abott (1961) have attempted to relate the physiology of the eyes of selected Crustacea to the animals' behaviour and luminescence. Pacific species were used, *Euphausia pacifica, Meganyctiphanes norvegica* and *Nematocselis difficilis*. In general, a linear relationship between the log of the stimulating light intensity and the magnitude of the electro-retinogram

TABLE 5.4. *Intensity of luminescence of some pelagic animals.* (Nicol, 1958)

Animal group and species	Radiant flux, μW or μJ per cm^2 receptor surface		Recalculated	
	Measured flux ($\mu W \times 10^{-6}$ unless given in μJ)	Recording distance (cm)	Flux at 1 m distance in air ($\mu W \times 10^{-9}$ unless given in μJ)	Temperature (°C)
Radiolaria				
Cytocladus major and	0.2	5.6	0.6	22
Aulosphaera triodon	1.7	5.6	5.3	22
Hydromedusae				
Colobonema sericeum	0.8	9.7	7.2	8
	1.0	9.7	9.5	8
Crossota alba	0.02	13	0.4	13
Aeginura grimaldii	0.5	14	9.3	11
Siphonophora				
Vogtia spinosa	0.7	14	13.7	8
	10.6	17.4	320.9	21.8
V. glabra	19.4	7.8	120	8
Rosacea plicata	0.6	7	2.4	20
	1.2	10.7	13.7	20
Hippopodius hippopus	0.4	8	2.6	22
	0.7	7.7	4.2	22

Scyphomedusae				
Atolla wyvillei	0.1	5.7	0.3	24
	10.2	14	199.9	13
Ctenophora				
Beroë ovata	0.89×10^{-6} μJ	13.8	16.95	24
	118	26.9	8538.5	24.5
Mnemiopsis leidyi	0.5×10^{2}	50	12500	—
	$>0.75 \times 10^{2}$	50	18750	—
Crustacea				
Acanthephyra purpurea	0.23	9	1.9	9
	1.01	9	8.2	9
Euphausia pacifica	$1.6 \times 10^{3}–2 \times 10^{3}$	1	100–200	—
Tunicata				
Pyrosoma atlanticum	1.18	10.8	13.8	14
	17	28.3	1361.5	23
	$8 \times 10^{3}–4 \times 10^{4}$	1	800–4000	—
Teleostei				
Searsia schnakenbecki	19	9	150	11
	53	9	430	11
S. koefoedi	98	14.7	2117	12–15
	130	14.7	2808	12–15
Myctophum punctatum	0.1×10^{-6} μJ	9.5	0.925×10^{-9} μJ	16
	5.8	9.5	52.345	16

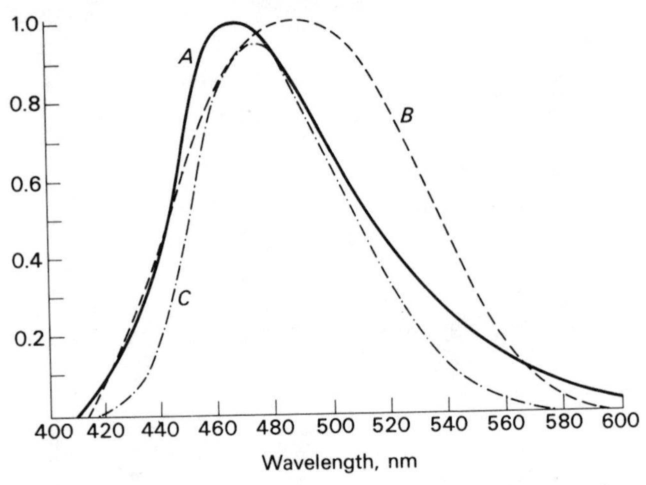

Fig. 5.13. The emission spectrum, *A*, of the luminescence of the lantern fish, *Myctophum punctatum* compared to *B*, the absorption (difference curve) of its retina; *C*, product curve based on *A* and *B*. (Nicol, 1969)

response was found and this allowed the spectral senstivity of the eyes to be measured. *E. pacifica* exhibited 2 peaks of sensitivity, one at 400 nm, the other at 490 nm. *M. norvegica* exhibited three peaks – at 515, 490 and 460 nm. *N. difficilis* has a bi-lobed eye in contrast to the normal single structure, and its upper lobe showed considerable sensitivity over the range 460–515 nm with a maximum showing most sensitivity at 490 nm. These animals vertically migrate but tend to inhabit dimly illuminated water (1.1×10^{-4} μW cm^{-2}). *E. pacifica* and *M. norvegica* luminesce, transmitting maximally at 476 nm with a second peak transmission at 520–540 nm. The intensity of their luminescence is of the order of 20 times that of transmitted daylight at their normal depth. The spectral sensitivity measurements obtained from the electroretinogram suggest that the visual pigment euphausiopsin (λ_{max} 460 nm) is in fact functioning in vision. The presence of peak sensitivities at other wavelengths may be related to the luminescence spectra of members of the same species, or of potential predators.

The nutritional cost and energetic aspects of bioluminescence has not been investigated in any detail, which is surprising in view of the way the subject might lend itself to quantitative treatment.

TABLE 5.5. *Percentage of benthic isopods with eyes, in relation to depth.* (Menzies & George, 1967)

	Peru–Chile Trench			North Atlantic (off Beaufort, USA)	
	% with eyes	No. of species		% with eyes	No. of species
Intertidal (Chile)	100	37	Intertidal	92	13
Shelf 5–200 m	76	34	Shelf 0–200 m	75	12
Trench wall 500–6000 m	0	54	Slope 200–2050 m	32	9
Trench floor 6000–6300 m	0	9	Continental rise 2000–3100 m	0	8
			Abyssal plain >4500 m	0	3

Fig. 5.14. A deep sea benthic fish without eyes, *Ipnops*. (Marshall, 1954)

LOSS OF EYES AND PIGMENTS

Many deep sea animals, chiefly invertebrates, lack pigment, and some lack eyes as well. Their similarity with cave-dwelling animals, many of which do not possess pigment, is striking. A quantitative study by Menzies and George (1967) provides some interesting data (Table 5.5).

The selective advantage of losing organs is a topic beyond the scope of this book although the deep sea provides a number of examples. A marked reduction in the size of the eyes occurs in the life time of *Ceratias holböelli* (Chapter 1) and other fish, whilst *Ipnops* (Fig. 5.14) provides an example of a fish which lacks eyes altogether in the adult stage. Perhaps the selective advantage of organ loss in deep sea animals lies in nutritional economy. It is important to establish the genetic basis for this form of loss. The eyes of fish develop poorly in permanent darkness (Blaxter, 1970), so it would be interesting to rear male *C. holböelli* or a more convenient animal in an illuminated environment to establish the inevitability or otherwise of its eye degeneration.

Deep sea sound

There are good physical and morphological reasons for supposing that water-borne sound waves are important to some deep sea organisms, particularly fish. Underwater sound consists of compression waves with concomittant displacement of the medium. Shallow water fish detect sound by means of otolith organs which respond to displacement as sound waves pass. Such waves are sometimes referred to as an AC stimulus in contrast to the DC stimulus provided by local water movements which are detected by lateral line organs.

The morphology of deep sea fish shows well developed auditory labyrinths and lateral line organs. Marshall (1971) has concluded from extensive studies that many bathypelagic fish have lateral line organs exposed to the environment, suggesting they are fully used to detect movements in the water. Mesopelagic or bottom living fish often have their lateral line organs enclosed in canals. The inner ear of deep sea fishes shows a full complement of structures which are variously emphasised according to species. There is no reason to doubt that the sacculus–lagena complex functions in the same way as it does in shallow water fish, and we may assume detection of underwater sound takes place at considerable depths. In order to discuss the special adaptations of the auditory sensory system that great depth may or may not require, it is necessary to consider the nature of underwater sound in general and the significance of hydrostatic pressure and distance in particular.

THE STIMULUS

Underwater sound travels at a velocity of about 1500 m sec^{-1}. More precisely, in seawater at 5 °C at 1 atm pressure and of salinity 35 ‰, the velocity of sound is 1471.0 msec^{-1} (Cox, 1965). Sound velocity is increased by factors which decrease the density or compressibility of the medium through which it travels. In

TABLE 5.6. *Sound velocity in the sea.* (McLellan, 1965)

Increase in	Change in compressibility	Change in density	Approximate net effect on sound velocity (compared to 1500 m sec^{-1})
1. Temperature	$<$	$<$	$>$ by 3 m sec^{-1} per degree C at the surface
2. Salinity	$<$	$>$	$>$ by 1.3 m sec^{-1} per unit ‰
3. Pressure	$<$	$>$	$>$ 1.8 m sec^{-1} per 10 atm

the sea, salinity, temperature and pressure are the important variables and Table 5.6 sets out their influence on sound velocity. It will be seen that compressibility is the dominant property, but it only changes the velocity of sound slightly in natural conditions.

A more important result of the natural distribution of these factors, particularly temperature, is that the acoustic impedance of the sea changes with depth. Sound becomes refracted and in certain circumstances confined to sound channels.

The intensity of sound diminishes with its progress through the water. First, there is the geometrical effect of sound spreading out from a point source. Second, sound is absorbed by viscous and thermal effects, readily imagined if we recall that sound is made of a succession of compression waves which will do mechanical work on the system through which it passes. A third cause of attenuation is absorption by solutes, particularly Mg^{2+} and SO_4^{2-} ions whose massive hydration shell interacts with sound and bulk water in a manner not fully understood (Horne, 1969). Both types of sound absorption act preferentially at higher frequencies. Finally, sound weakens because it is scattered by particles in suspension. As a sensory or communicating stimulus, sound is well suited to underwater conditions because its attenuation is gradual. The additional pressure and reduction in the number of particles in the deep sea benefit rather than diminish its potential as a sensory stimulus there. In contrast to light whose attenuation coefficient has been discussed, sound travels long distances in water. A 5 kHz sound beam, for example, diminishes several thousand times less rapidly with distance than light (Schevill, Backus & Hersey, 1962). The brightest bioluminescence at depth is a feeble and local event compared to the noises which pelagic organisms create. Because of the stratification of the deep water masses the horizontal transmission of sound is likely to be facilitated in the deep sea. Vertical transmission may be poor.

Whilst sound travels far and fast in the oceans it only involves small transient changes in the microstructure of the medium through which it passes. The transduction stage in the sensation of underwater vibrations involves remarkably small physical changes which may be affected by conditions at depth. Bauer (1967) has pointed out how an intense sound at 1000 Hz may only involve a pressure wave of 100 dyn cm^{-2} and a relatively slow movement of 10 Å in hypothetical particles in the sound transmitting medium. Such small displacements are the basis of sound detection and the question arises, does high pressure impose important constraints on the process? Further, is the noise level in the deep sea, against which any adequate stimulus has to stand out, significantly different from that prevailing in shallow seas?

BACKGROUND NOISE IN THE DEEP SEA

Background noise in the sea comes from water movements (waves, currents on the sea floor), thermal noise of high frequency and, rarely, from terrestrial sources such as earthquakes and turbidity currents.

Lomask and Frassetto (1960) used the bathyscaph *Trieste* to measure the change in noise level with increase in depth. In a flat calm they found no change in noise level with increase in depth but in a slight sea, sea state 2, low frequency noise declined to depths of 3000 m (Fig. 5.15). The bathyscaph could not be used in rougher conditions, although for most practical purposes sea state 2 is a rare and calm condition in which to work. The evidence suggests that the deep sea is indeed quieter than the surface waters, but the level of background noise is still determined by the weather.

TRANSDUCTION OF THE SOUND SIGNAL

Sound is detected in fish by the response of otoliths whose displacement stimulates hair cells. Similar structures in the lateral line organs respond to local water movements. A number of factors determine the response of the hair cells to displacement, namely the density of the otolith, the stiffness of the stereocilia and the viscosity of the medium surrounding them. The fluids of the

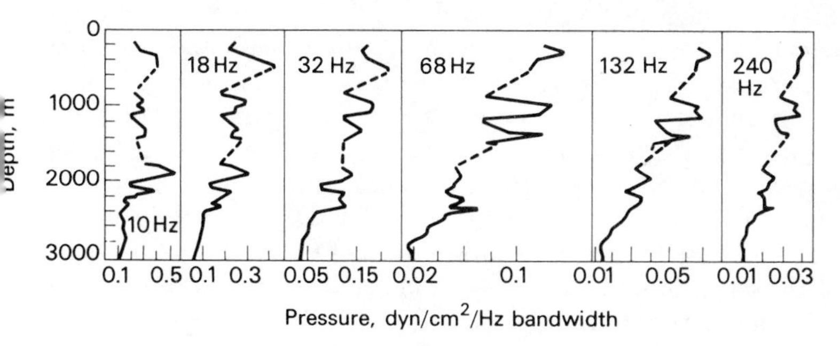

Fig. 5.15. Ambient noise levels as a function of depth in sea state 2. (Lomask & Frassetto, 1960)

sacculus and cupula may have pressure-sensitive properties. For example, the viscosity of the cupula fluid, said by Sato (1962) to be a polysaccharide gel, may require adjustment to work at extreme pressures.

In fish which utilise the swimbladder as a transformer in sound reception, some change in performance might be expected when the density of the gas contents rises significantly. The main constituent of the gas mixtures found in the swimbladders of deep sea fish is oxygen whose density increases by a factor of 320 over the pressure range 1–400 atm (Kunkle, Wilson & Cota, 1969). Swimbladders containing highly compressed gas will be somewhat reduced in their efficiency as transformers, but more important perhaps, changes in gas density during long range vertical movements will affect the acoustic characteristics of the swimbladder. In surface fish this effect is discussed by Sand and Hawkins (1972).

Elasmobranchs lack swimbladders but possess quite low hearing thresholds so there is no reason to suppose that fish in general at great depth need be insensitive to sound.

SOUND EMISSION

Fish produce noise by stridulating devices or by 'drumming' their swimbladders by means of special muscles. Sound producing fish, often the males of the species, are prevalent on the continental slope, suggesting that sound plays a part in mating behaviour there. However, Marshall (1971) notes examples of certain midwater fish which lack the necessary musculature to make sounds.

Marshall suggests that the ability of fish to produce sound correlates with life near the sea floor so it is conceivable (but mere speculation) that fish also locate the sea floor acoustically.

Griffin (1950) has described an extraordinary case of sound emission in the open ocean which illustrates the power of underwater sound. In March 1949 acoustical studies were being conducted by scientists from the Woods Hole Oceanographic Institution aboard a ship some 170 miles north of Puerto Rico. The depth of the ocean in that locality was estimated at 5110 m from nearby soundings. At around 1500 hours a hydrophone, suspended 20 m below the surface of the sea, picked up a series of paired short notes at a frequency of 500 Hz. Each note lasted from 0.3 to 1.5 sec. The first note in each pair normally exceeded the background noise level by a factor of three giving a sound pressure level of 85 db above 0.0002 dyn cm^{-2}. The second note in each pair was at background noise level but discernible to the human ear. Two important features of these paired notes were first, the mean interval between members of a pair was 1.58 sec, with a maximal variation of −6.9 per cent, +12.1 per cent, and second, the ratio of the amplitudes was 0.405. Thus the paired notes possessed a somewhat variable time interval and were fairly similar in sound level.

Griffin has argued that the sounds came from a marine animal but the possibility that they came from a creature producing paired notes with the observed precision and amplitude ratios cannot be eliminated. However, Griffin considered the explanation that the paired sounds were a primary note and an echo, and that some animal was rhythmically emitting a single loud sound pulse. The interval of 1.58 sec implies a sound path difference of 2370 m between the primary note and the echo, assuming sound travels at 1500 m sec^{-1}. The high amplitude ratio suggests the echo travelled a distance similar to that travelled by the primary note.

The question is then, what caused the echo? Sound channels probably played no part because multiple echoes were entirely absent. This leaves two possibilities. Either the sound was reflected from the sea surface or the sea floor. The sea surface is an unattractive possibility. The proximity of the hydrophone to the surface means the source of sound would have to be several times greater than 2370 m from the hydrophone. A source of sound capable of travelling such distances would be too intense to come from a marine animal. The sea floor is a more plausible reflecting

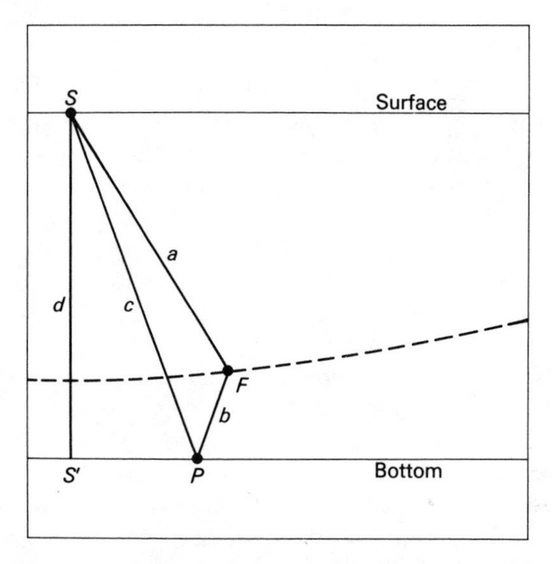

Fig. 5.16. Diagram of the relationship between location of hypothetical fish. *F*, hydrophone, *S*, and the point on the ocean floor, *P*, from which echo is reflected. See text. (After Griffin, 1958; see also Griffin, 1950)

surface. Fig. 5.16 illustrates the geometrical problem which Griffin so elegantly solved. The path difference of 2370 m between the primary sound path and the echo may be expressed

$$b+c-a = 2370 \text{ m}$$

Griffin argued that the low frequency sound will attenuate mainly as a result of geometrical spread. The amplitude of each note is related to the distance travelled and the efficiency with which the sea floor reflects the echo. The high ratio of 0.405 implies efficient reflectance in which the angle of incidence and reflectance are nearly equal. Calculations indicate that the source of sound was very loud and probably between 3925 and 1250 m below the surface. An animal capable of emitting such sounds could detect echoes from the sea floor with normal hearing ability and could therefore, in principle, locate its distance off the bottom. It is conceivable the animal was a marine mammal and not, as originally contemplated, a fish (Griffin, personal communication). The value of Griffin's argument lies in its illustration of how sound may be used in the deep sea.

SOME CONCLUSIONS

If we view sound as a 'good' signal for the underwater environment then it becomes even better with increase in depth. First, because its transmission is improved in the clear, compressed deeper water, and second, because other alternative signals such as light or chemical sense cannot compete for range or speed. However, it is not reasonable to take this engineering approach too far. Organisms evolve highly complex and specific relationships with their environment and in some cases sound is important and in others it may be unimportant.

There is no reason to suppose that deep sea conditions introduce fundamental limitations on the primary sound transducing mechanism, although gas bladder transformers may be slightly and deleteriously affected by high pressure. There is no firm evidence to show that long range acoustic devices have evolved to cope with the huge living space of the deep sea, although Griffin's observation (p. 232) remains intriguing. Both these conclusions must be regarded as highly provisional and will surely require revision when deep water listening buoys are deployed and when sound thresholds of deep sea fish are determined.

Deep sea chemoreception

In fish, chemoreceptors are found in olfactory organs, in taste buds in the mouth, gill cavities and elsewhere, and in the form of free nerve endings over the general body surface (Hara, 1971). Olfactory receptors are highly sensitive and typically respond to substances which are not detected by the taste buds or free nerve endings, although some overlap exists. The free nerve endings are the least sensitive of receptors; they function at close range in contrast to the long range olfactory organs.

A similar classification of chemoreceptors based on sensitivity applies in marine invertebrates (Laverack, 1968), which underlines the widespread importance of chemoreception in marine animals generally. Sensing food, the presence of other members of the species and potential predators are examples of the roles of chemoreceptors in marine life. That this sensory mode is of undiminished importance in the deep sea is apparent from the following experimental and morphological observations.

Experiments consisting of lowering a large can of dead fish to the sea floor and filming the animals subsequently attracted to it have demonstrated chemoreception in benthopelagic fish and Crustacea at 2000 m depth. This comparatively simple technique (simple now the deep sea camera is sufficiently developed) has considerable potential both for chemoreceptor and population studies at depth (Fig. 1.20).

The olfactory organs of bathypelagic fish have been studied by Marshall (1967) who concludes that, in general, males are typically macrosmatic (large olfactory organ) and females microsmatic. A typical case is illustrated in Fig. 5.19, which shows the olfactory system in *Cyclothone microdon*. In contrast, mesopelagic fishes only rarely exhibit sexual dimorphism in their olfactory organs. *Cyclothone* is also typical of bathypelagic fish in possessing large females and small males. In thirty specimens Marshall (1967) found that males averaged 3 cm in length and the females 5.5 cm.

What can be said of chemoreceptors which function at high pressure and in such a large living space as the deep sea? At the level of the receptor mechanism there is the possibility that pressure may affect the way molecules interact with receptor sites. It is, for example, conceivable that certain molecules are unsuited to serve as chemical signals in a high pressure environment if the activation volume of adsorption or complex formation is high, but this is pure speculation. The sensitivity of chemoreceptors in general is impressive when expressed in terms of the concentration of a substance which may be detected. *Anguilla* can detect phenylethyl alcohol in an experimental situation at a calculated threshold dilution of 3.5×10^{-19} M (Hara, 1971). Invertebrates may be equally sensitive. However, unlike sound, the effective range of chemical stimuli is obviously dominated by local conditions. Deep sea currents can be quite strong; on the other hand large near-stagnant regions of water probably exist through which chemical stimuli can only diffuse slowly. Stratification of the water column may influence the range of chemical signals (Chapter 1).

The prevalence of microsmatic male fish is evidence of the significance of chemical attractants in their sexual reproduction. Shoals of the bathypelagic fish *Cyclothone* appear capable of emitting chemical stimuli sufficient to attract males, but in this case Marshall suggests the population density of females is high with only a few metres distance between individuals. Perhaps the

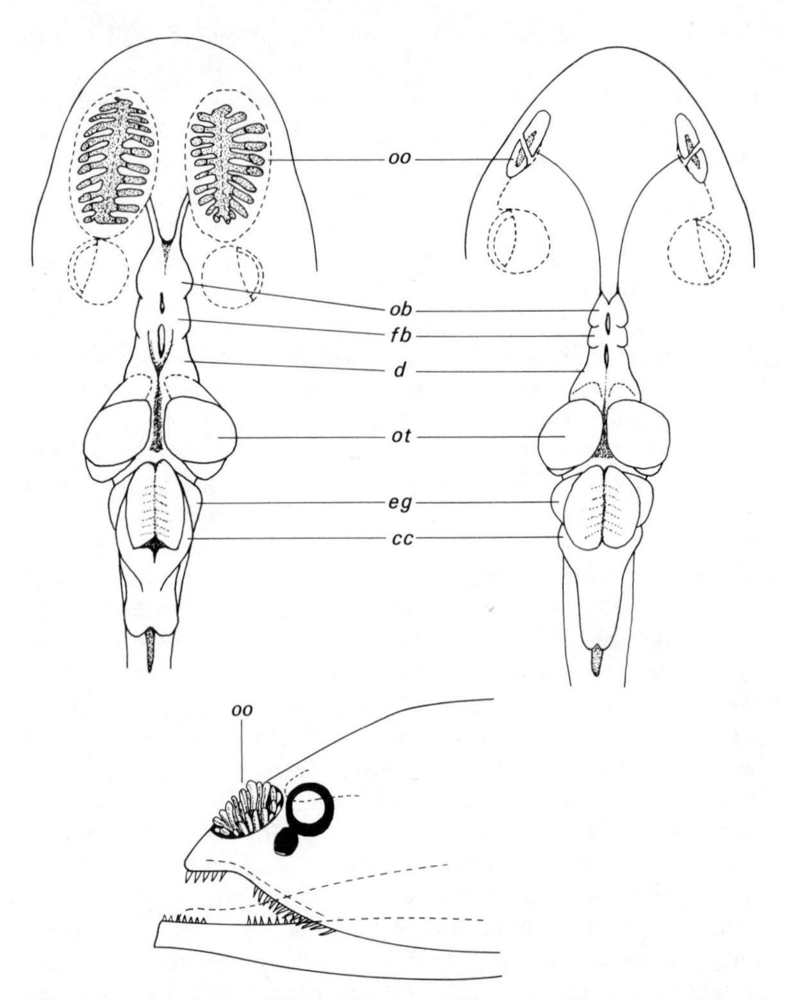

Fig. 5.17. Olfactory organs and brain of *Cyclothone microdon*. Left, male (× 16); right, female (× 16); bottom, head of male (× 16). *oo*, olfactory organ; *ob*, olfactory bulb; *fb*, fore-brain; *d*, diencephalon; *ot*, optic tectum; *eg*, eminentia granularis; *cc*, corpus cerebellum. (Marshall, 1967)

role of the chemical attractant is to achieve synchrony in breeding and it may be just one of several sensory modes used to hold the population of fish together. The ability of male angler fish to find females with which to establish a parasitic union is described on p. 21.

Sense of depth

Animals which can sense their ambient pressure possess a sense of depth. Fish with swimbladders which are used to achieve neutral buoyancy are capable of detecting a change in pressure, and so are certain littoral crustaceans (Blaxter & Tytler, 1972; also Chapter 4). As mentioned on p. 233, fish may determine their depth acoustically. Nothing seems to be known about depth sense in deep sea animals.

Locomotion and buoyancy in the deep sea

INTRODUCTION

The resistance to the movement of a macroscopic animal through water is determined by its size and velocity, and the density and viscosity of the water. The density of seawater increases steadily with increase in pressure whereas viscosity changes in a biphasic manner, reaching its lowest value at about 500 atm. Neither the viscosity nor density of seawater changes more than 4 per cent as a result of naturally occurring pressures so deep water offers virtually the same resistance to macroscopic animals as does shallow water.

In the case of the movement of microscopic animals and moving parts such as cilia, it is conceivable that pressure may affect the behaviour of water at their surfaces. The small size and the complex shape of microscopic animals, especially small planktonic Crustacea, further complicates the issue. For example, in the swimming of the shallow water copepod *Labidocera trispinosa* Vlymen (1970) has argued that in the 'hop and sink' progression typical of copepods, the bursts of rapid acceleration are not an energetically expensive form of progression. In fact, it is argued that the energy cost of movement through the water is negligible in the animal's total energy budget. Vlymen points out how the animal derives benefit from rapid movement in escaping predators without paying a high cost. It is difficult to imagine such a state of affairs in macroscopic animals where rapid acceleration involves a mass of muscle which undoubtedly requires significant energy from the animal's metabolism. Some bathypelagic fish appear to have relinquished powerful muscle and live a quiescent and economic life.

To a first approximation then, the deep sea does not present mid-water animals with special mechanical problems of locomotion. This is also true of those animals which crawl over, or burrow in, deep sea sediments where the nature of the sediment determines the problems of locomotion. This somewhat unexciting observation is worth making if only because the floor of the deep sea, covering twice the area of the terrestial environment, is populated by such creatures. Nevertheless, these same benthic organisms may well be interesting in respect of their buoyancy.

Animals tend to sink because protein-based tissues and skeletal materials are denser than seawater. Lipids, certain body fluids, and gases in bulk are lighter than seawater and are deployed in ways which vary in degree of refinement to bring animals close to neutral buoyancy. At depth, buoyancy devices have to generate expansion against significant hydrostatic pressure, and it will be argued here that uplift is obtained in one of two ways. The increase in partial molar volume of a solute (or its displacement of water) either involves a phase change or a low density arrangement of water around solute particles. Because the problem of deep sea buoyancy forces us to think of molecular events, it is worth noting the molecular forces which act to make tissues denser than seawater in the first place. The case of skeletal salts is simple; these are dense because of their atomic constituents and close packed arrangement. What of proteins? In Chapter 2, mention was made of how a protein, β-lactoglobulin, changed in volume and density on dissolving in water. In the dry crystalline state the specific volume (reciprocal of density) is 0.802 which diminishes to 0.751 in aqueous solution. Thus the solid state protein is less dense than the dissolved substance. Part of the change in density is attributed to the internal pressure of water and part to the electro-striction of the water by charged groups on the protein. β-lactoglobulin is probably typical of many proteins. Ovalbumin in the dry state has a density 1.2655 (Neurath & Bull, 1936), casein 1.318, horse serum globulins 1.279–1.312, and horse serum allumin 1.27–1.28 (Chick & Martin, 1913). When the last three proteins enter aqueous solution at room temperature their density was found to increase by 5–8 per cent, doubtless due to the forces already mentioned.

So far as is known, buoyancy devices in marine animals appear to compensate for heavy tissues as if protein-based tissues have a

TABLE 5.7. *Buoyancy mechanisms in some deep sea animals*

	Species	Depth (m)
	A. Bulk phase mechanisms	
Gas		
(*a*) Gas-filled swimbladder, examples from deep water with well-developed swimbladder presumed functional at depth (Marshall, 1972)	Bony fish	
	Bassogigas profundissimus	5610–7160
	Bassozetus toenia	4570–5610
	Nematonurus armatus	2600–3600
	Lionurus carapinus	> 2000
(*b*) Gas-filled shell (Denton, 1961, 1971; Denton & Gilpin Brown, 1971)	Cephalopods	
	Spirula	< 1000
Liquid		
(*c*) Lipids, regression of swimbladder in adult (Capen, 1967)	Bony fish	
	Lampanyctus mexicanus	Mesopelagic
(*d*) Hydrocarbon squalene in liver (Corner, Denton & Forster, 1969)	Elasmobranchs	
	Centroscymnus coelolepis	~ 1500
	B. Aqueous solutions	
(*e*) Ammonium-rich body fluid (Denton, 1971)	Squid, e.g. *Heliocranchia pfefferi* and many others	(Shallow water)
(*f*) Absence of SO_4^{2-} (Denton & Shaw, 1962)	*Beröe* sp.	(Shallow water)
None		
(*g*) Reduced heavy tissues (Denton & Marshall, 1958; Blaxter, Wardle & Roberts, 1971)	Bony fish	
	Gonostoma elongatum	~ 1000

density of 1.3 compared to the density of normal seawater which is 1.026. Buoyant proteins are an intriguing possibility however.

Neutral buoyancy is likely to be of advantage to an animal for two reasons. It can save energy and it can allow the animal to hover inconspicuously. The energetics of neutral buoyancy have been quantified in a most illuminating way (Alexander, 1972), but the value of being able to hang silently in the water is less easily quantified. Table 5.7 summarises a number of buoyancy mechanisms found in marine animals.

DEEP SEA SWIMBLADDERS

The physiology of the swimbladder of deep sea fish embodies such a range of physiological and physical phenomena that it could well be the centre-piece of a book on deep sea physiology. For two good reasons the topic is only dealt with in summary here; first, because the author has no practical experience in this field and second, because the subject is well dealt with in the literature, having attracted the attention of some brilliant experimenters and lucid writers. Accordingly, for the sake of completeness there follows only a brief description of the structure and function of the swimbladder of shallow water fish and a few comments on the special problems of its function at great depth. Those who wish to take this subject further must refer to papers by Enns, Douglas and Scholander (1967), Kuhn *et al.* (1963), Scholander and Van Dam (1954), Scholander (1954), Denton (1961), Alexander (1966), Steen (1970), and other papers referred to therein.

Fig. 5.18 shows a diagrammatic swimbladder. In some fish it connects by a duct to the alimentary canal, but in all deep sea fish it is a closed sac.

The gas gland is perfused by capillaries which form a long counter current rete which is normally buried in the swimbladder wall. The gas gland is the point from which gas leaves solution as a gas phase. The oval (Fig. 5.23) is another perfused patch which permits dissolved gas to be transported away from the bladder. Note the absence of a rete there. Gas exchange between oval and bladder is determined by muscular control over the extent to which the oval is exposed to the gas contents. The wall of the swimbladder is highly impermeable to gases and compliant. Its volume in relation to the body it buoys is fairly constant, for example in shallow water fish it lies in the range of 3–6 per cent of the body volume. The relative constancy is, of course, due to the approximately uniform densities of fish and seawater. In fish which lay down a lot of lipid during their life span, the swimbladder may be relatively small (Table 5.14).

Marshall's (1972, 1960) thorough studies and reviews of deep sea swimbladders have yielded a number of important conclusions. About one third of all mesopelagic species of fish possess a swimbladder of a size suited for buoyancy purposes. Others, including such common animals as *Cyclothone* spp. and

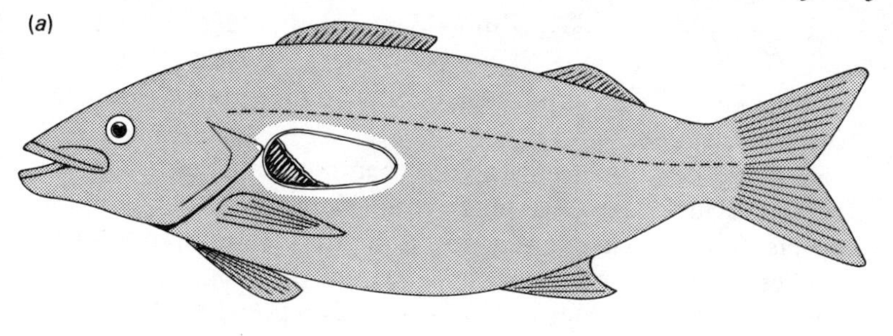

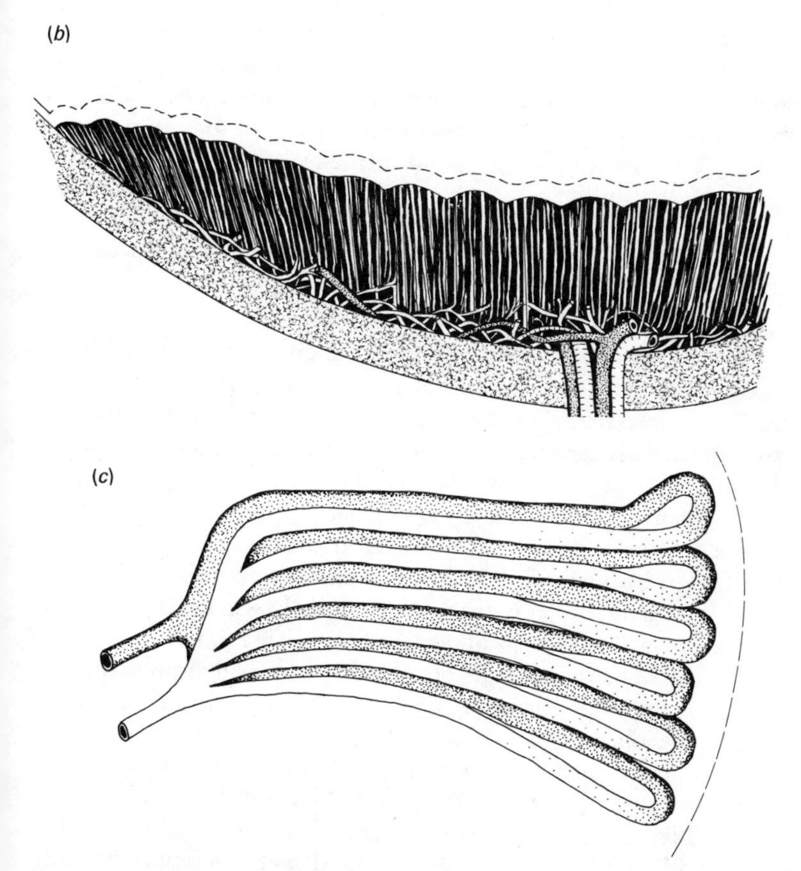

Fig. 5.18. Diagram of a swimbladder. (*a*) shows position within fish body, (*b*) the gas gland and (*c*) the rete. See text.

Stomias spp. do not possess a functional swimbladder in the adult state.

Bathypelagic fish living deeper than 2000 m do not possess gas filled swimbladders.

Fish living near the deep sea floor (benthopelagic) possess a gas filled swimbladder irrespective of depth. Marshall states that of the 16 species known to live in this way at depths greater than 3500 m, 11 possess an apparently functional swimbladder.

Benthic fish generally do not have a functional swimbladder, irrespective of depth. In this context the celebrated case of *Basogigas profundissimus* (Nielsen & Munk, 1964) is often quoted. A single specimen was retrieved from a trawl which collected sedentary animals from the floor of the Sunda trench (7160 m, 1.5 °C). On retrieval it was found to have an undamaged swimbladder, which is difficult to reconcile with the suggestion that it contained sufficient gas at a pressure of 700 atm to give it useful buoyancy. Further, there is no evidence that the individual was actually caught at the greatest depth at which the trawl was deployed and finally, no evidence to show that the animal was even alive when it was caught. With these points in mind we can provisionally regard it as the deepest fish with a functional swimbladder.

One might expect at least two pieces of evidence to establish the maximum depth at which swimbladders function as true buoyancy organs. It is essential to know that the fish in question actually possesses a fully developed swimbladder and not much lipid, and secondly it is important to observe the animal floating at depth. The present evidence suggests that useful upthrust might be obtained from a swimbladder in fish at depths to about 5000 m. The rat tail *Coryphaenoides* has been filmed 'floating or swimming slowly', presumably at a depth of 2000 m (Phleger, 1971).

Pressure is not the most important limiting factor in the distribution of gas filled swimbladders; an adequate food supply appears an equally important prerequisite to their inclusion in the repertoire of fish physiology.

The density of atmospheric gases at deep sea pressures and temperatures is less than that of the liquid hydrocarbons or lipids. Oxygen, the major constituent of the gas mixtures in the swimbladders found at depth, has a density of approximately 0.45 g cm^{-3} at 400 atm (Table 5.8*A*). The upthrust obtained from a given volume of swimbladder gas measured at environmental pressure declines markedly in the bathypelagic zone. Obviously the region where change in depth affects a given gas volume most is the top few hundred metres (Table 5.8*B*).

Functioning of the gas filled swimbladder at depth

Before considering some of the special features of swimbladder function at abyssal depths a brief description of the situation in shallow water is given. Fig. 5.18 shows the rete through which blood flows in counter current fashion. The solubility of atmospheric gases physically dissolved in the blood and the association of oxygen with haemoglobin decreases as the blood enters the efferent capillaries where a higher lactate level prevails. The partial pressure rises in accordance with Henry's Law, causing gas to diffuse across to the afferent rete and, by a continuous process, builds up a high partial pressure of gas in the blood at the swimbladder end of the afferent rete. This high partial pressure sustains that of the gas in the swimbladder.

Taking the rate of blood flow and the trans-rete diffusion of gas into account, an equation defining the partial pressure of gas in the blood at the gas gland end of the afferent capillaries has been derived by Enns, Douglas and Scholander (1967) following earlier papers by Scholander (1954) and Kuhn, Ramel, Kuhn and Marti (1963). In the notation used by Enns *et al.* (1967):

$$P_L = P_0 \exp\left(\frac{K}{F} L \frac{\Delta\alpha}{\alpha}\right) \qquad (2)$$

where P_L = partial pressure of stated gas at the gas gland end of afferent capillaries, P_0 = partial pressure of stated gas at the entrance to afferent capillaries, K = trans-rete diffusion of gas (Q (ml min^{-1} length^{-1} partial pressure difference^{-1}), divided by α the solubility coefficient of the stated gas in distilled water), F = blood flow rate (ml min^{-1}), L = length of rete, and

$$\frac{\Delta\alpha}{\alpha} = \frac{\text{gas solubility in distilled water} - \text{solubility in solution}}{\text{solubility in distilled water}}$$

The important point to note is that K is a very large figure (in other words trans-rete diffusion is fast), and it dominates the exponential term; $\Delta\alpha/\alpha$, or the salting out effect, is a small figure.

TABLE 5.8*A*. *Some properties of substances which contribute upthrust in marine animals* (*all data at atmospheric pressure unless otherwise stated*)

Substance†	Specific gravity (see Table 5.8*B*)	Source	Upthrust ($g\ ml^{-1}$)	Volume giving 1 g upthrust (ml)
Cod liver oil[1]	0.925	Cod liver	—	—
Triglycerides[2,3]	0.91–0.92	Fish body oil, *Euphausia*	0.106	10
Diacyl glycerol ethers[2,3]	0.89–0.91	Elasmobranch liver, e.g. *Dalatias, Squalus, Acanthias*	—	—
Squalene[1]	0.856	Elasmobranch liver, e.g. *Centrocymnus*	0.166	6
Wax esters[2,4]	0.857	(Albatross stomach oil)	—	—
Pristane[5]	0.78	Bony fish, whales, *Latimeria*, small amounts only, no buoyancy function demonstrated	0.246	4
n-pentane[6]	0.626	Not found in organisms, for comparison only	—	—
Oxygen‡ at 400 atm[7]	Density approx. $0.457\ g\ cm^{-3}$	Swimbladders	0.706	1.4
Dry oxygen at 2 atm total pressure 0 °C[1]	Density approx. $0.0028\ g\ cm^{-3}$	—	—	—
Ammonium-rich body fluids (isosmotic)[8]	Density $1.010\ g\ cm^{-3}$	Body fluid, *Heliocranchia*	—	—
Hypotonic body fluid of bony fishes[9]	1.013	Watery body fluid, *Gonostoma*	—	—
Seawater lacking SO_4^{2-} (isosmotic)[10]	1.022	*Beröe*, diatoms	—	—

† Specific gravities of lipids normally refer to the substance at room temperature compared to water at 4 °C.
‡ The critical temperature of oxygen is −118.8 °C, of nitrogen −147 °C and of argon −122 °C. Carbon dioxide has a critical temperature of 31 °C and a critical pressure of 72.8 atm. At its critical temperature and pressure its density is $9.46\ g\ cm^{-3}$ (reference 11).

References for Table 5.8A

1. *Handbook of Physics and Chemistry*, 49th ed. (1968), Chemical Rubber Co.
2. Lewis (1970).
3. Malins and Barone (1970).
4. Nevenzel (1970); Benson and Lee (1971).
5. Blumer, Mullin and Thomas (1963).
6. *International Critical Tables* (1926).
7. Kunkle, Wilson and Cota (1969).
8. Denton, Gilpin-Brown and Shaw (1969).
9. Denton and Marshall (1958).
10. Gross and Zeuthen (1948); Denton and Shaw (1962).
11. Glasstone (1964).

TABLE 5.8*B*. (After Cox, 1965)

Pressure above atmospheric (bars/atm)	Specific gravity of seawater (35‰, 0 °C)
0	1.028
100	1.032
500	1.050
1000	1.070

Plausible figures for $K = Q/\alpha$ used by Enns *et al.* for nitrogen are $65/0.02 = 3250$ ml min^{-1} cm^{-1} at 5 °C.

Taking $\Delta\alpha/\alpha$ as 0.00153 (a value based on the afferent–efferent difference in lactate in shallow water fish), F as 1 ml min^{-1}, and L as 1 cm, then P_L equals 112 atm when P_0 is 0.78 atm as it is in the case of nitrogen. Thus the generation of 100 atm of nitrogen is broadly accounted for, provided leakage through the swimbladder wall is small, which it is in shallow water conditions. It therefore appears that a shallow water swimbladder only requires a long rete if it is called upon to buoy a fish at considerable depth. To reinforce that over-simplifying assertion we can consider the permeability of the swimbladder wall of shallow water fish to very steep partial pressure gradients. Table 5.9 (after Kutchai & Steen, 1971) shows the permeability of the wall of two swimbladders from shallow water fish. The low permeability of the conger eel bladder is attributed to the presence of guanine crystals and is unlikely to be upset by a high ambient pressure (Denton & Shaw, 1962). Kutchai and Steen argue that oxygen will leak through the wall of the conger swimbladder at a rate equal to a plausible rate of secretion

TABLE 5.9. *Oxygen and carbon dioxide permeabilities of swimbladder walls in ml* $(cm^2\ min\ atm\ \mu m^{-1})^{-1}$ *compared with values reported for connective tissue and for water. Readings at STP.* (After Kutchai & Steen, 1971)

	O_2 permeability	CO_2 permeability
Swimbladder of common eel[a]	0.0106	0.308
Swimbladder of conger eel[b]	About 0.001	About 0.03
Connective tissue from frog rectus abdominus muscle	0.115	—
Central tendon of dog diaphragm	—	2.65
Water	0.34	9.75

[a] See Kutchai and Steen, 1971.
[b] See Denton, Liddicoat and Taylor, 1970.

when the partial pressure of the gas equals 250-fold that of the ambient water, i.e., at 250 atm or 2500 m.

The equilibrium pressure of dissolved gas rises with increase in hydrostatic pressure in an exponential manner. This somewhat surprising state of affairs has already been discussed in Chapter 4. The experimental verification of the thermodynamic arguments which yield this conclusion is limited, but in the absence of data to the contrary we will assume, with Enns *et al.* (1967) that dissolved gas activity rises 14% per 100 atm all the way to the greatest depths. These workers make the point that if we regard the sea as equilibrated with nitrogen at normal atmospheric pressure and consider a swimbladder being inflated with nitrogen, then at both 3000 m depth and a hypothetical depth of 13300 m, the ratio of P_{N_2} in the bladder to P_{N_2} in solution in the sea will be 246. The situation with oxygen is more complicated (p. 248) but it is clear that the exponential rise in gas equilibrium pressure with hydrostatic pressure helps rather than hinders the maintenance of gas in a swimbladder at great depth.

Special features of deep sea swimbladders – the rete

L, in equation (2) is increased in deep sea fish. Table 5.10 shows data selected from Marshall (1972) to illustrate this point and the original papers should be consulted to verify the generalisation. The rete in Table 5.10 are uncorrected for body size or swim-

TABLE 5.10. *Length of swimbladder rete in deep sea fish.* (Marshall, 1972, simplified)

Species	Length of rete (mm)	No. of rete	Range of depths where most commonly found (m)
Bassogigas profundissimus	15	1	
Bassozetus taenia	25	2	4570–5610
Nematonurus armatus	25	5	2600–3600
Lionus carapinus	25	6	> 2000
Coryphaenoides guntheri	20	4	1000–2000
Synaphobranchus kaupi	10	2	800–2000
Hymenocephalus italicus	6	2	300–800
Malacocephalus laevis	4	2	150–600
Vinciguerria attenuata	0.8	—	150–600
Myctophum punctatum	2	—	150–600

bladder size but the correlation with living depth is real and clearly connected with the working of the countercurrent multiplier.

The energy required to maintain the high partial pressure of gas within the swimbladder has been estimated by Alexander (1972). If the partial pressure of gas (oxygen or nitrogen) in the ambient seawater is P_w and the partial pressure in the swimbladder is P_s, then the work of compressing the gas is $RT \ln (P_s/P_w)$; R and T are the gas constant and absolute temperature respectively. Diffusional losses must be matched by the secretion of gas to maintain a steady state. Let D represent the number of moles of gas diffusing out of the swimbladder, a quantity which can be computed from permeability data and measurements of a particular swimbladder. Thus the steady state power required to maintain a swimbladder volume constant at depth is $DRT \ln (P_s/P_w)$. Using permeability data from the conger eel swimbladder and arbitrarily assuming that gas secretion proceeds with an efficiency of 5 per cent, Alexander calculates a 1 g fish would expend energy equivalent to respiring 35 $cm^3\ O_2 kg^{-1} hour^{-1}$ at 100 atm pressure to maintain its swimbladder full of oxygen (Fig. 5.19). This is 17 times the amount of energy required at 10 atm pressure, and would appear to be rather a high rate of expenditure to apply to a real fish.

The efficiency of gas secretion in even shallow water fish may be much higher than 5 per cent. Alexander's argument shows that

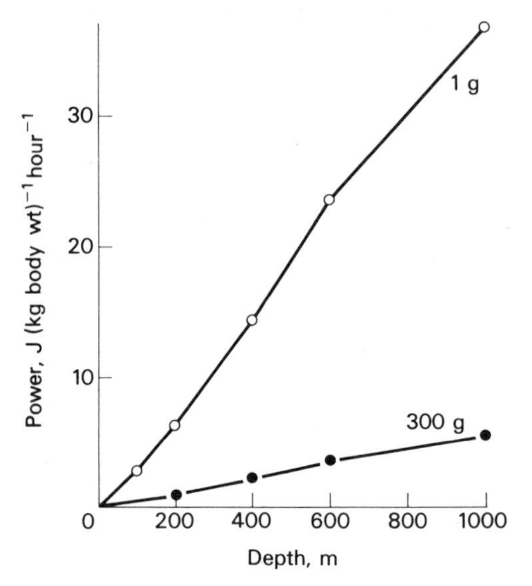

Fig. 5.19. Graphs of estimated power required to replace oxygen diffusing out of the swimbladder, against depth. Separate graphs are presented for a 1 g and a 300 g fish. An ambient P_{O_2} at the following depths is assumed: 0.2 ATA down to 200 m, 0.02 ATA at 400–1000 m (approximately). (Alexander, 1972)

swimbladders may become metabolically very demanding at great depth and we might reasonably anticipate adaptations in deep sea swimbladders to improve their efficiency.

Composition of the gas in deep sea swimbladders

Oxygen is the major gas in deep sea swimbladders, but nitrogen and argon are usually present at partial pressures in excess of that prevailing in the ambient seawater. A small amount of carbon dioxide is also present. Fig. 5.20 shows the results of Scholander and Van Dam (1953) who analysed the swimbladder contents from many different species. These results are consistent with a gas secreting mechanism which works in a physical rather than a selective chemical way. High concentrations of oxygen were also found in the swimbladders of hatchet fish retrieved from the oxygen minimum layer at a depth of about 800 m (Table 5.11). Under these conditions the P_{O_2} of the gas was 10000 times the

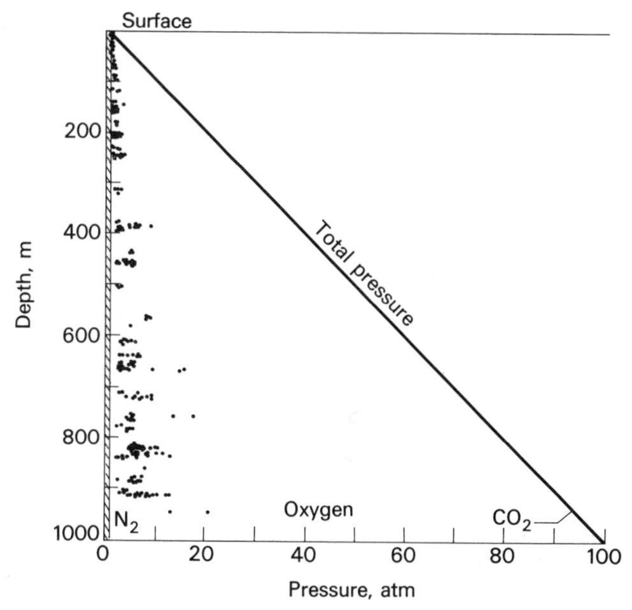

Fig. 5.20. Composition of gas in the swimbladders of deep sea fish. The partial pressures of nitrogen, oxygen, and carbon dioxide have been calculated from the composition of the gas mixtures and the depths of origin. The shaded area is equal to the partial pressure of the nitrogen in the seawater. The partial pressure of nitrogen in the swimbladder is to the left of each point, that of oxygen to the right. The partial pressure of the carbon dioxide is usually less than that represented by the thickness of the diagonal line. The over-all picture suggests a linear increase in the partial pressure of the nitrogen with depth. (Scholander & Van Dam, 1953)

ambient P_{O_2} which may still be accounted for by the 'salting out' mechanism of gas secretion. Note also the volume of gas present in these animals.

Deleterious effects of high partial pressures of gases

At partial pressures in excess of about 1 ATA oxygen is toxic to intact animals and impairs the function of many *in vitro* preparations.

Nitrogen at partial pressures of about 4 ATA exerts a depressive action on the mental performance of men, and at slightly higher pressures narcotises a number of physiological systems. The gas gland of the swimbladder works at elevated partial pressures of both these gases, and in particular, the gas gland of deep sea fish

TABLE 5.11. *The volume and oxygen content of the swimbladder gasses from oceanic fish.* (After Kanwisher & Ebeling, 1957)

	% O_2	Swimbladder gas as % volume of inflated fish at the surface
Surface pelagic species		
Exocoetidae		
Exocoetus monocirrhus	33	3.5
Cypselurus xenopterus	23–25	
Hirundichthys sp.	33	3.1
Hemiramphidae		
Hemiramphus sp.	29	
Deep water, fishes taken at the surface		
Myctophidae		
Myctophum evermanni	74–91	3.2–5.0
M. spinosum	77–91	4.0–5.5
M. affine	90	
Fishes taken at depth (all below 100 m)		
Sternoptychidae		
Argyropelecus lychnus	72–92	32–65
A. affinis	67–93	42–55
Sternoptyx obscura	71–88	
Vinciguerria lucetia	78–90	32–52
Ichthyococcus sp.	77–83	
Melamphaidae		
Melamphaes macrocephalus	73–87	37–65
M. nigrofulvus	67–77	56
M. opisthopterus?	71	
M. nycterinus	54–93	48–54
M. cristiceps?	61–81	
Myctophidea		
Lampanyctus omostigma	89	53
Lampadena bathyphila	85	

functions at a P_{O_2} of 100 or more atmospheres. D'Aoust (1969) has brought this into sharp focus by subjecting the epipelagic fish *Sebastodes miniatus* to a partial pressure of oxygen which only its gas gland would normally encounter. Fig. 5.21 demonstrates the fact that the gas gland functions normally in a P_{O_2} lethal to the whole animal. It is not yet clear if a gas gland which tolerates a few atmospheres partial pressure of oxygen would remain unaffected by a P_{O_2} of 100 atm. Intuitively, one might suppose further adaptive changes would be necessary. In this connexion pressure

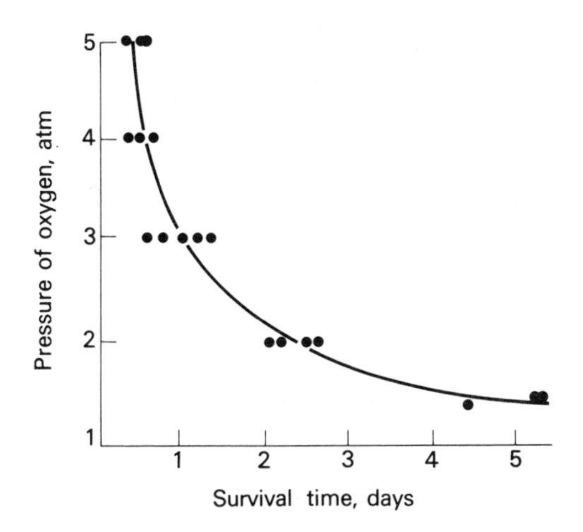

Fig. 5.21. Effects of an elevated P_{O_2} on *Sebastodes miniatus*. Survival time (horizontal scale) is plotted against oxygen pressure. In order to rule out the effects of hydrostatic pressure, the lowest partial pressure of oxygen (1.5 ATA) was also administered to fish in the chamber as compressed air so that the total pressure was 7.5 ATA. There was no significant difference in the survival time of fish treated with these two gas mixtures. (D'Aoust, 1969)

itself may play a role in modifying the potency of the gas (Chapter 7).

The sensitivity of gas glands to high pressures of nitrogen has not been studied. In general, there is at present little evidence to show that they are exposed to partial pressures much greater than 20 atm, but such pressures are, of course, mildly narcotic in some systems. Hydrostatic pressure can enhance or antagonise the narcotic action of nitrogen in a number of preparations (Chapter 7).

In a shallow water species, gas bubbles have been seen in a live gas gland. Lipids rich in cholesterol and phospholipids have been found within the swimbladder of *Coryphaenoides acrolepis* and *Antimora rostrata* (Phleger & Benson, 1971, Patton & Thomas, 1971), and it is conceivable that these substances are connected with the secretory process. The role and metabolism of lipids both inside and outside swimbladders of deep sea fish is obscure.

$\Delta\alpha/\alpha$ *at great depth*

A generalised salting out of dissolved gases could give rise to a multiplying effect in the rete sufficient to generate gas pressures in excess of those needed in the deep sea. In addition, a shift in the equilibrium $HbO_2 \rightleftharpoons Hb + O_2$ could also cause oxygen to leave the blood and enter the swimbladder. From solubility measurements carried out by Enns *et al.* (1967) the ratio $\Delta\alpha/\alpha$ for argon and nitrogen can be computed for different molarities. It would be interesting to know how hydrostatic pressure affects these data (Table 5.12).

If lactate released into the efferent capillary blood decreases the solubility of nitrogen in a manner comparable to NaCl then quite high partial pressures of nitrogen may be slowly generated. Lactate also shifts blood pH which might affect the equilibrium shown above. Indeed Scholander and Van Dam (1954) have shown that acidified blood will dissociate off oxygen at quite high pressures (Root effect), but probably not at a P_{O_2} of more than 40 atm. Perhaps other substances can release oxygen from oxyhaemoglobin against higher partial pressures. Fig. 5.22 summarises a recent view of the gas secreting process. Oxygen may be driven from the aqueous and haemoglobin components of blood by two different mechanisms. A pH effect on the oxygen binding capacity of haemoglobin dissociates the oxygen off into solution. A salting out effect drives it from solution to diffuse into the swimbladder. The first mechanism is limited to a P_{O_2} of 40 atm, the second is not limited by a naturally occurring pressure.

The problem of the rate of gas secretion is important. For example, Enns *et al.* (1967) suggest that a hypothetical 100 g fish at 200 atm might take 174 days to fill its swimbladder with oxygen. In short, in swimbladders working at great depth the equilibrium situation is broadly explicable. The rate at which buoyancy equilibrium is achieved in deep sea fish is not known but extrapolation from those systems which have been studied suggests that it may be achieved only very slowly unless supplementary mechanisms exist.

BUOYANCY BY A GAS FILLED, RIGID SHELL

Certain cephalopods achieve neutral buoyancy by means of a gas filled shell (Denton & Gilpin-Brown, 1971, Denton, 1971). The pressure within the shell is atmospheric or less and the shell withstands the ambient pressure. *Nautilus* and *Spirula* are mesopelagic animals living at pressures of less than 100 atm. Their coiled and chambered shells fail catastrophically at pressures of 60 and 170 atm respectively. The mechanism by which the shell, which is

TABLE 5.12. *Solubilities of nitrogen and argon in salt solutions at 20.48 °C.* (Enns, Douglas & Scholander, 1967)

	Nitrogen			Argon		
NaCl conc. (mole cm^{-3})	Solubility (cm^3 cm^{-3} atm^{-1})	$\Delta\alpha/\alpha^a$	(cm^3 $mole^{-1}$)	Solubility (cm^3 cm^{-3} atm^{-1})	$\Delta\alpha/\alpha^a$	(cm^3 $mole^{-1}$)
0	0.01545	—	—	0.03375	—	—
0.0953×10^{-3}	0.01495	0.0324	3.40×10^2	0.03271	0.0308	3.23×10^2
0.1594×10^{-3}	0.01454	0.0589	3.69×10^2	0.03195	0.0533	3.34×10^2
1.784×10^{-3}	0.00827	0.464	2.60×10^2	0.01901	0.437	2.45×10^2
5.259×10^{-3}	0.00293	0.810	1.54×10^2	0.00722	0.786	1.495×10^2

[a] $$\frac{\Delta\alpha}{\alpha} = \frac{\text{(solubility in distilled water)} - \text{(solubility in solution)}}{\text{solubility in distilled water}}$$

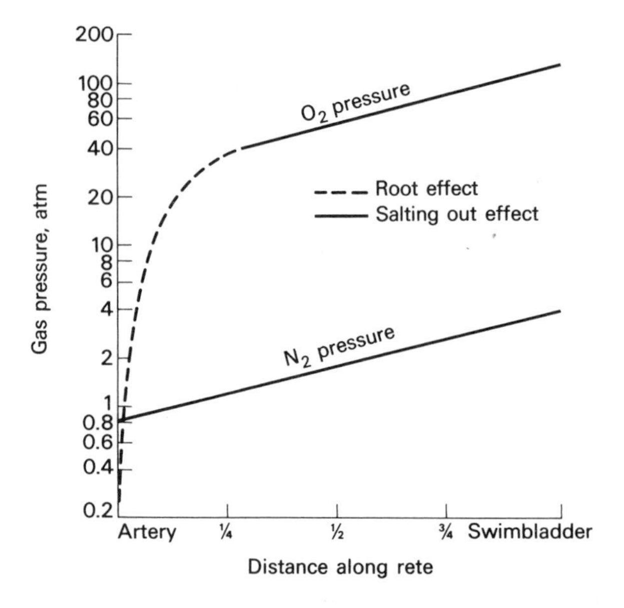

Fig. 5.22. Summary of the development of gas pressure along the swimbladder rete. (See text and Enns, Douglas & Scholander, 1967)

initially filled with water, is evacuated during the growth of these animals, is not fully understood; nor is it clear how the siphuncle, the perfusing stem running the length of the shell, withstands the ambient hydrostatic pressure and remains waterproof. An osmotic pressure gradient maintained by the active transport of salts appears one of the forces which holds back the water pressure, but such a mechanism in its simple version cannot account for the way in which *Spirula* withstands hydrostatic pressures greater than the osmotic pressure of its blood (about 20 atm).

LIPIDS

Some inactive myctophids (lantern fish) show an interesting reduction in their swimbladders and an accumulation of lipid material during their life (Barham, 1971). Capen (1967) has studied examples of this progress in some detail and found both a 'cottony' tissue growing inside and an oily tissue outside the swimbladders of, for example, *Lampancytus mexicanus*. Fig. 5.23 shows the swimbladder regressing and the lipid tissue increasing,

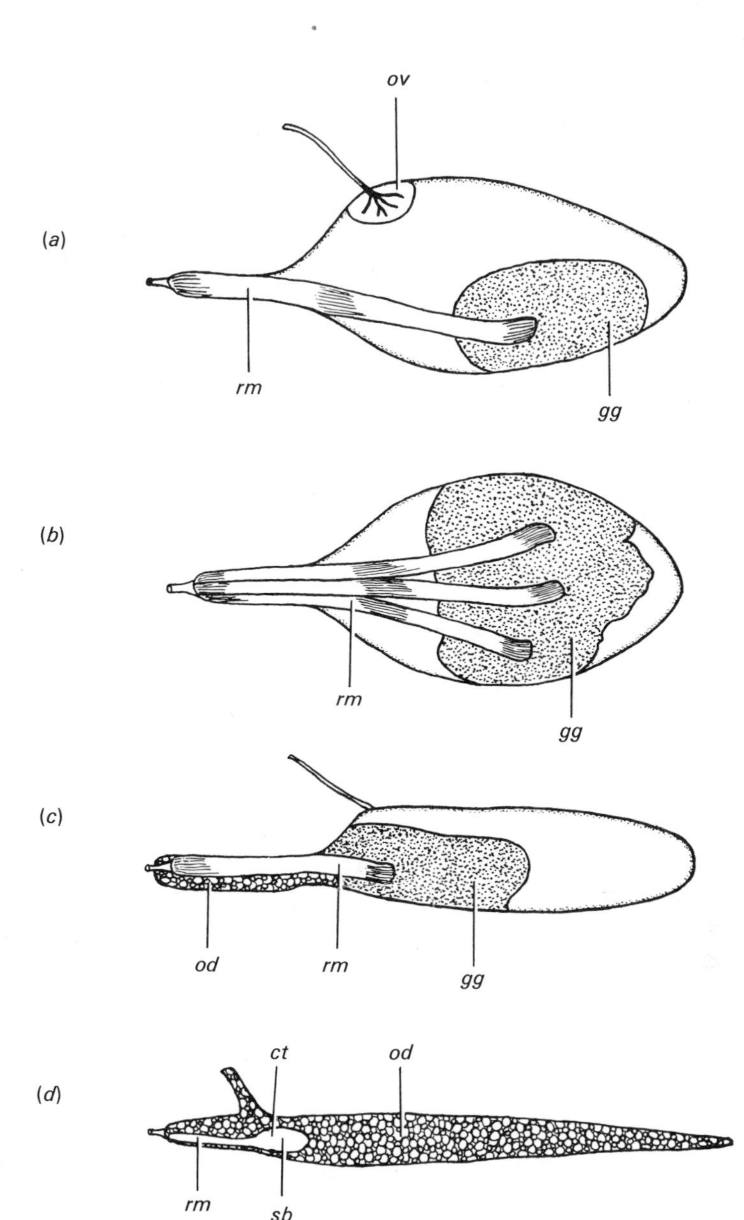

Fig. 5.23. Growth stages in oil-filled tissue surrounding the *Lampanyctus mexicanus* swimbladder. Partially diagrammatic. Rete mirabile and gas gland shown as if swimbladder wall were transparent. (*a*) and (*b*), 22 mm fish, left lateral view and ventral view, respectively; (*c*) 25 mm fish, left lateral view; (*d*) 60 mm fish, leftlateral view; in (*d*) the swimbladder, *sb*, is filled with cottony tissue, and the oval is not apparent. (*a*) and (*b*) enlarged 21½ times; (*c*) 23, and (*d*) 7½ times. *ov*, oval; *rm*, rete mirabile; *gg*, gas gland; *od*, oil droplets; *sm*, swimbladder; *ct*, cottony tissue. (Capen, 1967)

and Table 5.13 shows the relationship between the volume of the swimbladder volume and size of the whole animal during growth. Table 5.14 summarises data showing how swimbladder volume (not necessarily a good guide to gas volume) varies in a number of deep sea fish. The adults of a number of mid-water fish of the deep scattering layer may be slightly negatively buoyant in sea-water and it may be that gas filled swimbladders are too slow to adjust during vertical migrations.

Other myctophids including the genus *Lampanyctus* sp. contain interesting lipids (Nevenzel, Rodegker, Robinson & Kayama, 1969). Certain species contain large amounts of wax esters which probably provide significant upthrust (Table 5.15, e.g. *Stenobrachius*); others, e.g. *Diaphus*, contain negligible amounts. Presumably, the proportion of these substances present in an animal is related to its buoyancy and nutritional economy.

An outstanding example of a liquid buoyancy system is seen in a number of deep sea sharks whose livers contain a large quantity of the hydrocarbon squalene. Its specific gravity is 0.86 and when present in sufficient amounts it is capable of giving appreciable lift (Table 5.8*A*). Heller, Heller, Springer and Clarke (1957) found the liver oils in *Dalatias licha* and *Centrophorus uyato* to contain at least 70 per cent squalene, whereas in many other species, *Squalus* sp., *Carcharhinus* sp. and *Scyliorhinus* sp., squalene constituted less than 1 per cent of the liver oils. Corner, Denton and Forster (1969) observed that sharks such as *Dalatias licha* floated in surface seawater and subsequently demonstrated that the squalene was responsible for the animals buoyancy. This was first seen qualitatively in dead animals which were caused to sink by the removal of their livers. A more precise demonstration involved corrections for the differences in salinity, temperature and pressure between the deep sea water where the animals live and surface seawater in which they were weighed. Table 5.16 shows the example of *Centroscymnus coelolepis* which, in Plymouth seawater, required the addition of 18 g to achieve neutral buoyancy. At the animals normal depth (about 1500 m) the salinity was greater than that of Plymouth water and a corresponding correction was made by Corner *et al.* in the following way. If the volume of the animal is V, and the surface seawater density in which the animal floated with an upthrust of 18 g is d_a, then the change in that upthrust will be $V(d_a - d_b)$ g when the animal is transferred to a different

TABLE 5.13. *Swimbladder size in* Lampanyctus mexicanus *throughout growth.* (Capen, 1967)

Standard length (mm)	Swimbladder Minor axis (mm)	Major axis (mm)	Calculated volume (mm^3)
22	1.3	2.1	1.86
25	0.7	2.1	0.54
26[a]	2.1	3.4	7.84
29	1.0	1.9	0.99
29[a]	1.0	1.9	0.99
31	0.6	1.3	0.24
31[a]	0.9	1.9	0.80
31[a]	1.6	2.4	3.22
32[a]	1.3	2.6	2.30
35	0.8	1.4	0.47
36[a]	1.0	2.3	1.20
44	0.5	1.3	0.17
47[a]	0.8	1.6	0.54
49[a]	1.2	2.3	1.73
54	0.6	1.3	0.24
60	0.6	1.0	0.19
62	0.6	1.2	0.23
64	0.8	1.4	0.47
69[a]	1.0	1.8	0.94

[a] Fresh specimens.

TABLE 5.14. *The swimbladder as a percentage of total fish volume in a number of deep sea fish.* (Capen, 1967)

Species	Range of swimbladder percentage of total fish volume
Lampanyctus leucopsarus	0.001–3.04
L. mexicanus	0.032–6.03
L. omostigma	3.62–6.07
L. parvicauda	1.65–2.04
Diaphus theta	0.050–2.24
Myctophum aurolaternatum	0.57–0.91
Argyropelecus pacificus	0.68–0.81
A. lychnys	0.53[a]
A. hawaiensis	1.02[a]
Merluccius productus	0.61–1.44

[a] Only one specimen examined.

TABLE 5.15. *Lipids of Myctophidae* (Nevenzel, *et al.*, 1969)

Species	Mean standard length (cm)	Mean wt (g)	Tissue	Lipid (%)		Wax esters in total lipid (%)
				Fresh wt	Dry wt	
Hygophum reinhardtii	3.9 (3)[a]	0.58 (3)[a]	Whole fish[b]	3.34	14.0	10 (est.)[f]
Symbolophorus evermanni	6.5 (2)	3.7 (1)	Whole fish[c]	3.1	13.4	10 (est.)[f]
			Muscle[d]	1.16 ± 0.3	5.1 ± 0.7	—
			Viscera[e]	3.1	15.1	—
S. californiensis	No data	1.3 (1)	Whole fish	4.3	19.8	Trace[f]
Tarletonbeania crenularis	No data	0.43 (2)	Whole fish	2.1 ± 0.3	11.4 ± 1.8	5 (max.)
Diaphus theta	No data	2.1 (1)	Whole fish	15.8	57.0	Trace[f]
Stenobrachius leucopsarus	7.0 (17)	4.0 (17)	Whole fish	15.6 ± 1.4	56.4	90.9
			Muscle	18.5	54.7	90.5 ± 1.7
			Viscera[e]	14.4	63.9	90 (est.)[f]
Triphoturus mexicanus	6.0 (4)	1.85 (4)	Whole fish	14.5 ± 0.8	54.1	82.2
			Muscle	16.9	57.8	74.5
Lampanyctus ritteri	7.8 (5)	5.2 (5)	Whole fish	14.2 ± 0.3	50.6	58 (min.)
			Muscle	11.8	51.1	86.7
			Liver	—	—	30 (est.)[f]
			Viscera[e]	9.5	35.4	—

[a] No. of fish in parentheses [b] Three adult females [c] One male adult [d] Nine pooled individuals [e] Includes gut contents
[f] From thin-layer chromatograms.

TABLE 5.16. *The buoyancy balance sheet of* Centroscymnus coelolepis (*fish* ♂). (Simplified from Corner, Denton & Forster, 1969)

	Weights in air (g)
Whole animal	5260
Liver	1550
Weight of liver, % total	29%
Weight of oil in liver, % liver	82%
	Weights in Plymouth seawater (g)
Whole animal (18 g upthrust)	−18.0
Correction for salinity difference between Plymouth and deep sea water (whole animal)	−9.3
Correction for temperature difference between Plymouth water and deep sea water:	
(i) liver oil	+11.5
(ii) animal minus liver oil	+4.1
Correction for pressure difference between Plymouth water and deep sea water:	
(i) liver oil	+2.6
(ii) animal minus liver oil	—
Total net correction	+8.9
Calculated weight of fish in its natural environment	−9.1 g

salinity seawater of density d_{D}. The effects of temperature and pressure were allowed for separately. The cooler deep sea water caused a change in the upthrust produced by the squalene which was related to the temperature coefficient of expansion of oil and seawater. The upthrust in grams of oil floating in water changes with temperature according to αW, in which W is the weight of oil and α is $d_{\mathrm{B}}/d_2 - d_{\mathrm{A}}/d_1$ where d_{A} and d_{B} are seawater densities at high and low temperatures and d_1 and d_2 are densities of oil at the respective temperatures. Oils are caused to contract a lot more than seawater by a temperature decrease and in the buoyancy balance sheet it is seen that this is the largest single adjustment.

The negative buoyancy of the animal lacking its liver is affected by the cooler deep sea water to the extent indicated in Table 5.16. Corrections for the effect of pressure on the buoyancy of squalene and the negative buoyancy of the animal lacking a liver are both small. The compressibility of squalene has not been directly determined but it probably lies in the range typical of oils generally which is slightly in excess of that of seawater. It would be interesting to know the melting points of squalene and lipids at high pressure and low temperatures. The buoyancy correction is given by $10^{-5} \times (P_B - P_A) \times$ (density of seawater) $\times$ (volume of oil) where P_A and P_B are 1 atm and the normal ambient pressure of the animal respectively. The difference between the bulk compressibilities of protein and seawater is even smaller and Corner *et al.* quite reasonably discounted it in the buoyancy balance sheet (see Fig. 4.23, p. 160). The balance sheet for the individual given in Table 5.16 is quite representative and shows that squalene, the dominant liver oil with no apparent metabolic function serves a buoyancy role in life.

How much swimming effort does a shark have to make in order to move its huge squalene-filled liver through the water? Indirectly this question is answered by Alexander's theoretical studies (1972) which show that, to a first approximation and in small fish, the energy cost of moving a mass of lipid through the water is much less than the cost of providing an equivalent amount of hydrodynamic lift in the absence of the lipid. In many sharks the pectoral fins generate uplift and Corner *et al.* (1969) showed that in neutrally buoyant sharks the pectoral fins are significantly smaller than the fins of similar animals which are negatively buoyant.

The metabolic cost of synthesising squalene has been estimated at 0.7 cal g^{-1} compared to 0.5 cal g^{-1} for oleic acid, but more upthrust is obtained from squalene.

The control of the synthesis of squalene is presumably connected with any control the animals might have over their buoyancy. Figures from Corner *et al.* (1969) suggest that the mass of squalene would have to be controlled to within 1 per cent in order to keep the animal within 0.1 per cent of neutral buoyancy.

Malins and Barone (1970) have demonstrated a possible basis for the regulation of the upthrust provided by the liver of deep water sharks. They used *Squalus acanthias* whose liver does not contain large amounts of squalene. Although the whole animal is not neutrally buoyant, the liver does contain significant amounts of diacyl glyceryl ethers and triglycerides. 3 kg animals which were loaded with 100 g weights showed a change in the mixture of liver lipids over 2 days, with diacyl glyceryl ethers present in higher proportions. The specific gravity of the diacyl glycerol ethers is lower than that of the triglycerides so the changed ratio would to some extent, offset the effect of the weights.

The naturally occurring lipid which might provide most buoyancy in animals, pristane, has a specific gravity of 0.78 but is not found in sufficient quantities to exert much effect. Blumer, Mullin and Thomas (1963) have shown that starving *Calanus* do not utilise pristane, which only constitutes 1–3 per cent of the animal's lipid. The energy cost of negative buoyancy in copepods may well be negligible in the overall economy of the animal anyway.

NO BUOYANCY DEVICE

A condition of near neutral buoyancy is reached in some bathypelagic fishes by a reduction of the skeleton and muscles (Denton & Marshall, 1958). The buoyancy balance sheet of *Gonostoma elongatum* for example reads as follows. The weight of protein in water is 1.1 g per 100 g of fish, and the weight of skeletal material is also 1.1 g per cent. Upthrust is provided by the dilute body fluids, 1.2 g per 100 m fish, and from lipid to the extent of 0.5 g per cent. Thus without bulk lipid or gas buoyancy devices this fish weighs only 2.2–1.7 g or 0.5 g per 100 g in seawater, which approximates a fifth of the weight in water of a normal shallow water fish with its swimbladder removed. This ingenious economy is probably closely related to the animal's nutritional economy and feeding habits.

AQUEOUS SOLUTIONS

Certain marine animals, representatives of which may live in the deep sea, reduce their weight in water through the regulation of their ionic contents. The coelomic fluid in a variety of squid and a few Crustacea is isosmotic but lighter than seawater due to the presence of ammonium ions (Denton, 1971). *Gigantocypris mülleri* is one of the deep sea crustaceans which buoys itself in this way.

Other invertebrates appear to exclude SO_4^{2-} ions, thereby reducing weight (Denton & Shaw, 1962). The partial molar volume of SO_4^{2-} in aqueous solution is large and negative; it is an intensely hydrated ion and notably heavy (Duedall & Weyl, 1967). The prevalence of these buoyancy devices in the deep sea is not known. Only a limited uplift is obtained from a large bulk of 'light' fluid.

SOME CONCLUSIONS ABOUT BUOYANCY

The act of generating upthrust either involves motive power or molecular expansion. The deep sea has little effect on the resistance of water to muscular propulsion but exerts a variety of effects on buoyancy mechanisms. Some of the buoyancy devices employed in shallow water are adaptable to great depths. The buoyancy chamber of the cephalopods *Spirula* and *Nautilus* appears severely limited, and a theoretical study of its depth limitations would be an interesting sequel to the full elucidation of the buoyancy mechanism in these animals.

Other buoyancy systems work less well at increasing depths but are not subject to such an abrupt limitation by pressure as is the case in *Spirula*. Bulk phase buoyancy (Table 5.7), utilising gas or lipid molecules, is achieved by a phase change and the accompanying expansion of the system is due, in the case of lipids, to the formation of hydrophobic bonds (Chapter 2). The energy required to generate the gas pressure required in deep sea swimbladders is not particularly demanding, but the work required to sustain adequate buoyancy against diffusional losses, or to vary buoyancy daily, seems high.

Aqueous systems are different as no phase change is seen. Uplift is obtained by the selective maintenance of ions in which the process of molecular expansion has already taken place. Heavy and heavily hydrated ions are excluded. Although little upthrust is obtained by light aqueous solutions, deep sea animals may well be found to make extensive use of this type of buoyancy device.

The energy cost of generating upthrust at depth has yet to be investigated. Alexander's theoretical approach points the way. Making plausible assumptions, Alexander (1972) calculates that in fairly shallow water upthrust by muscular propulsion at constant depth uses about 12 times the energy which upthrust from a swimbladder would cost. Upthrust obtained from buoyant lipids is intermediate in its energy cost. At the 1000 m (100 atm) depth level the energy saving advantage of a buoyant swimbladder seems greatly reduced, but the widespread occurrence of swimbladders at depths greater than this suggests otherwise.

As we descend into the depths Nature has increasing difficulty in providing neutral buoyancy to save energy which, in turn, is becoming increasingly scarce. A more satisfactory description of these interacting forces and solutions to them can only come from physiological study of specific cases. Barham (1971) has suggested how two types of mesopelagic myctophids may be distinguished. There is an active type, exemplified by *Myctophum*, which have muscles, and in those cases which possess a swimbladder it is well developed and perhaps capable of rapid gas exchange. It is these active myctophids which migrate all the way to the surface at night and are members of the deep scattering layer in daytime. Inactive myctophids like *Lampanyctus mexicanus*, with a reduced swimbladder, high lipid content and weak body musculature, migrate over a lesser range and many have been seen to hang vertically and immobile in the water during daylight hours (Fig. 5.24). Barham has also made the interesting observation that these suspended animals may obtain upthrust from their opercular pump mechanism.

The biochemistry, physiology and behaviour of mesopelagic fish have a lot to tell us about how different design-solutions to a common physical environment are arrived at. The deep sea elasmobranchs, far less numerous than myctophids but rather more robust as experimental animals, may prove good bathypelagic animals to study.

Summary

Deep sea animals, like all others, use sensory information to maintain contact with their species, to find food, and to avoid being eaten. Deep sea sensory physiology is quite clearly distinctive. There is a fair measure of understanding of the physics of the sensory stimuli at depth. A certain amount is known of the gross processes involved in vision but experimental studies of hearing or chemo-reception in deep sea animals are almost entirely lacking. Experiments *in situ* are required to provide data on real, as opposed to laboratory, threshold sensitivities and the range of communication at depth.

Depth perception and orientation have yet to be investigated in deep sea animals. Deep sea fish, some of which possess the maximum practical sensitivity to light, may also exhibit extreme capabilities in these areas.

The sensory physiology of deep sea animals may be viewed as a hugely complicated response to the special nature of the environ-

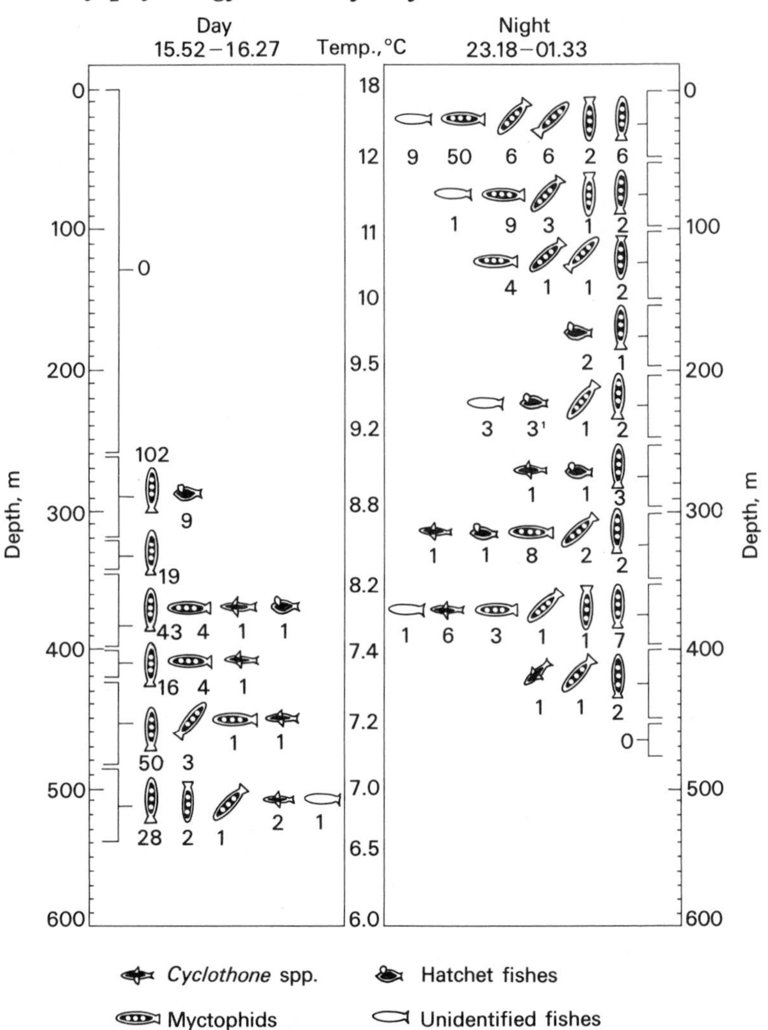

Fig. 5.24. A diagram showing numbers and posture of fishes observed from a deep submersible vehicle during two dives, 7 hours apart, in the San Diego Trough. Temperatures measured during the day-time dive are shown between the columns. (Barham, 1971)

ment. It has evolved in a context which requires economy of energy and has demanded maximum inventiveness with the minimum of raw materials. In the next chapter we will consider the supply of food energy in the deep sea in some detail.

6 DEEP SEA NUTRITION

Introduction

Life in the deep ocean depends on photosynthesis (primary production) which is restricted to a shallow sunlit layer of sea. Net synthesis of organic material is confined to layers in which the amount of organic matter respired is less than that synthesised. This layer rarely extends to 100 m depth and normally only to 30–50 m. Some of the food which is used by deep sea animals derives from land but this is a minor contribution and a source from which shallow seas benefit as well. The organisms inhabiting the deep sea, to a maximum of about 1100 m depth, derive most of their food from the surface sunlit layer. Fig. 6.1 illustrates the contrast between a productive shallow sea, such as the North Sea, and a typical ocean. The flux of light energy into the sea varies with latitude and season but Fig. 6.1 shows how similar inputs may be dissipated through vastly different water masses. The contrast between deep and shallow seas suggests that the deep sea is a sink into which energy, in the form of organic material, leaks from the surface layers. Both the food material and the animals which feed on it are liable to be enormously diluted in the depths.

A discussion of the nutritional physiology of the organisms inhabiting this region requires knowledge of the quantitative distribution of all kinds of organic material in the deep sea. This, in the terminology of marine ecology, is the standing stock, and its magnitude at any one time and place is related to the rates at which it is being consumed and produced. Measurements of standing stock reveal nothing about the turnover of non-living constituents nor do they indicate the rates at which organisms metabolise or feed, but they do provide an inventory of potential food which is interesting and important to know. The inventory may be classified as follows:

(1) organic material in solution
(2) particulate non-living organic matter

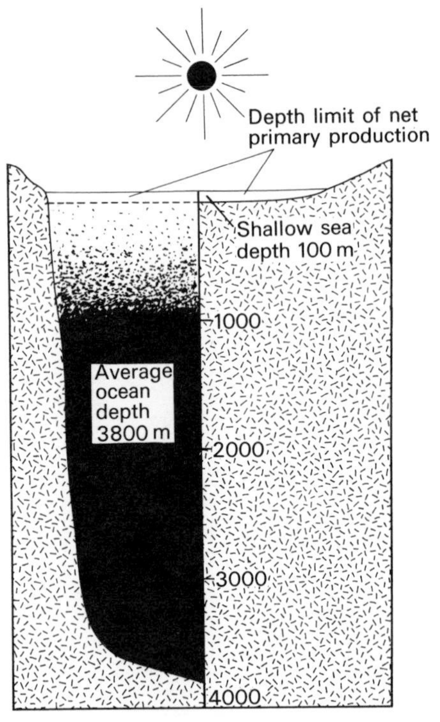

Fig. 6.1. Diagram showing the extent to which photosynthesis is confined to the surface of the oceans. A typical ocean depth is compared to that of a shallow sea.

(3) micro-organisms
(4) plankton (small organisms of limited mobility)
(5) nekton (larger, swimming animals).

For convenience we will deal first with pelagic and secondly with benthic material.

The standing stock

DISSOLVED ORGANIC MATERIAL

According to Duursma (1965), dissolved organic material includes substances which are not precipitated at very low pH or with $Al(OH)_3$, nor are they readily separated by ordinary filters or centrifugation. The significance of such dissolved organic material lies primarily in its potential use in heterotrophic processes. It arises by the leakage of material from living and dead organisms, some of which may be of terrestial origin. A comprehensive list of organic compounds found dissolved in bulk sea water is to be found in Duursma (1965). In summary, carbohydrates are found in quantities in the mg–$\mu g\,l^{-1}$ range, free amino acids in amounts of $\mu g\,l^{-1}$ or less, and carboxylic acids in concentrations between the two. Vitamins, B_{12} for example, are also present and are measured in concentrations of less than 5 $ng\,l^{-1}$ even at considerable depth (Carlucci & Silbernagel, 1966).

The profile in Fig. 6.2, an example of the distribution of dissolved organic carbon (DOC) in the deep water of the Atlantic, shows that an average of 0.60 $mg\,l^{-1}$ DOC occurs throughout the water column down to a depth of 5000 m. It is separated from sea-water samples by ultrafiltration and its concentration is usually determined by combustion to yield CO_2. Local variations in its concentration seem to be mainly attributed to hydrographic or seasonal effects. The identity of the deep water solutes is not known in any great detail but those referred to in Fig. 6.2 underwent little change when concentrated in the presence of marine bacteria and were therefore presumed to be of little value to micro-organisms in natural conditions. Dissolved organic material is by far the largest category of organic material in the sea; Sutcliffe, Baylor and Menzel (1963), and more recently Williams (1971), suggest the representative data in Table 6.1. DOC is slowly recycled through a number of routes. It may be degraded to its constituent parts, it may be concentrated into particles by physical chemical processes and then utilised by living organisms, or it may be concentrated directly into metabolising cells.

PARTICULATE NON-LIVING ORGANIC MATERIAL, POC

Like dissolved organic material, a particulate fraction is distributed throughout the water column. In practice POC is usually distinguished from DOC by filtration, a pore size of 0.45 μm conveniently retaining the former. Fig. 6.3 shows three vertical profiles obtained by Gordon (1970*a*, 1970*b*) in the North Atlantic. Particle concentrations of 10–50 $\mu g\,l^{-1}$ POC are typical of deep water but considerable variations in concentration arise, probably through differences in primary production and water transport processes. These data therefore require earlier ideas of the uniform

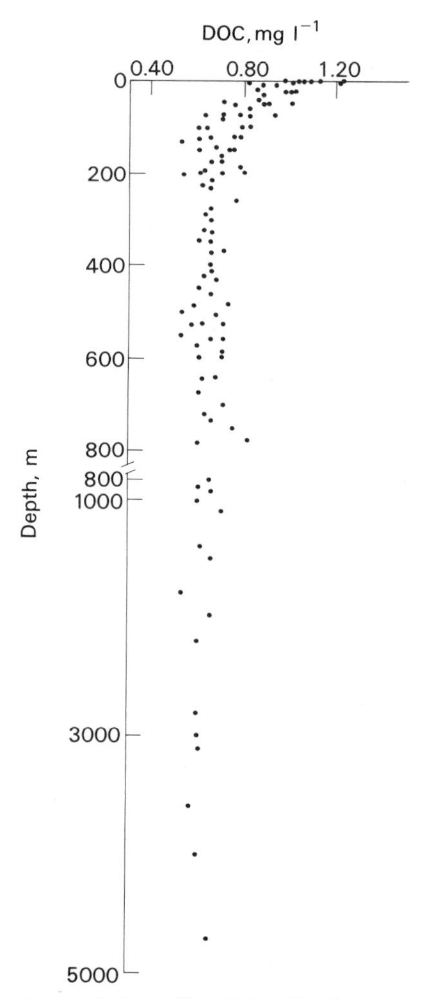

Fig. 6.2. Composite vertical profile of the distribution of dissolved organic carbon in the western North Atlantic Ocean on a section from 40° 43′ N and 69° 11′ W to 35° 25′ N and 67° 17′ W. (Barber, 1968)

distribution of POC with depth to be revised (Menzel, 1967). Non-living particles are of at least four types and these are listed in Table 6.2. Deep water organic particles contain little phosphorus and the carbon:nitrogen ratio increases with depth (Hobson & Menzel, 1969). It is interesting that lipids are absent. Gordon's classification refers to particles larger than 5 μm and excludes

TABLE 6.1. *Organic material in ocean water.*

Sutcliffe, Baylor & Menzel, 1963	(kg per m² of seawater column)
Dissolved organic material	2.4
Particulate non-living organic material	0.5
Living organic material	<0.05
Williams, 1971	
Above 300 m depth, POC $\frac{1}{10}$ of total organic material	
Below 300 m depth, POC $\frac{1}{50}$ of total organic material	

TABLE 6.2. *Classification of particulate organic carbon in the sea.* (After Gordon, 1970*a*)

Particle	Composition determined by histochemical tests	Form
Aggregate	Carbohydrate, some protein	Rounded, includes some phytoplankton and bacteria, also crystals
Flakes	Protein, some carbohydrate	Thin scales, some bacteria inclusions
Fragments	Carbohydrates only	Decaying organic tissue, typically cellulose
Unclassified	Miscellaneous	

faecal pellets and individual crystals. It also excludes particles of clay found in a layer of water known as the nepheloid layer because of its light scattering properties; this layer may extend several hundred metres up from the ocean floor and continental slope (Ewing & Thorndike, 1965). The physical properties of deep water particles are not well known but those found in shallow water show electrophoretic mobility and carry a negative charge (Neihof & Loeb, 1972).

A recent estimate of the amount of particulate organic material in the oceans has been made by Williams (1971) who suggests that

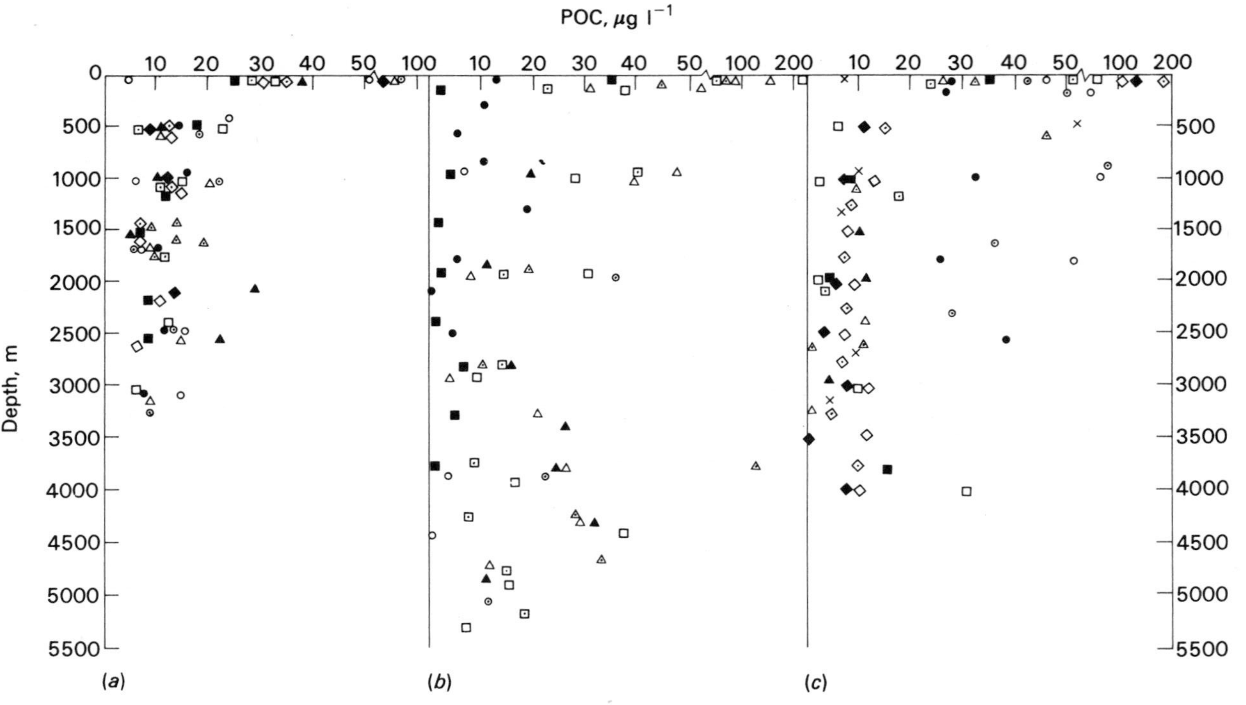

Fig. 6.3. The vertical distribution of particulate organic carbon. Each point refers to a single analysis and each symbol to a locality. (*a*) the Irminger Sea, (*b*) the temperate North Atlantic and (*c*) the Sargasso Sea. (Gordon, 1970*b*)

of the total organic material, POC constitutes about one tenth in waters of less than 300 m depth and only one fiftieth in deeper waters (Table 6.1).

PELAGIC MICRO-ORGANISMS

Eukaryotes and prokaryotes make up a mixed population of micro-organisms which is thinly distributed throughout the oceans. The following terms are sometimes used in the literature: micro-plankton (cells 50–500 μm), nanoplankton (10–50 μm), and ultra-plankton (0.5–10 μm) (Strickland, 1965).

Although the existence of bacteria and flagellates in the nano-plankton category has been known for a long time, macroscopic nekton and plankton tend to dominate our thinking about pelagic life. As judged by metabolic turnover the ultraplankton appears to far outweigh any other category of organism in the shallow seas (Pomeroy & Johannes, 1966). Less is known of the extent of the microbial metabolism going on in deep water but a certain amount of evidence points to the existence of a variety of heterotrophic organisms in the aphotic depths.

Before considering the quantitative distribution of deep sea micro-organisms some consideration has to be given to the methods used to determine the numbers present. The two main methods involve counting cells which are seen with a microscope, and growing cells in suitable nutrient media. Jannasch and Jones (1959) have compared various methods under these two headings.

Counts with a microscope can be made on seawater which has been concentrated by filtration, a procedure which usually involves fixing the cells with 1 per cent formaldehyde. A complicated concentrating procedure known as the Cholodmy method is sometimes used (Table 6.3). The concentrated suspension of bacteria and other particles is then transferred to a microscope slide with a calibrated pipette, and the bacteria in a known volume of seawater are counted. An experienced observer may discriminate between the inanimate particles we have just considered and bacteria, but not between stationary, living bacteria and dead bacteria. Stains have been used to distinguish between living and dead cells but they cannot distinguish between actively metabolising and 'resting' cells.

Culturing methods are also used to enumerate bacteria in

TABLE 6.3. *A comparison of the results obtained from different methods used to determine the numbers of bacteria in seawater (all data compared to the pour plate method).* (Jannasch & Jones, 1959)

	Growth methods			Direct counts		
	Pour plate method (1 and 2)	Extinction dilution method (3)	Macrocolonies on membrane filter (4)	Microcolonies on membrane filter (5)	Direct counts on membrane filter	Direct counts after filtering by the Cholodmy method
Mean	1	21.8	1.16	32.1	147	2100
Range	—	0.3–35	0.2–4.0	1.6–34	13–840	108–9700
% error	—	19.2	17.1	12.7	8.5	20.7

seawater and these, like the microscopic procedures, will be familiar to most biologists. Jannasch and Jones compared five methods (Table 6.3):

1. The agar pour plate method, in which a diluted seawater sample is mixed with cooling agar (40 °C) in a petri-dish. Individual cells are thus dispersed and reveal themselves by growing into a colony of cells which is easily seen with the naked eye.
2. A similar method may be used using silica gel which can be prepared to set at any temperature of biological interest.
3. The extinction dilution method. A seawater sample is divided into 10-fold dilutions down to 10^{-6} or so and each incubated in nutrient broth. The most probable number (MPN) of individual cells in the original seawater sample may be computed from the number of tubes yielding a turbid suspension after a few days.
4. The macrocolony method. Seawater is concentrated by filtration through a millipore filter which becomes inoculated with bacteria. The filter is transferred to cellulose pads which are impregnated with nutrients and colonies grow up after a few days. Each colony is assumed to come from one cell.
5. A similar method may be used with colony counts carried out by microscopic methods. This attempts to cut down the incubation period and eliminates errors due to growth of one colony over another. A more realistic low concentration of nutrients may also be used.

All these methods which improve bacterial growth are subject to uncertainties and the resultant 'count' is highly selective. Microscopic counts generally yield higher figures than growth methods.

The range of these results is typical of that found by other workers who have counted aquatic bacteria and should always be kept in mind in any consideration of bacterial populations in the sea. The presence of clumps of cells, nutritional and other factors which affect growth, such as pressure and temperature, all distort one or other methods. It is hard to avoid the conclusion that a lot of marine bacteria do not grow in the media provided and without exhaustive tests it is impossible to know the precise number of cells and their physiological state in a particular water sample. As recently as 1965 MacLeod discussed the evidence for the existence of 'special' marine bacteria, analogous to marine animals, and concluded that bacteria found in the sea are typically Gram-negative motile rods, psychorophilic, and only distinctively marine in their ability to survive in the sea. According to MacLeod less than 1 per cent of the marine bacteria seen by microscopy can be grown in the laboratory.

Bathypelagic bacteria

The number of cells in deep seawater is very low, typically less than 10^2 ml^{-1} as determined by culture methods. According to Kriss (1960) the number of bacteria is generally related to the accumulation of organic material at interfaces between water layers of differing density, although Gordon (1970*b*) found no particular correlation between bacteria and particulate material. Kriss and co-workers have carried out extensive deep water surveys of bacterial numbers and Figs. 6.4 and 6.5 reproduce some of their results. Cell numbers were determined by macrocolony counts of filtered seawater, the filters being incubated on peptone agar at 25–30 °C for 3–4 days at normal atmospheric pressure. However these data have been criticised by Sieburth (1971) and Bogoyavlenskii (1964) for a variety of reasons, including quantitative errors caused by the contamination of samples and the use of a rich culture medium.

Johnson, Schwent and Press (1968) have isolated bacteria from deep water in the Indian Ocean. Their findings are summarised in Table 6.4. The survey reveals strains of *Spirillum* and *Bacillus* to be exclusively of deep water origin; further work is desirable to substantiate this in view of the obvious influence of hydrography on the distribution of bacteria in the sea.

Other cells

Plant-like cells in the deep sea at least to depths of 400 m were first investigated by Lohmann over 50 years ago (Lohmann, 1920). Fournier (1970) has recently described two types of cell which were obtained from depths around 400 m in the western Atlantic. The most abundant cell type lacked organelles and was 2.5–15 μm in diameter. Smaller and less common cells contained organelles which resembled a chloroplast and a nucleus. These cells are indistinguishable by light microscopy from those lacking

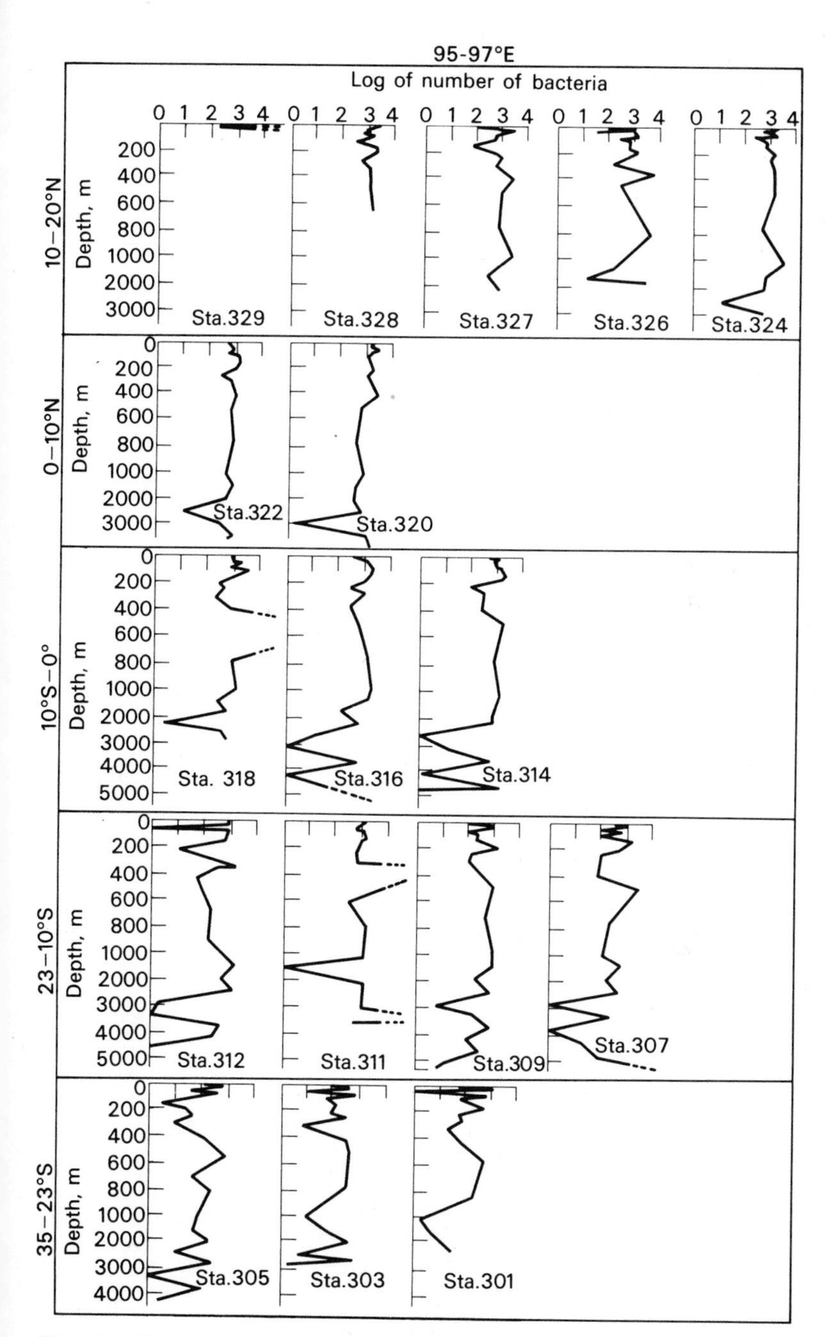

Fig. 6.4. Vertical distribution of bacteria (and similar cells) in the Indian Ocean between 20° N and 35° S. Horizontal scale, log of the number of bacteria per 40 ml volume of water. See text. (Kriss, Labedeva & Mitzkevich, 1960)

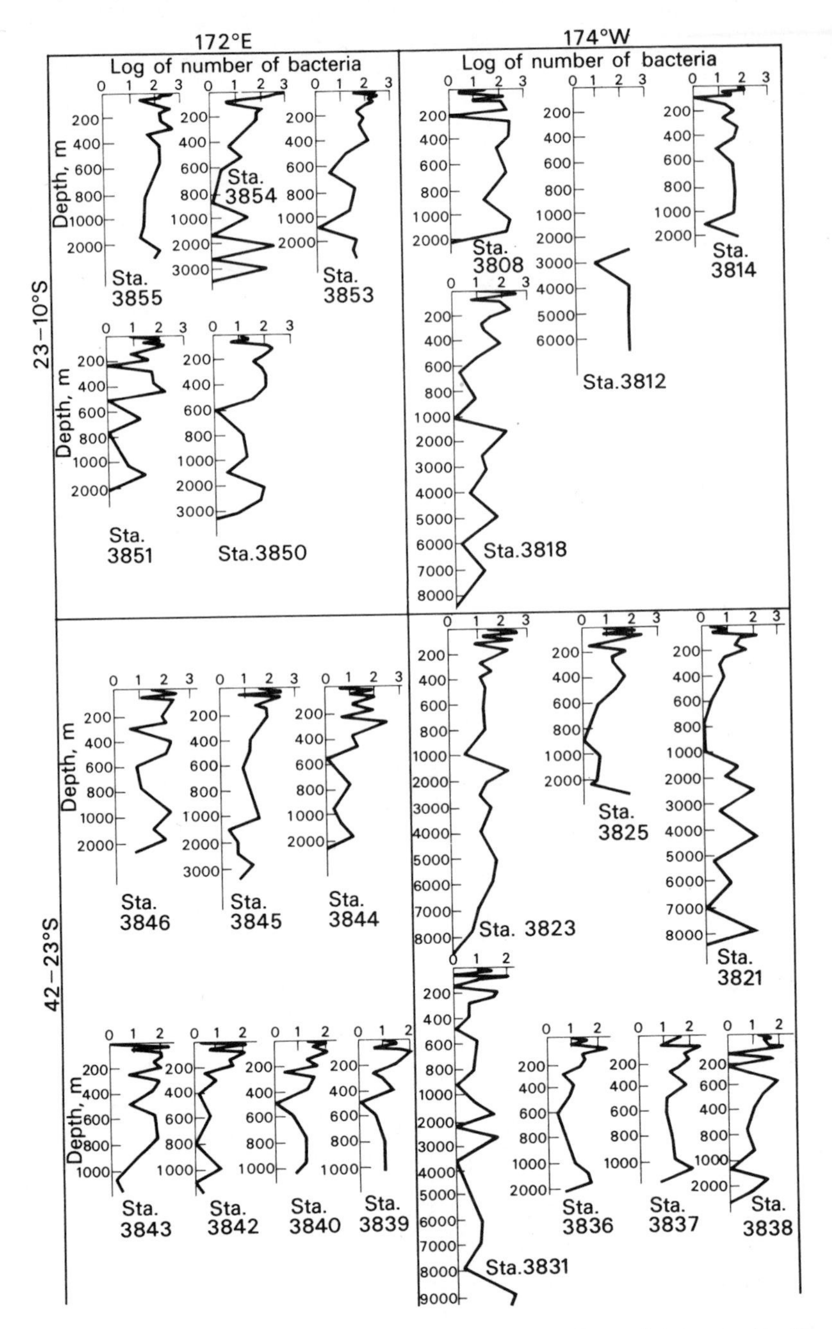

Fig. 6.5. Vertical distribution of bacteria (and similar cells) in the Pacific Ocean between 10° S and 42° S. Horizontal scale, log of the number of bacteria per 50 ml volume of water. See text. (Kriss, Abyzov & Mitzkevich, 1960)

TABLE 6.4. *Types of bacteria from deep water in the Indian Ocean.* (Johnson, Schwent & Press, 1968)

Name	No. of strains	Depth of origin (m)	
Achromobacter	2	200–3000	Pelagic
Flavobacterium	—	200–3000	Pelagic
Pseudomonas	—	200–3000	Pelagic
Alcaligenes	2	200–3000	Pelagic
Vibrio	2	200–3000	Pelagic
Spirillum	2	> 1000	Pelagic
Corynebacterium	—	200–3000	Pelagic and benthic
Bacillus	3	> 1600	Pelagic and benthic

organelles and are probably included in the counts of cells which were filtered from seawater samples from depths to 5000 m in a number of ocean stations. Fig. 6.6 shows the vertical distribution of this heterogeneous population of cells, the so called 'olive green cells' (Fournier, 1971). A recent view is that they are present at depth in numbers related to the productivity at the surface (Fournier, 1972). So far, convincing evidence that these particular cells are living is lacking, but even if they are not they may provide a little food for deep sea animals. Normally, they contain carbon equal to less than 10 per cent of the POC at depth. Their sinking rate is difficult to determine but Fournier suggests a rate of 2.6 cm day^{-1} for a single cell and 100-fold that for small clumps.

Whilst Fournier finds cell concentrations of the order of 5×10^3 cells per litre, Bernard (1958) reports the presence of eukaryote cells in the deep sea at somewhat higher concentrations. Indeed, the claim is made that flagellates may be more numerous at depth than in the sunlit shallow waters. *Cyclococcolithus* sp., *Coccolithus* sp., flagellates and representatives of the Myxophyceae and Dinophyceae occur at depths in excess of 2000 m (Fig. 6.7). As with all quantitative distribution data, we have to consider the methods employed in collecting and counting the organisms. The cells were allowed to settle and then observed with an inverted microscope. No evidence is provided about the condition of the cells and it is clearly of importance to establish whether or not they were alive at their depth of collection. Indeed Fournier (1968) argues that a lot of the cells counted by Bernard were in fact

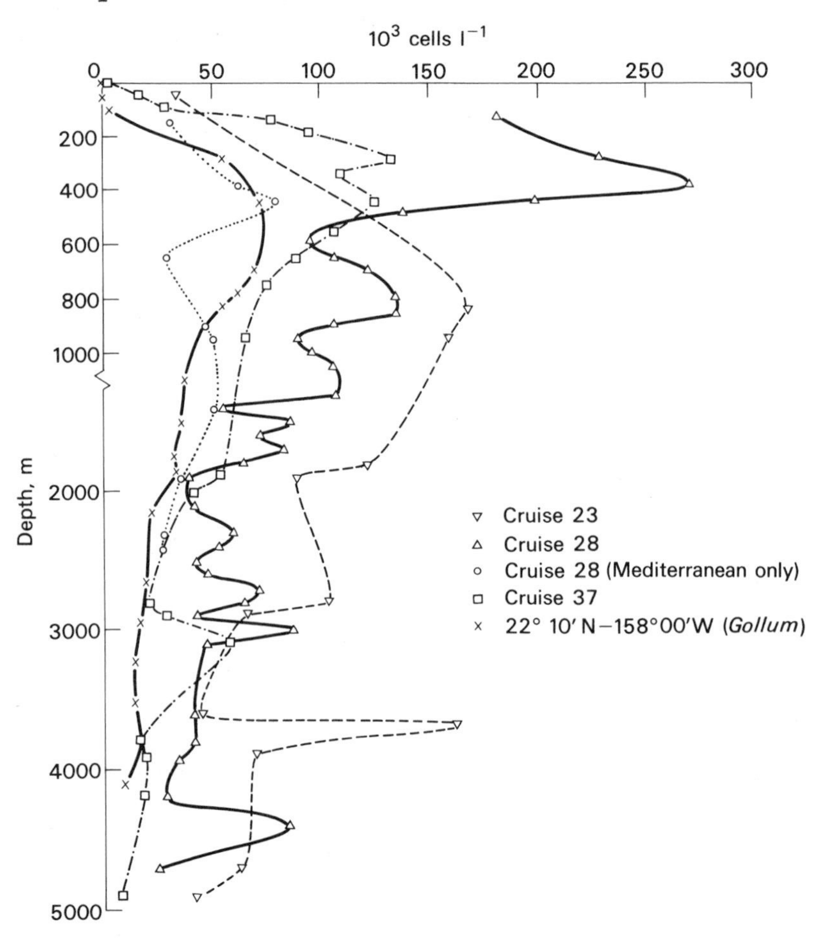

Fig. 6.6. The vertical distribution of olive green cells. Data have been averaged at 50 m intervals in the upper 1000 m and thereafter at 100 m intervals. Cruise 23, tropical Atlantic; cruise 28, North Atlantic to Mediterranean, at approximately 37° N; cruise 37, western Atlantic 20–40° N, and *Gollum* station, subtropical Pacific. For station details see original text. (Fournier, 1971)

inorganic particles, or less likely, cell remnants. Two reasons for Fournier's criticism are first, the density of some of the cells which Bernard calculated from sinking rates exceeds a plausible value, but approximates the density of calcite, secondly, the number of *Cyclococcolithus fragilis* reported by Bernard in deep water off Algeria would yield particulate carbon concentrations equal to the

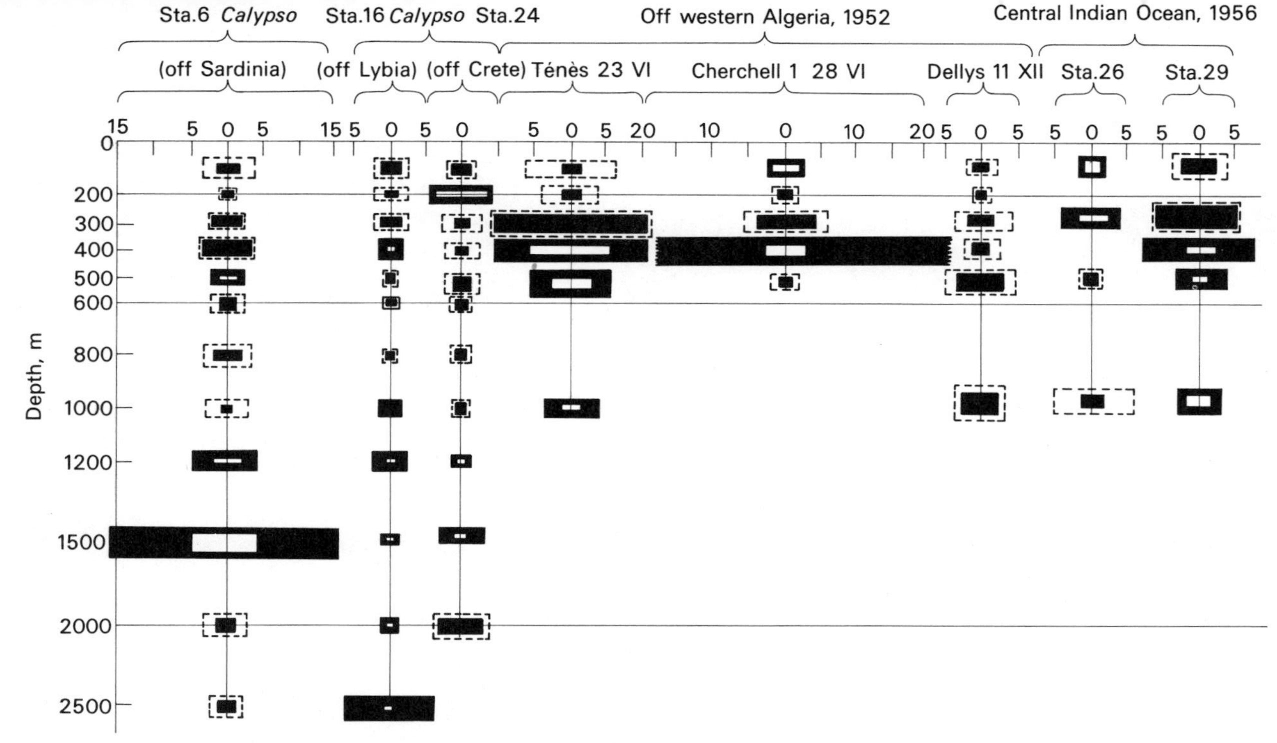

Fig. 6.7. Vertical distribution of *Cyclococcolithus* (palmelloid cells, black rectangles) and naked flagellates (dashed rectangles) in the Mediterranean Sea and Indian Ocean. Horizontal scale, hundreds of cells per ml. See text (Bernard, 1963)

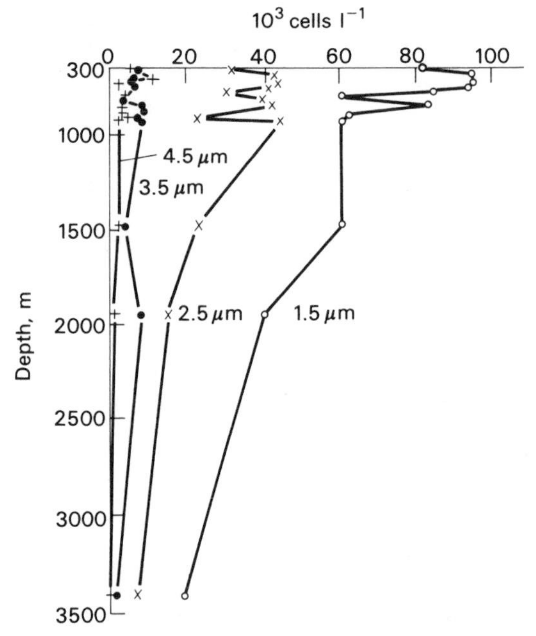

Fig. 6.8. The vertical distribution of various green cell sizes at position 31° 45′ N, 120° 30′ W in the Pacific Ocean. Size indicated refers to estimated cell diameter; cell numbers were obtained by direct microscopic counts. (Hamilton, Holm-Hansen & Strickland, 1968)

total POC values determined in those waters by Fournier. In view of the undoubted presence of other substantial sources of POC in the water Fournier is led to the suggestion that many of the 'cells' do not contain carbon and are not cells at all.

Hamilton, Holm-Hansen and Strickland (1968) have isolated spherical cells from the Pacific not unlike those described by Fournier and in similar concentrations. The cells are 1.5–4.5 µm in diameter and green or yellow in colour. Those from below 2000 m have been successfully cultured in the laboratory, thus demonstrating their potential activity in the sea. At depths between 1000 and 3500 m off California, cell densities of $1–60 \times 10^3$ cells l^{-1} were determined by direct microscopic counts of filtered cells (Fig. 6.8). Water samples were also analysed for their content of ATP which may be converted to 'living organic carbon'. 'Living' carbon thus determined was three times the amount calculated from the viable bacterial counts made by filtering the water sample, incubating the filters for 10 days and counting colonies. Carbon from the larger coloured spherical cells contributed more than 20 per cent of the portion of the 'living' organic carbon at greater depths. In all samples the living organic material represented between 1 and 6 per cent of the total particulate matter at depths below 1000 m. This is in line with the data in Table 6.1.

Yeast and fungi also occur in the mid-water but in very small numbers.

ZOOPLANKTON

Zooplankton contributes a small fraction of the total metabolism being carried out in shallow water where the dominant organisms are the μ-flagellates. In deep water, zooplankton is relatively more important although its absolute numerical density falls off steeply with depth. The existence of deep water planktonic animals poses a number of rather basic nutritional questions, such as, can plankton populations sustain themselves at all depths or are certain deep regions of the oceans 'sterilised'? and what effects do seasonal fluctuations have on the deep sea animals whose food supply is ultimately derived from the photic layer?

Quantitative distribution studies of deep water zooplankton have been carried out since the nineteenth century. A variety of methods have been used, each of which produces selective data. The technicalities of plankton sampling in deep water will be discussed in Chapter 7, here it is only necessary to mention the necessity for sampling known depths using nets or other devices which filter a known volume of water.

Vinogradov (1962*b*) working in the Pacific on the Russian ship RV *Vitjaz* has measured the concentration of zooplankton at depths down to nearly 8000 m. Data are shown in Fig. 6.9. A marked decline in the biomass (which was measured as wet weight and expressed per cubic metre) is seen with increased depth. This suggests a downward translation of material through food chains with progressive diminution in the standing crop at each level. The concentration of deep water plankton is clearly related to the standing crop and productivity at the surface (Fig. 6.9).

Quantitative studies in the Atlantic Ocean have yielded the biomass figures shown in Table 6.5 (Yashnov, 1961). At depths

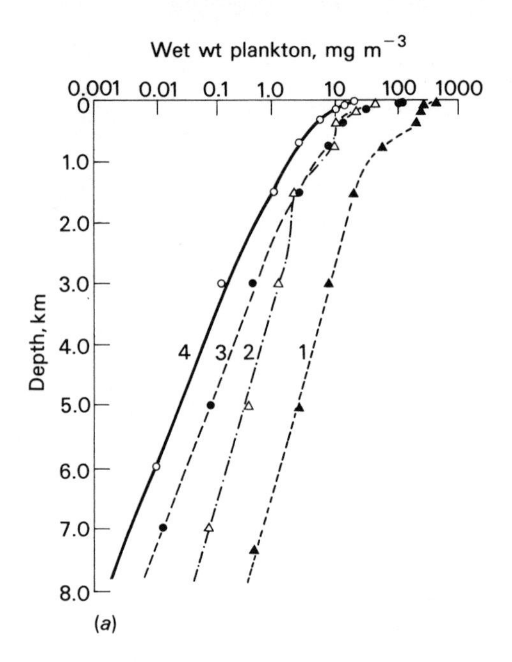

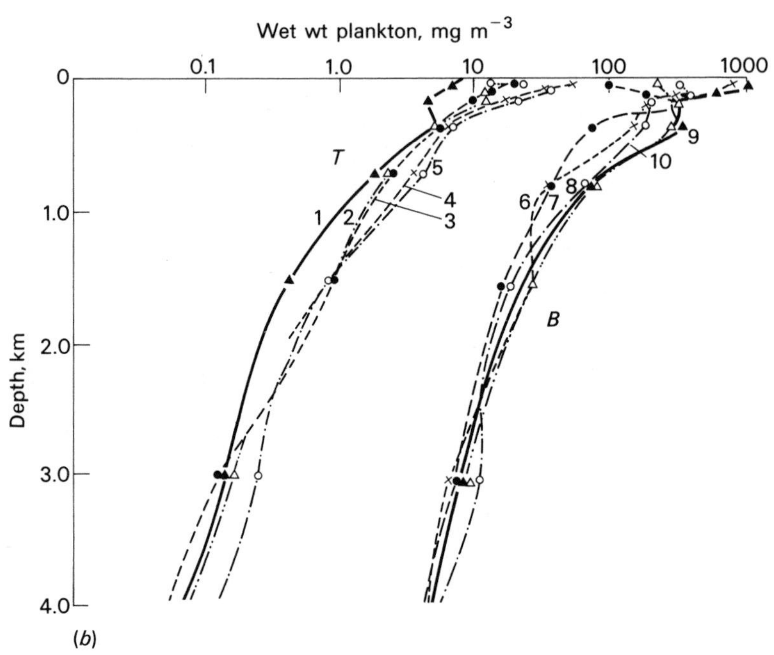

Fig. 6.9. For legend see opposite page.

around 4000 m in the Sargasso Sea zooplankton seems to exist at concentrations of 0.1 mg wet wt m^{-3}, as in the Pacific. The Sargasso Sea is not highly productive but is accessible and has been studied by a number of other workers. Menzel and Ryther (1961) measured the seasonal fluctuations in deep water biomass (Table 6.6) and obtained dry weight data in general agreement with the wet weight data of Yashnov (1961); see Table 6.5. In June the biomass at 2000 m depth is six times the August level and twice that seen in February. Physical conditions may be constant at these depths but seasonal changes at the surface make themselves felt.

The mean annual total number of planktonic animals found in given depth intervals is given in Table 6.7 from Deevey and Brooks (1971). These authors report copepods and ostracods to be the most abundant but at depths of only 1500–2000 m the total number of animals to be found is typically 5 in each cubic metre. Grice and Hulseman (1965) found copepods distributed one to several cubic metres at depths below 3000 m in the north-east Atlantic.

It is to be hoped that these quantitative studies are not in serious error through the inclusion of intact but dead animals. Wheeler (1967) found copepod carcasses more numerous than living animals below a depth of 2000 m in the Atlantic.

It is evident from a number of observations that there is a micro-scale distribution of plankton in deep water. The bathyphotometer of Clarke and Breslau (1959) has revealed a discontinuous distribution of luminescence some of which may come from small animals, and Bernard (1958) observed from a bathyscaph how the numbers of small individuals per volume fluctuated approximately threefold over a vertical distance of 10 m at depths of 1200 m and 2100 m (see also Pere's observation, Chapter 7, p. 358). The small scale

---

Legend to Fig. 6.9

Fig. 6.9. The vertical distribution of plankton in the Pacific Ocean (*a*) in deep sea trenches, and (*b*) in the upper 4000 m in tropical (*T*) and boreal (*B*) regions of the Pacific. The numbers on the horizontal axes refer to wet weight of plankton filtered with a net comprising 14 meshes per cm. In (*a*) the trenches are: 1, Kurile–Kamchatka (average); Kermadec (sta. N 3829); 3, Bougainville (sta. N 3663); 4, Marianas (sta. N 3686–3689). In (*b*) station numbers are 1, N 3810; 2, N 3876; 3, N 3686–3689; 4, N 3804; 5, N 3824; 6, N 2208; 7, N 2119; 8, N 2120; 9, N 2218; 10, 2076. For details of stations see original text. (Vinogradov, 1962*b*)

TABLE 6.5. *Vertical distribution of the mass of zooplankton in various areas of the tropical region of the Atlantic Ocean. Values were obtained from volumetric determinations and are given in* $mg\,m^{-3}$. (Yashnov, 1961)

Depth (m)	Gulf Stream, northern profile, 40–43° N	Labrador Current 43 °N	Continental slope of North America, 38–41° N	Gulf Stream, southern profile, 35–37° N	Sargasso Sea, S.W. part, 20–37° N	Northern Equatorial Current, 16–19° N	Shore region of Africa, 15–22° N	Canary Current, 27–36° N
0–50	74.5	562.0	199.0	54.9	64.7	89.6	1115.0	78.6
50–100	42.3	270.0	94.0	35.9	59.9	80.7	214.0	55.0
100–200	26.2	240.0	35.1	23.3	32.1	30.6	94.7	27.0
200–500	25.4	61.0	30.8	11.1	10.2	15.9	59.9	15.6
500–1000	15.7	—	19.0	6.0	4.1	5.4	27.2	6.5
1000–2000	5.0	—	7.8	2.8	1.2	1.6	—	2.1
2000–3000	—	—	—	2.1	0.6	1.0	—	0.8
3000–4000	—	—	—	—	0.3	0.3	—	0.2
4000–5000	—	—	—	—	0.1	—	—	—

TABLE 6.6. *Seasonal variation in the vertical distribution of plankton in the Sargasso Sea (Atlantic Ocean).* (Menzel & Ryther, 1961)

Depth (m)	Dry weight ($mg\,m^{-3}$)		
	June	August	February
0–100	4.92	1.90	1.16
100–200	1.96	0.64	0.82
200–300	1.22	0.61	1.46
300–400	1.05	0.49	0.95
400–500	0.93	1.54	0.93
500–1000	1.55	0.69	1.21 (500–1250 m)
1000–2000	1.34	0.23	0.68 (1250–2000 m)

TABLE 6.7. *Numbers of planktonic animals in selected depth intervals in the Sargasso Sea. The numbers are A, mean annual total number per* m^3 *(or per* 10^3 m^3 *when in italics), B, mean percentages, both at stated depth intervals.* (Deevey & Brooks, 1971)

	0–500 m		500–1000 m		1000–1500 m		1500–2000 m		0–2000 m
	A	*B*	*A*[a]	*B*	*A*[a]	*B*	*A*[a]	*B*	*A*
Calanoids	89.8	42.1	12.0	43.1	5.4	45.7	2.8	48.8	27.5
Other copepods	60.9	28.0	10.4	37.7	4.2	38.3	2.0	36.4	19.4
Total copepods	150.7	70.1	22.4	80.8	9.6	84.0	4.8	85.2	46.9
Ostracods	14.8	7.0	2.2	7.7	*530*	4.8	*190*	2.8	4.4
Other Crustacea (including larval forms)	7.1	3.3	1.1	3.5	*370*	3.4	*270*	4.4	2.2
Total Crustacea	172.6	80.4	25.7	92.0	10.5	92.2	5.2	92.4	53.5
Tunicates	15.1	6.5	*185*	0.6	*20*	0.0	*20*	0.5	3.8
Chaetognaths	6.5	3.0	*312*	1.1	*25*	2.2	*35*	0.6	1.8
Coelenterates	6.8	3.1	*309*	1.0	*100*	0.8	*13*	0.4	1.8
Larval forms (noncrustacean)	4.4	2.1	*540*	1.9	*115*	1.0	*32*	0.6	1.3
Protozoa	5.1	2.6	*690*	2.5	*320*	3.4	*160*	5.0	1.6
Miscellaneous	4.9	2.3	*103*	0.9	*20*	0.4	0.0		1.3
Total no. per m^3	215.3		27.9		11.3		5.5		65.0
Total no. per m^2	107650		13952.5		5672.5		2750		130025

[a] Italicised numbers are $\times 10^{-3}$, i.e. no. per 1000 m^3.

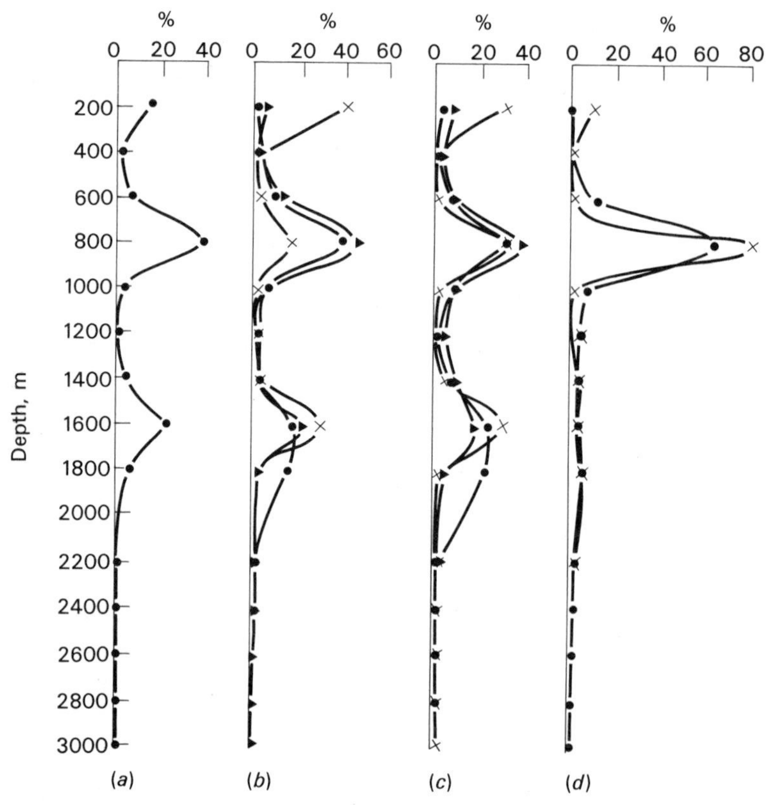

Fig. 6.10. The vertical distribution of plankton from one station in the Sargasso Sea expressed as (*a*) percentages of the total catch, (*b*) the percentage distribution of fish, ●, copepods, ▲ and euphausiids, and (*c*) the percentage distribution of decapods ●, chaetognaths, ▲, and salps, ×. (*d*) the percentage distribution of coelenterates, ●, and 'residue', ×. (Leavitt, 1938)

structure of water masses (Cooper, 1967) may allow plankton to aggregate in relative abundance. Leavitts' (1938) early work on the distribution of deep plankton in the north of the Sargasso Sea demonstrated two peaks of abundance which have yet to be explained (Fig. 6.10).

The vertical distribution of species is difficult to work out but Grice and Hulseman (1965) have obtained data for no less than 187 copepods in the north-east Atlantic (Fig. 6.11). A few species occur over a very large vertical range, notably *Spinocalanus abyssalis* (200–5000 m), *S. magnus* (200–5000 m) and *Mimocalanus*

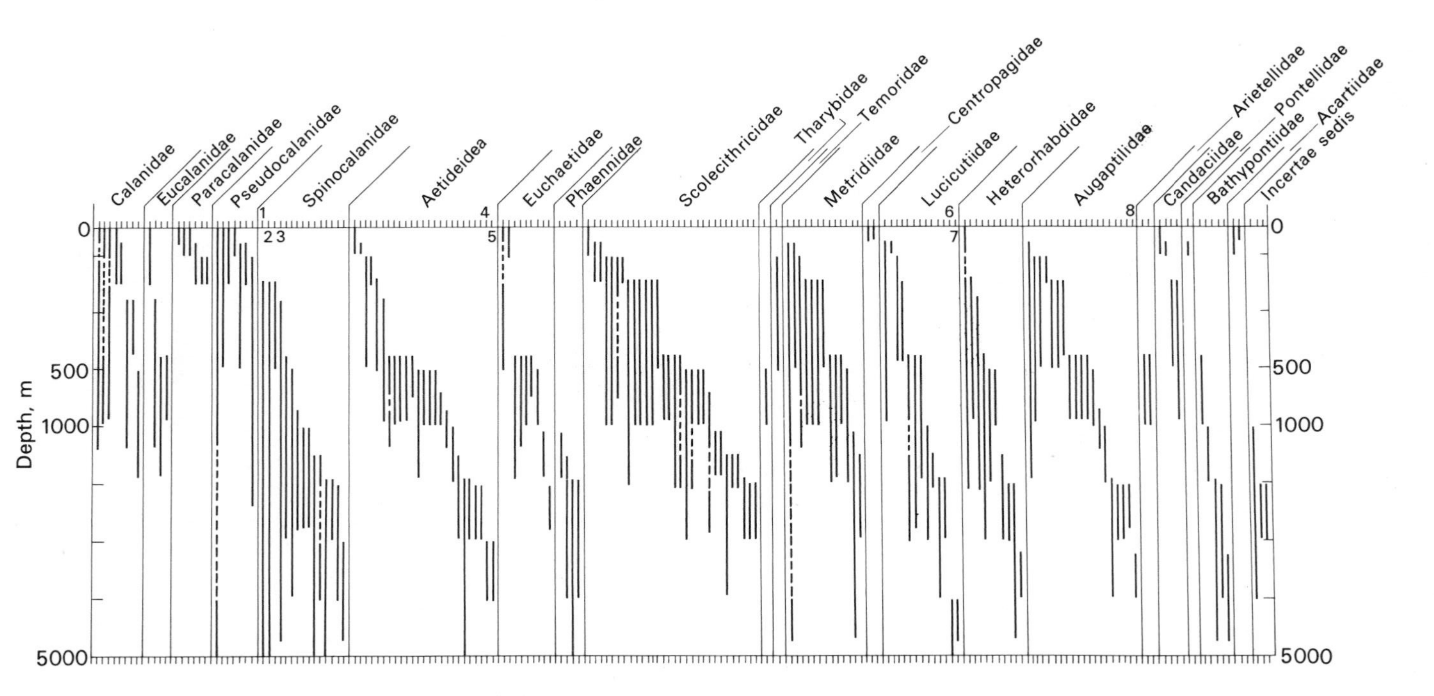

Fig. 6.11. Vertical distribution of species of copepods, grouped in families. Dashed line indicates species was found above and below the depth interval. The numbers correspond to species mentioned in the text; see original text for other species. Note change of scale at 500 and 1000 m. (After Grice & Hulseman, 1965)

cultrifer (300–4700 m), numbered 1, 2 and 3 respectively in Fig. 6.11. Certain species are confined to great depths, e.g. *Bradyetes inermis* (3000–4000 m), *Chiridiella subaequalis* (3000–4000 m), *Lucicutia curta* (4000–5000 m), *L. anomola* (4000–5000 m) and *Eaugaptilus gracilis* (3000–4000 m), numbered 4 to 8. Numbers of individuals per 5 cubic metres range from approximately 20 to 1. The last case can be envisaged as two copepods each about 1 mm long occupying a volume equivalent to a small room. It is difficult to imagine a population of small individuals maintaining themselves permanently in such a dispersed fashion.

A number of copepods not normally thought of as bathypelagic, penetrate to a depth of 2000 m in the Norwegian Sea during the winter. They include *Calanus hyperboreus*, *C. finmarchicus* and *Pseudocalanus minutus*, the latter two being most abundant (Østvedt, 1955).

Species of copepods from the hadal zone (6000–8500 m) have been obtained by Russian workers. *Spinocalanus similis*, var. *profundalis*, *Parascaphocalanus zenkevitchi* and *Metridia similis* comprised over 80 per cent of the individuals. Nine specimens of four species of amphipods were also collected (Wolff, 1960). These hadal animals, which are colourless, are assumed to be alive at their depth of collection.

Plankton which is seriously damaged by a net is difficult to collect and count. Siphonophores, jelly fish and similar soft bodied animals have therefore to be counted by other methods. A population of jelly fish and siphonophores has been observed by photographic means just above the sea floor at 2000 m depth off south California (Hartman & Emery, 1956).

NEKTON

Large actively swimming animals are certainly present in the deep sea in significant numbers but few quantitative studies have been carried out. The larger deep sea prawns reaching in excess of 15 cm in length are often recovered in nets in greater numbers than fish. We do not know if the large Crustacea outnumber the fish or if the latter evade the net in greater numbers. Perhaps squid, which are powerful swimmers are the most difficult of all to sample (Clarke, 1966).

The problem of counting even mesopelagic animals is a formid-

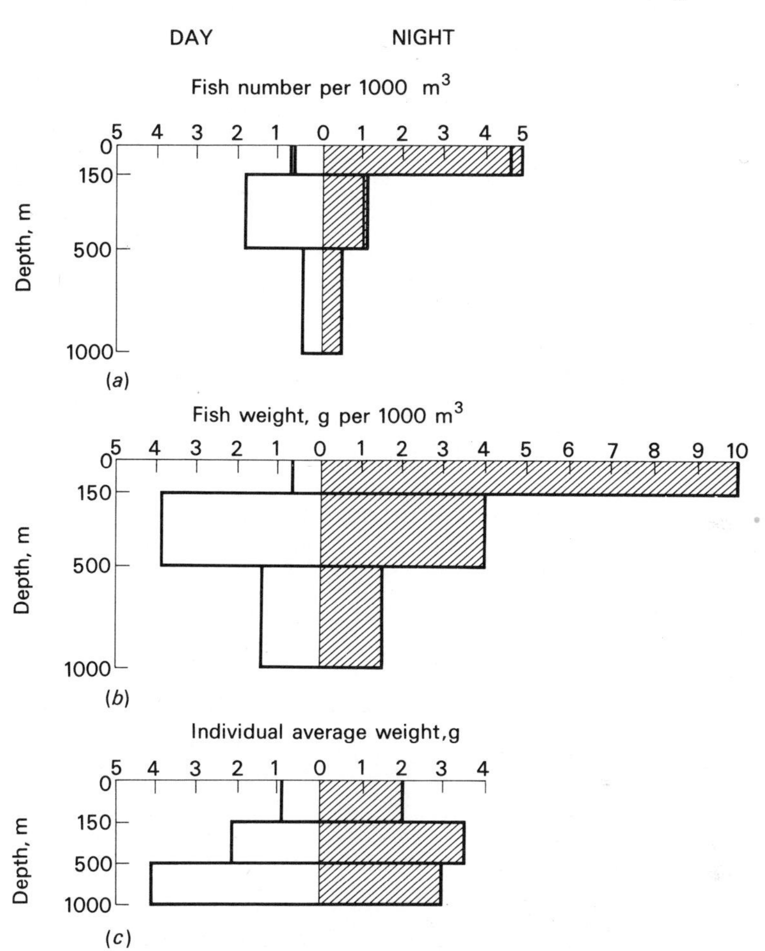

Fig. 6.12. Vertical distribution of deep sea fishes. Comparisons of the day and night catches for the three depth strata in (*a*) numbers per 1000 m^3, (*b*) grams wet weight per 1000 m^3, and (*c*) average individual weight. The spaces between the two outside lines in the upper histogram denote the number of larval fishes. (Pearcy & Laurs, 1966)

able one. In the case of fish, vertical migration has to be allowed for so quantitative mid-water trawling has to be carried out over a vertical range which exceeds that of the fishes under investigation. Errors in trawling techniques are discussed by Pearcy and Laurs (1966), and in Chapter 7. Data from Pearcy and Laurs (Fig. 6.12) shows the fish biomass in the top 1000 m of sea off the Oregon coast where the total water depth is 2000 m. More fish were caught in the total water column at night than in the day time. Comparison with Fig. 6.10 shows that, to a first approximation, fish are equal to 1×10^{-3} or 10^{-4} of the biomass present in the form of zooplankton at 1000 m depth. The commonest species of fish were *Lampanyctus leucopsarus*, *Diaphus theta*, *Tarletonbeania crenularis*, and *Tactostoma macropus*.

The sparseness of bathypelagic animals is one of their most interesting features. To date, the greatest depth from which a fish has been recovered, and at which it was assumed to be alive, is 7965 m (*Bassogigas* sp., Anon., 1970). A squid, of an unidentified species of the family Cirroteuthidae (Jahn, 1971) has been photographed at 5145 m where it looked alive. Both these animals are probably benthopelagic. In short, it is not possible to state with any certainty what the greatest living depth is for a population of midwater animals.

MACROSCOPIC BENTHOS

The commonest deep sea benthic animals are worms (Polychaeta), bivalves (Lamellibranchiata), Crustacea, and brittle stars (Ophiuroidea). The general picture of their distribution is one of decreasing biomass with increasing depth or distance from land. Russian workers have measured the benthic biomass in abyssal regions and found wet weight quantities of not more than several grammes per square metre, whereas shallow water benthic populations occur in densities as high as several thousand $g\,m^{-2}$ (Zenkevitch, Barsanova & Belyaev, 1960). Fig. 6.13 shows the distribution of material on the floor of the Pacific Ocean. In the tropics and well away from land a benthic biomass of 0.01 $g\,m^{-2}$ occurs. As a result, the sea floor at depths in excess of 3000 m supports less than 1 per cent of the total benthic biomass of the world's oceans (Table 6.10, p. 296). See also Belyaev and Vinogradova (1961).

Detailed analysis of the animals living along a transect crossing the floor of the Sargasso and nearby seas has been conducted by Sanders and Hessler (1969). Table 6.8 from Sanders, Hessler and Hampson (1965) summarises some of their findings. It is noticeable that the numerical density of animals (worms, crustaceans, bivalves, all of similar size) decreases by a factor of 25 from a sea

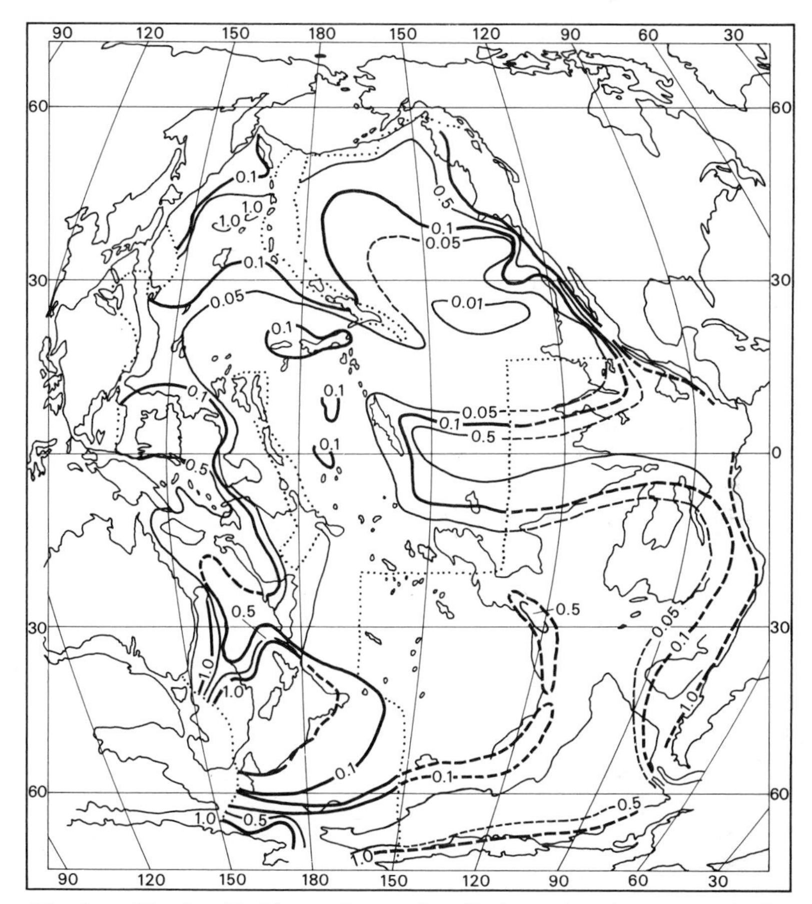

Fig. 6.13. The benthic biomass in g m^{-2} at depths greater than 2000 m in the Pacific and Indian Ocean, according to data of the expeditions on the ships *Vityaz* and *OB*, and in the Atlantic Ocean according to the data of the *Galathea* expedition. (After Menzies, George & Rowe, 1973; after Zenkevitch, 1969)

floor depth of 200 to one of 2500 m, along a 60 km transect down the continental slope. The productivity of the surface waters cannot account for this and neither can the nature of the sea floor. Is it conceivable that the downward flux of material through the water column involves increased losses at a critical depth such that only a small fraction of the shallow water biomass can be supported beyond 2000 m? The diversity of the benthic fauna has been the main interest of those who have studied the transect

TABLE 6.8. *Numbers of animals on the deep sea floor along a line between Bermuda and New England.* (Sanders, Hessler & Hampson, 1965)

Station (See original text)	Depth (m)	Latitude	Longitude	No. animals in sample	No. animals per m^2
55 (North)	75	40° 27.2′ N	70° 47.5′ W	3791	13073
C – 1	97	40° 20.5′ N	70° 47′ W	3082	5314
Sl. 2	200	40° 01.8′	70° 42′	6455	12910
Sl. 3	300	39° 58.4′	70° 40.3′	11907	21263
Sl. 4	400	39° 56.5′	70° 39.9′	4439	6081
D – 1	487	39° 54.5′	70° 35′	5115	8669
E – 3	823	39° 50.5′	70° 35′	3008	2979
F – 1	1500	39° 47′	70° 45′	997	1719
G – 1	2086	39° 42′	70° 39′	1120	2154
GH – 1	2500	39° 25.5′	70° 35′	365	521
GH – 4	2469	39° 29′	70° 34′	299	467
HH – 3	2870	38° 47′	70° 08′	636	748
II – 1	3742	37° 59′	69° 32′	—[a]	—[a]
II – 2	3752	38° 05′	69° 36′	391	1003
JJ – 1	4436	37° 27′	68° 41′	264	264
JJ – 3	4540	37° 13.1′	68° 39.6′	101	158
KK – 1	4850	36° 23.5′	68° 04.5′	113	92
LL – 1	4977	35° 35′	67° 25′	67	55
MM – 1	5001	34° 45′	66° 30′	27	33
NN – 1	4950	33° 56.5′	65° 50.7′	51	38
OO – 2	4667	33° 07′	65° 02.2′	58	126
Ber. 7	2500	32° 15′	64° 32.6′	91	120
Ber. 5	2000	32° 11.4′	64° 41.6′	89	189
Ber. 4	1700	32° 17′	64° 35′	217	271
Ber. 3	1700	32° 16.6′	64° 36.3′	126	274
Ber. 2	1700	32° 16.6′	64° 36.3′	189	215
Ber. 6	1500	32° 14.3′	64° 42′	208	178
Ber. 8	1000	32° 21.3′	64° 33′	326	729
Ber. 1 (South)	1000	32° 16.5′	64° 42.5′	243	528

[a] Sample excluded from quantitative analysis because of small size.

(Dayton & Hessler, 1972) and it is possible that the species composition of the community living there may influence the total numbers of individuals.

The quantitative distribution of deep sea benthic fish is not known although Grey (1956) records 800 species. Of these, 260 are found at depths greater than 2000 m and 43 species are regarded as confined to such depths.

THE HADAL REGION

The organisms which inhabit the ocean trenches differ from those in the abyssal regions in at least two respects. First, they live at much higher pressures, in excess of 600 atm. Secondly, and surprisingly, they live in an environment which may be better provided with food than some abyssal plains. Ocean trenches are normally close to islands and tend to trap terrestial material. They are also narrow; for example the Philippine Trench is only 1–3 km wide at its greatest depth which is 10 km, and it seems possible that organic material may be funnelled to the bottom. Whatever the mechanism of food supply, Nature has provided physiologists with relatively well fed high pressure organisms for study (Belyaev, 1972).

Wolff (1970) argues that the hadal region is ecologically distinctive and not a mere extension of that found in abyssal regions. Lists of the species and the relative numbers of individuals which live on the sea floor at depths greater than 6000 m have been compiled by Wolff (1970) and show that common hadal animals are sea cucumbers (Holothurioidea), worms (Polychaeta), bivalves (Mollusca), anemones (Arrthozoa) and Pogonophora. The relative numbers of individuals found in specific trenches are given in Fig. 6.14. Note, for example, that in the Japan trench worms and sea cucumbers and bivalves are the most numerous benthic animals. Note also that this statement is based on the results of 6 trawls which yielded a total of 2930 specimens. Some trenches have not been investigated as thoroughly and some of the results are heavily biased by a single rich haul which may not be representative. The pogonophores appear numerous in the Kurile–Kamchatka trench only by virtue of a single sample.

Quantitative sampling is difficult to do at these depths but the limited data available are very interesting. At depths between 8330 m and 9950 m on the floor of the Kurile–Kamchatka Trench a biomass of 0.3 $g\,m^{-2}$ is thought to be present, on the basis of a few grab samples. On the floor of the Banda Trench (6580–7270 m) a richer biomass of the order of 10 $g\,m^{-2}$ occurs (Wolff, 1960).

Some major groups of animals have not been found in the hadal zone, which is interesting and perhaps important. In Table 6.9 we see that most kinds of animals inhabit the trenches, but decapods (e.g. crabs) are noticeable for their absence at depths greater than 5160 m. It would be premature to regard decapods

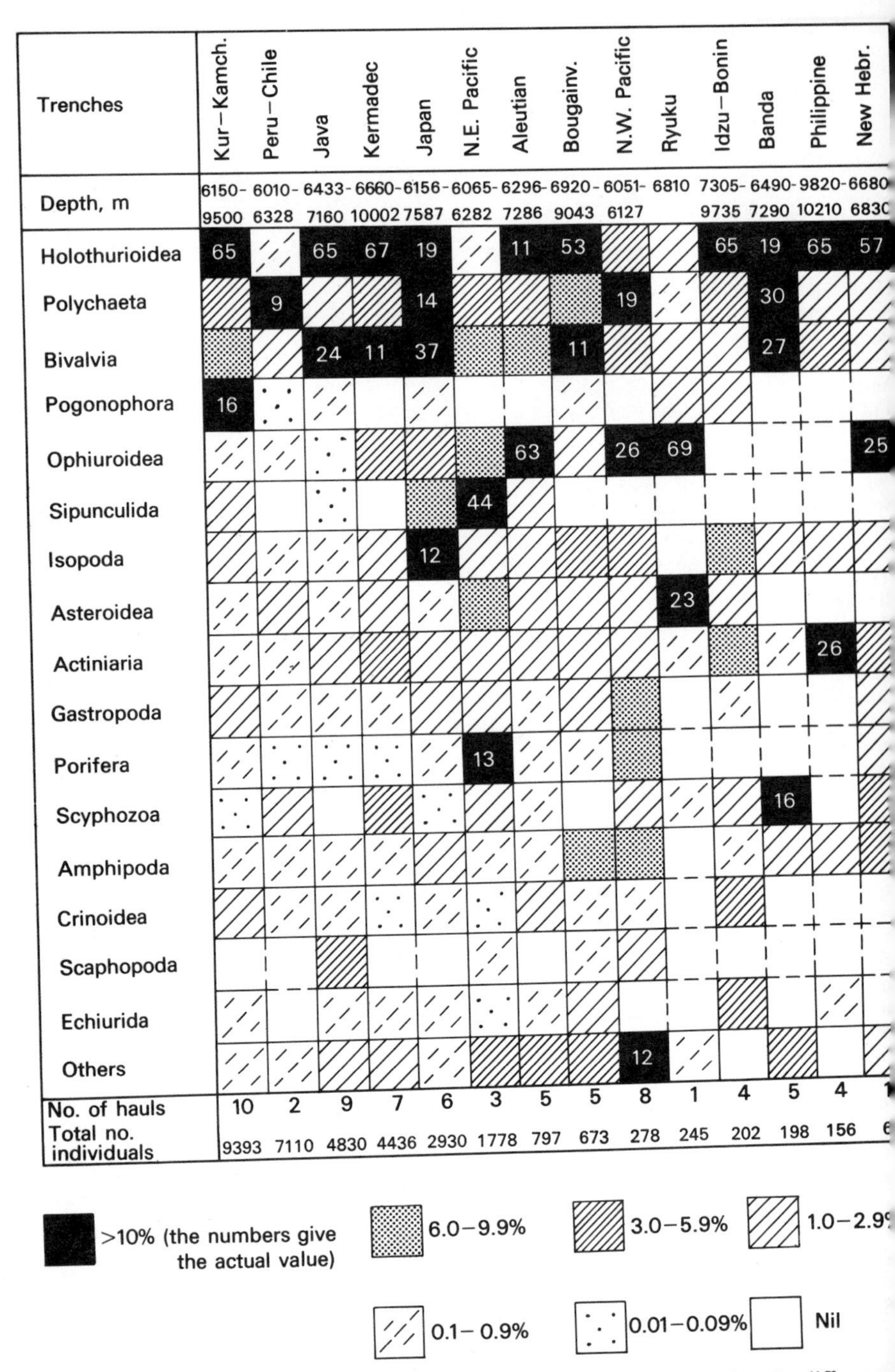

Fig. 6.14. Percentage of individuals of major groups of benthos in different trenches and basins. (Wolff, 1970)

TABLE 6.9. *Known maximum depths reached by benthos.* (After Wolff, 1970)

roup	Genus and species	Depth (m)	Trench
raminifera	*Sorosphaera abyssorum* Saidova	10415–10687	Tonga
rifera	*Asbestopluma occidentalis* (Lambe)	8175–8840	Kurile–Kamchatka
ydrozoa	*Halisiphonia galatheae* Kramp	8210–8300	Kermadec
yphozoa	*Stephanoscyphus* sp.	9995–10002	Kermadec
nnatularia	*Umbellula* sp.	6620–6730	Kermadec
ntipatharia	*Bathypathes patula* Brook	8175–8840	Kurile–Kamchatka
tiniaria	*Galatheanthemum* sp.	10630–10710	Marianas
urbellaria	1 sp.	6200	Milne Edwards
ematoda	1 sp.	10415–10687	Tonga
emertea	1 sp.	7210–7230	Kurile–Kamchatka
lychaeta	*Macellicephaloides* n.sp(?)	10630–10710	Marianas
hiurida	*Vitjazema* sp.	10150–10210	Philippine
punculida	*Phascolion pacificum* (Murina) and *P. lutense* (Selenka)	6860	Kurile–Kamchatka
iapulida	*Priapulus* sp.	7795–8015	Kurile–Kamchatka
rripedia	*Scalpellum* sp.	6960–7000	Kermadec
ysidacea	*Amblyops magna* Birstein and Tchindonova	7210–7230	Kurile–Kamchatka[a]
macea	? *Bathycuma* sp.	7974–8006	Bougainville
naidacea	? *Herpotanais kirkegaardi* Wolff	8928–9174	Kermadec
poda	*Macrostylis* sp.	10630–10710	Marianas
nphipoda	2 spp.	10415–10687	Tonga
capoda	*Parapagurus* sp.	5160	Celebes Sea
cnogonida	*Heteronymphon profundum* Turpaeva	6860	Kurile–Kamchatka
lacophora	*Chaetoderma* sp.	8980–9043	Bougainville
stropoda	1 sp.	9520–9530	Kurile–Kamchatka[b]
aphopoda	1 sp.	6920–7657	Bougainville
valvia	*Phaseolus* (?) n.sp.	10415–10687	Tonga
yozoa	*Bugula* sp.	8210–8300	Kermadec
inoidea	*Bathycrinus* sp.	9715–9735	Idzu–Bonin
lothurioidea	*Myriotrochus bruuni* Hansen	10630–10710	Marianas
teroidea	*Hymenaster* sp.	8185–8400	Kurile–Kamchatka[c]
hiuroidea	1 sp.	7974–8006	Bougainville
hinoidea	*Pourtalesia* sp. ? *aurorae* Koehler	7250–7290	Banda
gonophora	*Heptabrachia subtilis* Ivanov	9715–9735	Idzu–Bonin
cidiacea	*Situla pelliculosa* Vinogradova	8330–8430	Kurile–Kamchatka
ces	*Bassogigas* sp. (cf. *profundissimus* (Roule))	7965	Puerto Rico

[a] A mysid from 7246 m in the Aleutian Trench is still unstudied.
[b] Belyaev's record from 10415–10687 m in the Tonga Trench is not considered, as shells were empty.
The deepest published records are *Porcellanaster* n.sp. and *Hymenaster* sp. at 84–7614 m in the Volcano Trench. See original text for full references.

TABLE 6.10. *Distribution of the standing crop of benthic animals.* (Zenkevitch *et al.*, 1960)

Depth (m)	% area sea floor	% benthic biomass
0–200	7.6	82.6
200–3000	15.3	16.6
> 3000	77.1	0.8

(or any of the other apparently restricted groups) as incapable of adapting to hadal conditions, but there seems little doubt that they are not numerous in this zone.

Giantism in hadal isopods is a well documented phenomenon which is most intriguing. Whether this trait is due to hydrostatic pressure as suggested by Wolff (1960) remains to be seen. It seems unlikely.

The outstanding insignificance of the hadal biomass in the nutritional budget of the world's oceans is apparent from Table 6.10. 0.8 per cent of the total benthic biomass lives below 3000 m and we may safely conclude that the hadal biomass constitutes no more than 1 per cent of that.

BENTHIC MICRO-ORGANISMS

The existence of bacteria in deep sea muds at a depth of some 5000 m was first demonstrated by Certes in 1884. Table 6.11 from ZoBell (1942) shows not only the concentration of bacteria on the surface layers of sediments, but also the extent of the penetration of the bacteria into abyssal sediment. In slowly sedimenting deposits the deeper layers of bacteria have probably grown from a culture more than 10^6 years old (Morita & ZoBell, 1955). Microbial culturing procedures readily demonstrate a variety of morphological and physiological types of bacteria in these sediments. Aerobes were most numerous in the samples investigated at atmospheric pressure by ZoBell in 1938 (Table 6.12, and ZoBell & Morita, 1957; also Table 3.11) but anaerobes are an important part of the flora in the Philippine and other trenches. Note that the higher counts listed in Table 3.11 were obtained from cultures which were incubated at high pressure. All the cells experienced decompression during recovery from the sea floor.

TABLE 6.11. *Number of bacteria per g wet sediment in different strata of cores. Bacterial numbers determined by culture methods.* (ZoBell, 1942)

Core number	XIV-37	XIV-45	XIV-53
Station location	32° 26.4′ N 117° 41.3′ W	32° 36.4′ N 117° 27.8′ W	33° 03.3′ N 117° 25.5′ W
Water depth	1040 m	1190 m	471 m
Core depth (inches)	Bacteria per g	Bacteria per g	Bacteria per g
0–1	38000000	7500000	840000
1–2	940000	250000	102000
4–5	88000	160000	63000
9–10	36000	23000	19000
14–15	2400	8700	1500
19–20	400	2100	2200
29–30	180	600	370
39–40	330	200	190
59–60	250	300	210
79–80	130	100	140
99–100	290	150	140

The nature of the sediment has an important influence on the numbers of bacteria, coarse sand supporting fewer cells than fine silt. The most important single factor seems to be availability of organic material which is readily adsorbed onto finer particles. Two very distinctive deep sea sediments are red clay and Globergerina ooze, and both contain little organic material. Table 6.13 shows the numbers of bacteria which Morita and ZoBell (1955) found in deposits at several thousand metres depth in the Pacific. Concentrations of 10^1–10^4 viable bacteria per gramme of wet sediment were found by the extinction dilution method on the surface of Globergerina ooze, and slightly higher concentrations on red clay.

The concentrations of organic carbon in deep sea sediments are important to know. From the floor of the Sargasso Sea Sanders *et al.* (1965) obtained sediment containing 0.2–1.2 per cent organic material, and from the South Atlantic Hobson and Menzel (1969) obtained samples containing 0.4–0.7 per cent organic material. Bacterial carbon in deep sea trenches has been estimated by ZoBell and Morita (1959) at 0.2–2.0 mg per litre of surface sediment.

TABLE 6.12. *Relative number of different physiological types of bacteria found in the topmost 3 to 5 cm strata of three representative marine sediment samples, the number being expressed as the reciprocal of the highest dilution in which the differential media showed their presence. A plus sign (+) indicates that the bacteria were demonstrated but their abundance not estimated.* (ZoBell, 1938)

Sediment sample no.	8160	8330	9309
Station location	32° 51.2′ N 117° 28.3′ W	33° 25.9′ N 118° 06.5′ W	33° 44.2′ N 118° 46.1′ W
Depth of overlying water	780 m	505 m	1322 m
	Bacteria per g sediment (wet basis)		
Total aerobes (plate count)	930000	31000000	8800000
Total anaerobes (oval tube count)	190000	2600000	1070000
Ammonification: peptone → NH_4	100000	1000000	1000000
Ammonification: nutrose → NH_4	10000	1000000	100000
Urea fermentation: urea → NH_4	100	+	1000
Proteolysis: gelatin liquefaction	100000	10000000	1000000
Proteolysis: peptone → H_2S	10000	1000000	100000
Denitrification: NO_3 → N_2	100	10000	10000
Nitrate reduction: NO_3 → NO_2	100000	10000000	10000
Nitrogen fixation	0	0	0
Nitrification: NH_4 → NO_2	0	0	0
Sulphate reduction: SO_4 → H_2S	1000	1000	10000
Dextrose fermentation	10000	100000	1000
Xylose fermentation	10000	+	10000
Starch hydrolysis	10000	100000	10000
Cellulose decomposition	1000	+	1000
Fat hydrolysis (lipoclastic)	1000	+	+
Chitin digestion	100	+	+

While heterotrophic bacteria dominate the microbial population of deep sea sediments other types of micro-organisms coexist. Diatoms in a sediment sample from the Weber Deep (depth 7400 m) were examined by Wood (1956) who found them to contain cytoplasm. Some species grew at 500 atm pressure (Table 6.14). Foraminifera and fungal spores also apparently survive in deep sediments in unknown but probably very low cell densities.

TABLE 6.13. *Bacterial numbers in the sediment at the top and bottom of cores collected at different Mid-Pacific (MP) stations. Bacterial numbers determined by dilution technique.* (Morita & ZoBell, 1955)

Station number MP (see original text)	Location of station		Water depth (m)	Length of core (cm)	Type of sediment	Bacterial titre[a]	
	Latitude N	Longitude W				Top	Bottom
3–1	20° 51′	127° 09.0′	4390	747	Red clay	3	0
5–3	14° 22.1′	133° 06.8′	5300	40	Red clay	4	2
7–2	12° 47.5′	134° 26.4′	4758	106	Globigerina ooze	1	1
10–2	4° 37.2′	140° 00.3′	4365	89	Globigerina ooze	1	1
15–1	10° 43.5′	145° 53.2′	4987	92	Volcanic ash	4	2
17–2	14° 38.3′	151° 58.4′	5942	122	Red clay	3	3
20–2	20° 27.0′	154° 55.1′	3825	96	Red clay	4	2
21–2	20° 47.0′	159° 59.0′	4484	145	Red clay	4	4
25–E	19° 02′	169° 44′	1759	77	Globigerina ooze	4	4
27–2	19° 35′	171° 50′	3750	88	Globigerina ooze	3	3
30–1	18° 27′	173° 14′	3709	71	Red clay	2	2
33–H	17° 53′	174° 27′	1707	55	Globigerina ooze	2	1
33–L	17° 51′	174° 17′	1720	43	Globigerina ooze	4	1
35–1	19° 21′	174° 58′	4841	75	Red clay	4	4
35–2	19° 02′	174° 58′	3935	363	Red clay	3	3
36–1	16° 48′	176° 27′	5032	319	Globigerina ooze	4	3
37–1	17° 06′	177° 18′	5032	275	Globigerina ooze	4	2
38–1	19° 02′	177° 18′	4712	366	Red clay	3	0
40–1	15° 35′	177° 30′	4121	387	Red clay	3	0

[a] 0 = No viable bacteria demonstrated in 1.0 g of wet sediment
1 = At least 10 but <100 viable bacteria per g of wet sediment
2 = At least 100 but <1000 viable bacteria per gram of wet sediment
3 = At least 1000 but <10000 viable bacteria per g of wet sediment
4 = At least 10000 but <100000 viable bacteria per g of wet sediment

TABLE 6.14. *Diatoms in the sediments of deep sea trenches.* (Wood, 1956)

Station (see original text)	Location		Water depth (m)	Date collected	Trench	Diatoms present
	Latitude	Longitude				
418	10° 20′ N	126° 30′ E	10387	15 VII 51	Mindanao	
422	10° 49′ N	126° 01′ E	2010	24 VII 51	Mindanao	+
440	10° 25′ N	126° 40′ E	10610	14 VIII 51	Mindanao	+
463	10° 16′ N	109° 51′ E	7214	3 IX 51	Sunda	+
492	5° 31′ S	131° 01′ E	7445	20 IX 51	Weber	+
496	5° 36′ S	131° 06′ E	7465	23 IX 51	Weber	+
497	—	—	>9000	24 IX 51	Weber	+
517	6° 31′ S	153° 58′ E	9020	11 X 51	Solomons	+
517?	6° 31′ S	153° 58′ E	9255	13 X 51	Solomons	
608	44° 31′ S	167° 50′ E	390	18 I 52	Milford Sound	+
645	35° 16′ S	178° 40′ E	8515–8425	13 II 52	Kermadec	
650	32° 20′ S	176° 54′ W	6794	16 II 52	Kermadec	
658	35° 51′ S	178° 31′ W	7837–7901	21 II 52	Kermadec	+
677	38° 38′ S	175° 53′ W	6370	4 III 52	Kermadec–Tonga	
678	28° 30′ S	175° 53′ W	9437	4–5 III 52	Kermadec–Tonga	
678 (water)	28° 30′ S	175° 53′ W	9335	4–5 III 52	Kermadec–Tonga	
686	20° 53′ S	175° 31′ W	10080	11 III 52	Kermadec–Tonga	

Dynamic aspects

The problem before us is to quantify the flux of organic material, and describe the pathways it takes through the deep sea and through the organisms inhabiting it. On the gross scale this is a job for the ecologist but at the level of the individual organism physiological, physical and chemical processes influence both the rate of flux and nature of the materials metabolised and so the physiologist and chemist have a contribution to make in this area.

The central question is simply – how does a given deep sea organism obtain and metabolise its food? We will consider only gross food needs because we are mainly concerned with bulk quantities. The detailed nutritional needs of deep sea organisms is not yet a subject for study. From the first question a number of other questions follow. How much energy does the organism in question require? How is this partitioned between maintenance

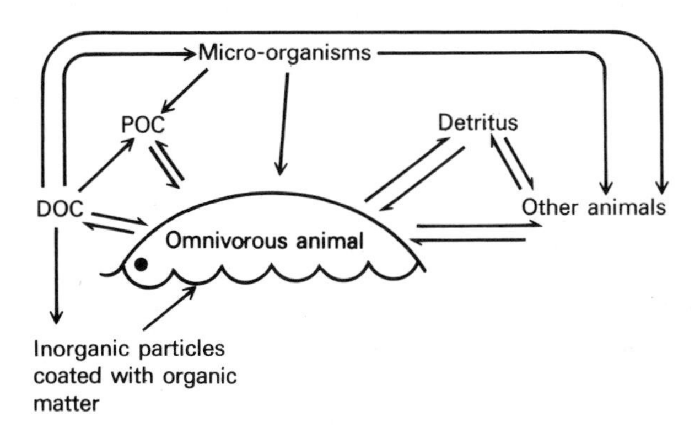

Suspended material frequently contains minerals such as calcium carbonate coated with an organic layer (Wangersky, 1968)

Fig. 6.15

and growth, and how do these compare with 'normal' marine organisms? Consider the hypothetical case of an omnivorous animal (Fig. 6.15). In the sparsely supplied deeps it may conceivably obtain food from five or six different sources. Each source requires special means of acquisition and is increasingly scarce as the dimensions of the particles of material increase.

Deep sea nutrition

Animals may feed on solutes, they may filter particles from suspension, or consume solid organic and inorganic material. They may also feed carnivorously. All these methods of feeding apply to pelagic and benthic animals, although the quantity of food available differs significantly between these two major habitats. The largest component of the organic material in the oceans is dissolved, so the ways in which this material becomes incorporated into animals will be considered first. As shown in Fig. 6.15, the major pathway is the flux of DOC into micro-organisms, that is, the heterotrophic nutrition of micro-organisms. Other routes will also be considered in turn.

MID-WATER MICROBIAL ACTIVITY

If we regard the bathypelagic region of the ocean as a gigantic but dilute broth culture, can the gross metabolic processes, such as the oxygen consumption of the cells, their metabolic turnover, growth rate and growth yield, be quantified? Information bearing on these questions is becoming available for shallow seas and there are even a few pieces of information about micro-organisms in deeper waters.

Parsons and Strickland (1962) measured the rate of uptake of glucose and acetate in seawater samples from 175 m depth and calculated rates of carbon uptake of the order of 0.003 $mg C l^{-1} hour^{-1}$ at *in situ* temperatures. Similarly, Vaccaro and Jannasch (1966) estimated an uptake of from 0.001 to 0.062 $mg C l^{-1} hour^{-1}$ by seawater micro-organisms from 100 m depth. K_m values of 10^{-7} M for the uptake process were apparent in the first study. Riley (1970), in a major paper on the fluxes of organic material in the oceans, quotes some results of experiments on seawater from 1000 m depth which produced uptake rates of 0.001 $mg C l^{-1} hour^{-1}$, equivalent to about 10 per cent of the available organic carbon per day.

Banoub and Williams (1972) have quantified the uptake of nutrients by shallow water micro-organisms and calculated that respiration used approximately 5 per cent and growth up to 30 per cent of the available substrate per day. Thus a large fraction of the DOC may be converted to cellular material each day under certain conditions, although in oceanic waters generally a much smaller total flux (1 per cent per day) has been measured by other workers.

TABLE 6.15. *Change in dissolved organic carbon (DOC) after concentration and incubation in the dark at 20 °C.* (Barber, 1968)

Location	Depth (m)	DOC (mg l^{-1})	Conc. DOC (mg l^{-1})	Conc. DOC + 1 mth inc. (mg l^{-1})	Conc. DOC + 2 mths inc. (mg l^{-1})
Woods Hole, Mass.	0	1.9	4.2	2.2	—
Woods Hole, Mass.	0	1.9	4.2	2.1	—
36° 54^{1} N, 68° 11′ W	50	0.8	3.6	3.6	3.7
36° 54′ N, 68° 11′ W	50	0.8	3.6	—	3.7
38° 47′ N, 69° 20′ W	1200	0.7	3.7	3.7	—
38° 47′ N, 69° 20′ W	1200	0.7	3.7	3.7	—
38° 47′ N, 69° 20′ W	1200	0.7	3.7	3.7	—
38° 55′ N, 67° 22′ W	4000	0.6	2.4	2.5	2.6
38° 55′ N, 67° 22′ W	4000	0.6	3.2	—	3.1
35° 26′ N, 67° 17′ W	4900	0.6	3.3	3.3	—

The organic solutes were concentrated by pressure dialysis using a diffusive membrane which retained molecules with a molecular weight greater than 500.

Samples of water from 1800 m investigated by Banoub and Williams (1972) possessed undetectable metabolic activity, that is, less than 1 per cent of that found at the surface.

These fragmentary data show wide variations in heterotrophic activity in the sea but are consistent with the expected low metabolic turnover in deep water. Despite the generally low level of metabolic activity seen in seawater, micro-organisms seem capable of a high growth yield (Williams, 1970), perhaps because of the wide spectrum of nutrients available.

Some of the DOC may be of little nutritive value to micro-organisms, as shown by Barber (1968) who concentrated seawater samples to boost microbial metabolism but found instead little or no utilisation over one or two months (Table 6.15). Unlike the author of Table 6.15 the present author sees no evidence therein that deep sea water alone contains DOC which is particularly resistant to microbial attack; some of the DOC in 50 m samples appears equally resistant. It seems generally agreed that there are at least two fractions of dissolved organic material in seawater, a relatively rapidly metabolised fraction and a slowly turning over pool whose steady state concentration is probably much larger. An example of apparently rapid metabolic activity at depth is provided by Wada and Hattori (1972) who demonstrated that the rate at which nitrate was reduced to nitrite at 400 m was similar to the rate at 4000 m.

According to Riley (1970) there is no serious discrepancy between plausible rates of oxygen consumption of natural deep seawater as determined experimentally, and the flux of oxygen from the air–water interface. The oxidation of organic material in the deep sea by microbial metabolism is not limited by the supply of oxygen which is generally plentiful.

The incorporation of dissolved organic material into mid-water animals (Fig. 6.15) probably takes place on a limited scale but there is no evidence that such animals acquire a net gain in food energy by this means. The situation with certain benthic animals is different and is described on p. 317.

THE FLUXES OF PARTICULATE MATERIAL

Dissolved organic substances can give rise to particulate material by processes which occur at the air–water interface and by different reactions within the water column. Neither phenomena is well understood. Interface processes involve the spread of surface-active material on a gas bubble and a subsequent condensation of the material when the bubble collapses. Salt particles can be formed from spray at the surface of the sea and may attract an organic coat on re-entry into the water (Sutcliffe, Baylor & Menzel, 1963). Whatever the mode of formation at the surface, particulate material has to be distributed into deep water before it is of use to the animals there. The sinking rates of particles which resemble balls rather than flakes are determined by their degree of aggregation in accordance with Stokes' Law, and according to Riley (1970) the fastest sinking rate under experimental conditions is 3 m day^{-1} with most particles sinking at rates of less than 1 m day^{-1}. Thus particles which are formed at the surface will be about ten years old when they enter the abyssal regions of the oceans unless vertical water movements greatly accelerate or retard their transport. Having concluded that particles form at the sea surface and sink into deep water, we have to await direct measurements of the flux of the surface-formed POC through the water column before deciding how important it is to the bathypelagic fauna.

Particulate material forms within filtered seawater by processes which may be accelerated by the presence of cells or a particulate nucleus, but which are otherwise regarded as physical processes. Concentrations of POC in the region of 0.1 mg C l^{-1}, which is slightly higher than seawater concentrations, have been produced experimentally in a few days. Concentrations of POC can give a false impression of their food value because only some 20 per cent by weight seems to be susceptible to hydrolysis by enzymes (Gordon, 1970*b*).

The net downward flux of POC (excluding macroscopic debris) is probably affected very little by mid-water animals because it is too sparsely distributed, although it is just conceivable that particles may be locally concentrated to a level sufficient to sustain filter feeders (see p. 308). Consequently, the benthic animals are probably the main beneficiaries of the downward flux which Riley (1970) estimates to supply the benthos at a rate of 1–5 mg C m^{-2} day^{-1}.

Small cells of the phytoplankton which sink from the eutrophic zone are rarely found intact below 500 m depth, probably due to the intense grazing in shallow water and rapid disintegration combined with a slow sinking rate. The tendency for mineral skeletal components to dissolve at depth may be reduced by the presence of an adsorbed protective organic layer which may have nutritional significance.

Fournier (1972) suggests that the population of olive green cells settles out at rates which are determined by the degree of clumping, and could supply the mid-depths with carbon at rates of up to 875 μg per m^2 of water column per day. In Table 6.16 Fournier has set out the sinking rates for clumps of cells of different sizes, and computed the corresponding flux of carbon to stated depths. The depth of the bottom envisaged in Table 6.16 is not stated but is probably about 4500 m. The data were calculated from the concentration of cells found in the sub-tropical Pacific which was approximately half that in other parts of the world. The significance of the flux of olive green cells into deep water is not very great as judged in bulk quantities. For certain species of animals however, the flux may be vital. Its accumulation on the bottom might be as high as one third of the minimum estimated biological demand, although probably much less (Fournier, 1972).

TABLE 6.16. *Amount of carbon delivered by olive-green cells, over a range of settling rates, to two mid-water depths and the bottom, plus (in parentheses), the percentage of the minimum estimated requirement this represents at those depths.* (Fournier, 1972; Riley, 1970)

	Settling rate	Carbon delivered ($\mu g m^{-2} day^{-1}$)		
	($m day^{-1}$)	2000 m	3000 m	Bottom
Single cells	0.02	2.5 (0.01)	1.7 (0.01)	1 (0.1)
Minimum aggregate	0.1	12.5 (0.07)	8.7 (0.05)	5 (0.5)
Cells clumped (20 μm)	2.0	250 (1.5)	174 (1.0)	100 (10)
Maximum aggregate	7.0	875 (5.2)	609 (3.5)	350 (35)

Other solid particles to be found in the mid-depths are exoskeletons of crustaceans, carcasses and faecal pellets.

The nutritive value of faecal pellets from *Paelaemonetes pugio*, a littoral prawn, fed on the diatom *Nitzschia* has been demonstrated by Johannes and Satomi (1966). The pellets consisted of 20 per cent organic carbon, most of which was bacterial protein synthesised from food left undigested by the animal. Faecal pellets from *Meganyctiphanes* and other species of euphausiids are to be expected to sink from the eutrophic zone at rates within the range 126–862 $m day^{-1}$, according to size and compactness and disregarding upwelling currents (Fowler & Small, 1972). There is evidence that such a downflux of material takes place off the Oregon coast. Isotopes with a short half life introduced at the surface over a bottom depth of 2800 m were found, by gamma-ray spectrometry, to enter sea cucumbers on the sea floor very rapidly. According to the investigators, Osterberg, Carey and Curl (1963), the isotope was distributed downwards in macroscopic particles and faecal pellets seem a likely means of transport.

The extent to which faecal pellets from Crustacea are suited to transport organic material to the depths is largely determined by the resistance of the outer cuticle to rupture or dissolution. Once the pellet loses its compactness its sinking rate decreases and the contents will progress little further.

The descent of larger particles such as eggs, exoskeletons, carcasses and plant debris have all been considered possible routes along which organic material can enter deep water. Crustacean

TABLE 6.17. *The rate of sinking in m day^{-1} of particulate material in seawater, laboratory conditions*

POC	0.1–10	Riley, 1970
Olive-green cells	0.02–7	Fournier, 1972
Foraminifera		
> 250 μm	1900	Berger & Piper, 1972
125–62 μm	250	
Copepod carcass	80–416	Vinogradov, 1961
Pteropod inside shell	910–2270	Vinogradov, 1961
Euphausiid:		
faecal pellets	126–862	Fowler & Small, 1972
eggs	132–180	Fowler & Small, 1972
exoskeletons	248–800	Fowler & Small, 1972
freshly killed animals	1760–3170	Fowler & Small, 1972

exoskeletons can occur in considerable numbers in the standing stock of material but their flux has not been investigated beyond the stage of laboratory studies of moulting. The copepod *Calanus finmarchicus* moults 1 per cent of the food it assimilates (Conover, 1968) but *Euphausia pacifica* loses a third of its net growth as exoskeleton (Lasker, 1966).

Some laboratory sinking rates of particulate material are shown in Table 6.17. It is interesting that in Lake Baikal small crustaceans which die at the surface sink 1500 m to the bottom where they are found to be intact. Vinogradov supports the notion that Crustacea larger than about 1 mm stand a good chance of sinking sufficiently fast in relation to their rate of decomposition to reach significant depths, whereas smaller carcasses are much more liable to be dissipated in shallow water. According to Vinogradov (1961) dead pteropods containing the remains of tissue occur at depths below 500 m in much smaller numbers than would be expected by their rate of sinking as determined in the laboratory (Table 6.17). However many pteropods are likely to start sinking from the surface with empty shells (Conover, personal communication).

Field experiments seem essential to determine the real sinking rates of all grades of particulate material into the deep sea. Midwater animals probably capture little of it, but when it reaches the bottom it becomes concentrated and more available to animals. As Riley has pointed out, the influx of the particulate fraction to the deep sea is given by $F = SC$, in which S is the sinking rate and C the particle concentration. Both S and C are complex terms which vary temporally and spatially.

HOLOTROPHIC NUTRITION IN MID-WATER

(i) Small free swimming animals, such as the copepods investigated by Grice and Hulseman (Fig. 6.11) seem to face an impossible task of acquiring small particulate food from their environment. Studies of how similar organisms feed in the eutrophic zone have shown that many filter feeders are not simple automata but capable of a degree of selection. Even so, they appear to require a concentration of particles on which to feed in the range of 2–50 $mg\,l^{-1}$, in order to meet the energy requirements of filtration. This is a thousand times the bulk concentration of particles in deep water (Jørgensen, 1966).

Knowledge of the ways in which small planktonic animals feed comes chiefly from a study of their mouthparts and gut contents. In the course of a most valuable review of Russian work, Vinogradov (1962*a*) argues that deep zooplankton is sustained by food chains and vertical migration rather than by sinking material. Bacteria in the mid-water are not an important source of food because they are too sparse, small and not noticeably related in their abundance to the zooplankton which might feed on them. Bacteria are probably consumed by Radiolaria which are of sufficient size to be subsequently filtered by pteropods and copepods.

According to Vinogradov, most animals of the deep water zooplankton are predatory feeders and many of those whose mouthparts appear suited to filter feeding migrate to the eutrophic zone where food is more concentrated. An example of this is seen in copepods which live at a depth of 1000 m, and deep water pteropods, both of whose gut contents were found to include diatoms along with Radiolaria.

Mysids have been studied in particular by Tchindonova whose work Vinogradov discusses. Animals which were caught at depths of 4000 m or more were found to contain diatoms and shallow water Crustacea in their gut (e.g. *Boreomysis curtirostris*). So the probability is that these animals fed in or close to shallow water. Carnivorous mysids found at considerable depth appeared to feed at that depth, for example *Petalophthalmus armiger* which sucks the contents of prey, a method of feeding thought to take place in a number of different groups. Other carnivores bite their prey and yet others (not mysids) swallow their prey whole. Filter feeding in deep water mysids appears restricted to two species.

Of 77 species of deep sea animals which Tchindonova examined, only nine were judged to be filter feeders, including the two mysids just mentioned, two pteropods and five ostracods. There is a tendency to regard such minority groups as unimportant, which they probably are as purveyors of energy through the ecosystem, but Tchindonova's conclusion is really an astonishing one and if correct implies that certain filter feeders have remarkable skill in locating food.

The decline in plankton biomass with depth (Fig. 6.9) has been regarded by Vinogradov as further evidence of the importance of a food-chain flux of material through the abyssal region. This is surely not so; falling particles could conceivably sustain a similarly distributed biomass and we need additional information to distinguish between these two processes. If we provisionally set aside all contributions from falling food then the flux of material through food chains might be explained as one layer of animals feeding on an upper layer, causing a decline in biomass with increase in depth as in Fig. 6.9. However we cannot entirely discount sinking food, nor can long range vertical migrations be ignored. The latter may be diurnal or once-a-lifetime as in the case of *Ceratias holböelli* (Chapter 1) and the copepod *Calanus cristatus*. This copepod grows in shallow water and descends to depths of 1000–4000 m where it matures on its food reserves, finally leaving eggs in an otherwise empty exoskeleton (Vinogradov, 1962*a*).

In view of the morphological evidence, small particles of food must be regarded as a minor component of the diet of deep sea zooplankton in general. Large particles, present in unknown but lesser numbers, may be preyed upon.

(ii) Large mid-water animals are all predatory, and in the case of fish exhibit striking morphological features.

(iii) The rate of flux of material into deep water through food chains must be in equilibrium with the food consumption of the animals. This can be measured experimentally but we have already seen that the flux proceeds at a rate sufficient to transmit marked seasonal changes in biomass to 2000 m depth (Table 6.6).

TABLE 6.18. *Relationship between depth of occurrence and the respiratory rate in some pelagic fishes and crustaceans* (Childress, 1971)

Depth (m)	Respiratory rate (mg O_2 (kg dry wt)$^{-1}$ min^{-1}) Range	Mean	s.d.	No. of individuals	Dry wt range (g)
0–400	6.2–15.3	12.6	±3.7	19	0.0041–0.104
400–900	1.46–10.0	4.5	±2.5	53	0.0049–2.22
900–1300	0.65–1.63	1.2	±0.3	8	0.132–2.02

The rate at which an animal respires oxygen provides a measure of its food requirements. The complications and uncertainties in converting Q_{O_2} to food intake will be considered shortly. Here we may note the basic stoichiometric relationships, namely, 5 calories are liberated from the respiration of 1 ml of oxygen (measured at NTP) when carbohydrate is oxidised, and 4.7 and 4.2 calories respectively when fat and protein are oxidised. As there is no lack of oxygen but a food shortage in the deep sea, it is worth noting that, on a weight basis, fats liberate more than twice as much energy as carbohydrate or protein when oxidised.

Childress (1971) has measured the oxygen consumption of crustaceans and fish trawled from depths down to 1000 m. The results are summarised in Table 6.18. Animals from the deepest layer included *Gnathophausia gracilis* (Mysidacea) and *Gigantocypris agassizii* (Ostracoda). Animals from the middle layer included *Gnathophausia ingens*, *Acanthephyra curtirostris* (Decapoda) and the fishes *Nectoliparis pelagicus* and *Melanostigmas pammelas*. Shallow water animals included *Euphausia pacifica*, *Hyperia galba* (Amphipoda) and *Sergestes similis* (Decapoda). Childress performed the experiments by placing individuals in a sealed chamber at 5.5 °C and measuring the depletion of oxygen continuously with an electrode over 12–24 hours at normal atmospheric pressure. Antibiotics were present, and controls run to measure the low level of microbial oxygen consumption which has been deducted from the data in Table 6.18. These data were obtained from animals which were allowed four hours to settle down after handling, and were taken from within the P_{O_2} range of approximately 0.1–0.04 ATA which is above the critical P_{O_2} of

TABLE 6.19. *Preliminary data for the rate of oxygen consumption of* Gigantocypris mülleri *at* 10 *or* 8 °*C, and selected pressures.* (*Macdonald, Gilchrist & Teal,* 1971; Macdonald & Teal, unpublished)

Duration of experiment and number of animals	Q_{O_2} (mm³ (gm fresh wt)$^{-1}$ hr^{-1})	Pressure and temperature
(i) 20 minutes 1 adult per experiment, 2 experiments	20 (average)	No effect up to 500 atm, 10 °C
(ii) 20 minutes 1 juvenile per experiment, 2 experiments	140	At 1 atm, 10 °C
(iii) Same as (ii)	5	After the application of 200, 400 or 50 atm, 10 °C
(iv) 1 experiment lasting 8 hours, involving 8 adults		
(*a*) First hour	168	At 1 atm, 8 °C
(*b*) 4 hours (P_{O_2} not less than 0.04 ATA)	105	At 50 atm, 8 °C
(*c*) Final 3 hours (P_{O_2} restored by replenishing seawater at constant pressure)	56	At 50 atm, 8 °C

the animals. According to Childress no effect of animal size was seen on the weight-specific Q_{O_2}.

The animals from around 1000 m depth respired at a tenth of the rate of the surface group. Thus a given dry weight of animals from deep water probably represents a food requirement one tenth of that of a similar dry weight of shallow water animals. The measurement of collected biomass by wet weight may underestimate the reduction of true dry weight biomass with depth if the animals in general become more watery. Thus the decline in biomass with depth shown in Fig. 6.9 represents an even greater attenuation of metabolism than the curves might suggest.

On a number of occasions the oxygen consumption of *Gigantocypris* has been measured at high pressure at sea. The first experiments were of twenty minutes duration but a longer experiment has confirmed that *Gigantocypris* consumes oxygen at a fairly low rate which, in adults, seems to be unaffected by pressures over the range it normally experiences (Table 6.19). The techniques used in the experiments are described in Chapter 7.

Deep sea nutrition

Some measurements of the respiration of marine animals at high pressure carried out by Teal (1971) and Teal and Carey (1967) may have been affected by the locomotor changes induced by pressure and produced results from which food requirements in nature should not be calculated. It will be recalled that moderate pressure exerts little effect on the oxygen consumption of *Marinogammarus*, a shallow water animal, provided accomodation to the excitatory effect of pressure is allowed to occur (Chapter 4).

Translating experimentally determined rates of oxygen consumption into food requirements in nature involves a series of assumptions, some of which may be important. The effect of confinement may seriously affect fish but it does not seem to alter the activity of Crustacea noticeably.

We are obliged to make certain reasonable assumptions about the orthodoxy of intermediate metabolism in these animals and assume that anaerobic metabolism is a negligible fraction of the total through-flow of material when oxygen is available. The anaerobic capacity of mid-water animals has yet to be studied but some preliminary observations (Macdonald & Teal, unpublished) show that whereas *Systellaspis debilis* fails to recover after a few minutes oxygen deprivation, other animals living in a similar P_{O_2}, e.g. *Gigantocypris*, recover activity after one hour in oxygen free water at normal atmospheric pressure.

The limited information we have on the oxygen consumption of mid-water animals probably reflects basal food needs. We need to know how much the animals can raise their metabolic rate during vigorous exercise and how active they are in nature. According to Teal and Carey (1967), actively swimming large Crustacea increase their Q_{O_2} threefold, and according to Vlymen (1970) small shallow copepods metabolise at a basal rate which masks variations in locomotion. *Astacus* increases its Q_{O_2} by 60 per cent during sub-maximal exercise. We can therefore only estimate real food requirements from rates of oxygen consumption when we have more information about locomotor activity in nature and the energy cost of such activity.

Gigantocypris appears a buoyant animal with a fairly steady level of swimming activity in the laboratory which does not appear to be greatly influenced by confinement. If its behaviour in nature is similar then the food which it would require to sustain its oxygen consumption (Table 6.19) would be about one large copepod per month. If we extend Childress' finding that deep water animals metabolise at 1/10 of the rate of shallow water species, then an animal such as *Certias holböelli* (a 'floating trap') may only require prey equal to 10 per cent of its own weight at three month intervals or longer.

The low rate of metabolism seen in mesopelagic animals may be the result of a reduction in the mass of tissue which is otherwise metabolically active, or it may be brought about by a reduction of cellular metabolism, or both. The proportion of dry weight to water content in some deep sea fish is less than normal (Blaxter, Wardle & Roberts, 1971). The question of how much the rate of metabolism of animals can be reduced during intervals between activity is interesting. *Ceratias*, it has been suggested, may only use its jaw muscles once every three months and we might speculate that such muscles have special properties. The muscle of the alimentary canal is probably active for a longer period of time.

The food requirements of animals may be determined by ways other than measuring oxygen consumption. Packard, Healey and Richards (1971) have extracted the electron transport system (ETS) from plankton homogenates and measured the activity of samples obtained from different depths. Plankton from 1000 m yielded activities of 1–20 per cent of the value obtained from plankton from 100 m depth. In view of the authors comment that 'the relationship of the *in vitro* ETS activity to the *in vivo* standard respiratory rate is not well known', it is difficult to assess the significance of this novel type of experiment. The results are by no means inconsistent with the respiration data in Table 6.18. Other methods of measuring the metabolic activity of plankton, such as the determination of phosphorous or nitrogen excretion, have not been applied to deep sea animals. Feeding and starvation experiments with deep sea animals will require a number of technological developments (Chapter 7) and will probably be first carried out on conveniently large sedentary benthic animals recovered by deep submersible vehicle.

(iv) *Long range vertical migration.* Examples of deep sea plankton which feed at the surface have been mentioned. The energy balance of such animals is clearly different from that of the permanent inhabitants of the deep sea. Large mesopelagic Crustacea such as *Acanthephyra purpurea* and *Gennades elegans* are known to migrate daily over several hundred metres (Waterman, Nunnemacher, Chace & Clarke, 1939) and these are more or less available for study. Consider one such organism; making the assumptions about metabolism which have already been mentioned, it is necessary to demonstrate that the extra oxygen consumption which is involved in swimming upwards is more than offset by the increased quantity of food captured. Assume the animal's Q_{O_2} is increased threefold in the effort of achieving an upward swimming velocity of two animal lengths per sec so that 1000 m will be covered in a few hours. The Q_{O_2} expressed on a daily basis will therefore be increased by an amount dependent on the duration of migration, but certainly by less than 50 per cent. Moving up 1000 m will increase the availability of food material several fold for perhaps 12 hours a day. Energetically, therefore, long range vertical migrations appear feasible.

There are as yet insufficient data on the migration, swimming rate, oxygen consumption and feeding rate of any one organism to describe its energy balance. Although fish may obtain nutritional advantage from long vertical migrations this may conflict with economies achieved by the uplift from a swimbladder (p. 252).

BENTHIC MICROBIAL ACTIVITY

It is well known that micro-organisms grow rapidly when surfaces are present on which nutrients become adsorbed. The local environment of a cell seems more important in determining its metabolic performance than the bulk concentration of nutrients. For this reason deep sea sedimentary bacteria may have a more active metabolism than mid-water heterotrophs.

There is not a great deal of information on the concentration of bacteria in deep sea sediments and there is even less on their turnover. ZoBell and Morita have shown (1959) that the division time of deep sea bacteria incubated in a nutrient-rich medium at 1000 atm and 3 °C varies from 2 to 20 hours. A standing crop of bacteria replenishing itself once a day would provide a daily amount of organic material for animal consumption equal to the standing stock of cellular material in the sediments. This might be between 1 and 5 $mg\,m^{-2}$ of surface sediment. The daily flux of organic material to the sea floor would have to be in excess of this

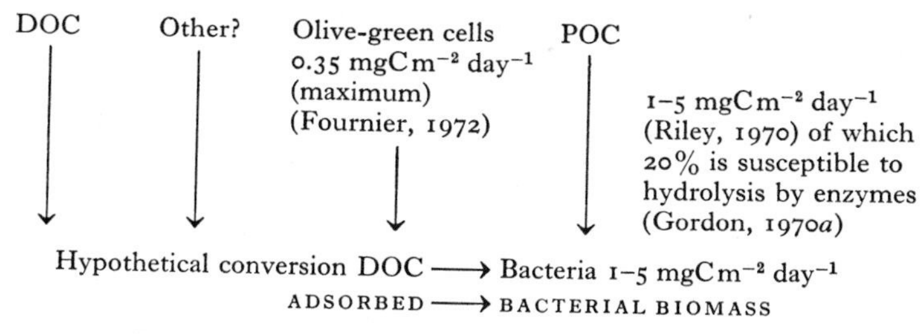

Fig. 6.16. Flux of organic material to the sea floor.

figure to account for the inefficiency with which bacteria convert soluble nutrients into cell material. Fig. 6.16 shows that the flux of POC to the sea floor is similar to the hypothetical conversion of DOC into bacteria.

The solid nutrient material which is available each day for animal consumption on the ocean floor may be itemised as 1–5 mgC m^{-2} bacteria, 0.2–1.0 mgC m^{-2} POC, 0.35 mgC m^{-2} olive green cells, and an unknown amount of organic material adsorbed onto non-nutritive particles. The flux of the latter may be significantly influenced by the flow of water at the sediment surface.

Experiments which attempt to measure sedimentary bacterial respiration *in situ* have been initiated by Pamatmat and Banse (1968) but no data from oceanic depths are available yet.

HOLOTROPHIC PROCESSES ON THE SEA FLOOR

Benthic animals may feed on other animals (carnivores), on material suspended in the water, on detrital organic material deposited on or in the sediments, and on soluble organic material.

According to Sokolova (1959) the benthic invertebrates in the Kurile–Kamchatka Trench consist of 20 per cent carnivores, 25 per cent suspension feeders and 55 per cent detrital feeders. On the abyssal plain of the Atlantic, Sanders and Hessler (1969) find about half the fauna to be detrital feeders. Menzies (1962) has examined benthic isopods from the abyssal plain of the West Atlantic in some detail and reports that most are detrital feeders. Gut contents typically included sediment, coccoliths and Foraminifera. On the assumption that the sediment contained 0.7 per cent organic material and the isopods grow with an overall gross efficiency of 30 per cent, 100 mg sediment would convert to

0.2 mg of animal tissue. As shallow water worms can ingest sediment equal to their own weight every four hours it is reasonable to assume benthic isopods can do the same. Thus a 1 mg isopod might take 16 days to 'process' 100 mg of sediment, synthesising 0.2 mg of new tissue. Let us assume a rich benthic fauna biomass of 1 g m^{-2}, of which half consist of detrital feeders similar to the isopods studied by Menzies. Such a population would be capable of removing about 20 $\text{mgC m}^{-2}\text{day}^{-1}$ from the sediment, ignoring organic material excreted by the animals. Comparing this with the maximal estimate of the carbon flux of 10.35 $\text{m}^{-2}\text{day}^{-1}$ from the sources indicated in Fig. 6.16 we see that if the metabolic rate of the isopods is halved their hypothetical food needs are met by the downflux of organic material available.

Of the many specific measurements which are required to confirm or refute this kind of speculative balance sheet, the energy cost of obtaining organic matter on the sea bed is one which might not attract the most attention. One would like to know the cost of feeding to a detrital feeder. Jørgensen (1952) has made estimates of the energy cost of filter feeding, the analogous problem for animals such as sponges and bivalves which made up 25 per cent of Sokolova's (1959) survey of trench benthos. Bivalves (in shallow water) filter water at a rate of approximately 15 l hour^{-1} and consume 1 ml of oxygen during the process. Thus, on the basis that 0.8 mg of mixed organic material is oxidised by 1 ml of oxygen, Jørgensen suggests that 0.05 mg organic material per litre would have to be available to sustain the effort of filtration. For growth to take place in addition, a higher concentration of organic particles than found by Gordon in deep water (Fig. 6.3) would be required. However, POC refers to particles larger than 4.5 μm and filter feeders can retain much smaller particles (Jørgensen, 1966).

A reduction in the metabolic rate of filter feeders may significantly affect the critical concentration of particles and it might favour survival in impoverished conditions up to some limit, after which, inexorably, the energy cost of filtration would exceed the energy content of the food. We can only conclude that sufficiently high local concentrations of nutrient particles do occur for the benthic filter feeders to live on, but we have little idea of extraction rates or the energy cost to the animals.

Oxygen consumption measurements have been carried out on only one type of abyssal benthic animal, the pogonophore *Sibo-*

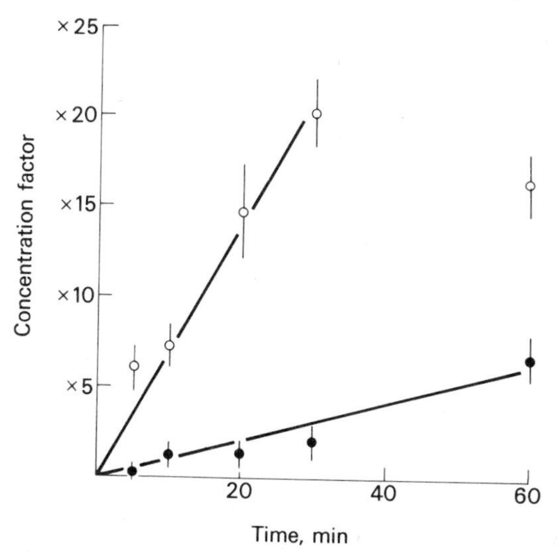

Fig. 6.17. The relation between quantity of [^{14}C]phenylalanine taken up, expressed as a concentration factor, and time, in *S. ekmani*. ●, animals in their tubes; ○, animals removed from their tubes. Vertical bars represent twice the standard error of the mean. Diagonal lines represent the initial rate of uptake. (Little & Gupta, 1969)

glinum atlanticum. A Q_{O_2} of 0.12 ml O_2 per gram wet weight per hour at 15 °C and normal atmospheric pressure was measured at sea on fresh animals recovered from 1300–1500 m in the Bay of Biscay by Manwell, Southward and Southward (1966). A Q_{O_2} of 0.06 ml O_2 per gram wet weight per hour at 5 °C was obtained from *Siboglinum ekmani*, which was trawled from depths of 240–680 m (Little & Gupta, 1969). These measurements suggest that deep sea benthic animals may not have a particularly reduced rate of metabolism.

HETEROTROPHIC ANIMALS

The possibility that marine animals may utilise dissolved organic substances as a food source has been realised for many years (Johnston, 1972). Bacteria, other micro-organisms, internal parasites and cells growing in tissues all acquire energy in this way. Free living animals, on the other hand, are at a disadvantage in this respect because they are relatively large and exposed to low concentrations of nutrients.

Deep sea nutrition

For an animal to acquire useful food from the environment it has to obtain more energy from metabolising the food than it spends in obtaining it. In view of the claims made about the uptake of labelled organic materials by marine animals, particularly invertebrates, it is important to be clear on the evidence needed to demonstrate a net gain in energy by the uptake of solutes.

Johannes, Coward and Webb (1969) have made this point in a paper describing the flux of free amino acids through the marine flat worm *Bdelloura candida*. The animals were fed 24 hours before the experiments and their microbial contamination minimised by the use of antibiotics, ultra violet light and washes in sterile seawater. They were then exposed to ^{14}C-labelled protein hydrolysate at a concentration similar to that found in natural conditions for a period of 20–120 min, washed and analysed for uptake of the labelled compounds. Other animals were incubated in seawater which was subsequently analysed for unlabelled amino acids. The rate of uptake of the labelled compounds was approximately one quarter of the rate of loss of unlabelled amino acids. This experiment demonstrates a net flux of amino acids from the animal into the environment, the reverse of heterotrophy, but does not preclude the possibility that useful uptake occurs in different conditions perhaps, for example, in conditions of near starvation. The minimum criterion for establishing heterotrophy is a *net* uptake of material sufficient to provide a significant contribution to metabolism in conditions which apply in nature.

In the deep sea the pogonophores are the only animals which appear to be heterotrophs. Lacking a gut they must take in small particles of soluble food through their body wall (Chapter 1). They are difficult animals to obtain in good condition for experimental work and of the limited number of experiments which have been performed on them most have utilised shallow water specimens.

The mechanisms by which these animals take in food are not fully understood but three lines of evidence have shed some light on the problem.

Morphological examination has revealed the presence of microvilli which penetrate the cuticular lining of the body. (Fig. 1.29, Chapter 1). These microvilli seem to be uniformly distributed irrespective of the thickness of the cuticle. They protrude into the cavity within the animal's tube and not into the true external environment.

TABLE 6.20. *Uptake of denatured Chlorella protein by* Siboglinum ekmani *(500 μCi l^{-1}, approx.* 1000 *μg l^{-1})*

Time (min)	Average wt of animals ± SE (mg)	Presence (1) or absence (0) of tubes	Concentration factor ± SE
10	0.50 ± 0.089	0	0.14 ± 0.09
60	0.40 ± 0.050	0	11.59 ± 2.5
60	0.55 ± 0.092	1	0.03 ± 0.01

A number of preliminary experiments which have attempted to measure the permeability of the body wall to likely nutrients have been described. Little and Gupta (1969) obtained specimens of *Siboglinum ekmani* from depths of 240–680 m in Norwegian fiords. Most of the animals were severed at the annulus, the 'posterior' portion being lost; despite this they survived well. They were exposed to [^{14}C]phenylalanine and [^{14}C]glycine solutions at 5 °C. The amount of label which was present in the tissues was subsequently determined by equilibriating the intact animal in 100 μl of 80 per cent ethanol for three hours. Counts were made of the radioactivity in a subsample of the ethanol extract and of the macerated and dried animal tissue. From the ethanol count the extent to which the labelled amino acid was concentrated in the animal was computed, and from the low counts obtained from the animal tissue it was evident that over 95 per cent of the labelled compound remained in an ethanol-soluble form. [^{14}C]phenylalanine was concentrated 20-fold from an external concentration of 2×10^{-4} M by those animals which were removed from their tubes, while those remaining inside their tubes concentrated the label 6-fold (Fig. 6.17). Similar results were obtained with [^{14}C]glycine at external concentrations of 6×10^{-4} M. Protein from *Chlorella* sp. was not concentrated at a comparable rate.

The same authors attempted to measure the influx of amino acids through the tube wall and demonstrated a significant permeability, although their exact figures seem to be open to question. Data in Table 6.20 indicate that proteins do not penetrate the tube to any useful extent.

Working independently, Southward and Southward (1970) have carried out similar uptake measurements on *S. ekmani* and

TABLE 6.21. *The permeability of the body wall of* Siboglinum ekmani *to labelled amino acids.* (Southward & Southward, 1970)

A. Uptake of [^{14}C]phenylalanine by *Siboglinum ekmani* as a function of time. (Animals removed from tubes; concentration 2.3×10^{-6} M at 6 °C)

Time (min)	No. of animals	Mean wet wt (mg)	Concentration factor mean ± s.d.	Rate of uptake mean ± s.d. ($\mu g\,g^{-1}$ per hr)	% 'in-soluble'
15	5	0.26	7.81 ± 5.89	13.22 ± 9.97	6.9
30	5	0.28	19.06 ± 5.81	16.15 ± 4.92	3.3
60	5	0.34	20.21 ± 5.39	8.55 ± 2.28	3.0
120	4	0.21	55.12 ± 14.92	11.86 ± 3.21	3.4
1080	5	0.17	229.44 ± 107.37	5.39 ± 2.52	8.1

B. Uptake of [^{14}C]phenylalanine by *Siboglinum ekmani*, as a function of concentration. (Animals removed from tubes: period 2 hr)

Concentration (M)	No. of animals	Mean wet wt (mg)	Concentration factor mean ± s.d.	Rate of uptake mean ± s.d. ($\mu g\,g^{-1}$ per hr)	% 'in-soluble'
1.0×10^{-5}	9	0.22	31.10 ± 10.34	37.03 ± 12.32	1.6
4.2×10^{-6}	5	0.30	67.00 ± 23.58	25.77 ± 9.07	1.9
2.3×10^{-6}	4	0.21	55.03 ± 14.89	11.86 ± 3.21	3.4
6.6×10^{-7}	5	0.20	159.33 ± 46.25	9.68 ± 2.81	2.9
2.1×10^{-7}	5	0.29	157.28 ± 97.72	3.09 ± 1.92	4.8

on a shallow water species *S. fiordicum*. Table 6.21 summarises their data from *S. ekmani*. The rate of uptake is related to the external concentration of the amino acid but the concentration factor achieved after two hours of incubation is lowest in the strongest amino acid solution.

The penetration of [^{14}C]phenylalanine into *S. fiordicum* differs markedly from that seen in *S. ekmani*. A lot of the amino acid becomes incorporated into an ethanol-insoluble fraction and the concentration factors are much lower.

The influx of [^{14}C]glucose from external concentrations of 10^{-7} M was demonstrated in both species. A significant portion of the labelled glucose entered an ethanol-insoluble fraction.

Chemical determinations of the change in the concentration of glucose in a solution bathing twelve specimens of *S. ekmani* established the *net* uptake of this substance at a rate of 34.2 μg per g fresh weight per hour from an ambient concentration of

TABLE 6.22. *The permeability of the body wall of* Siboglinum fiordicum *to palmitic acid and sodium butyrate.* (Southward & Southward, 1970)

A. Uptake of [^{14}C]palmitic acid by *Siboglinum fiordicum*, as a function of time. (Animals out of tubes: concentration 4.3×10^{-9} M)

Time (hr)	No. of animals	Mean wet wt (mg)	Concentration factor, mean	Rate of uptake, mean ± s.d. ($\mu g\ g^{-1}$ per hr)	% 'in-soluble'
2	3	0.84	50.60	0.0154 ± 0.0073	2.5
14	2	1.05	89.0	0.0077 ± 0.0008	3.0

B. Uptake of sodium butyrate by *Siboglinum fiordicum* as a function of time. (Animals in tubes: concentration 8.9×10^{-6} M)

Time (hr)	No. of animals	Mean wet wt (mg)	Concentration factor, mean ± s.d.	Rate of uptake, mean ± s.d. ($\mu g\ g^{-1}$ per hr)	% 'in-soluble'
2	3	1.02	0.43 ± 0.16	0.2158 ± 0.0824	4.4
5	3	0.88	0.80 ± 0.65	0.1615 ± 0.1312	5.3
18	2	0.98	6.77 ± 2.52	0.3712 ± 0.1386	1.7

9.2×10^{-5} M (Southward & Southward, 1970). Glucose was concentrated 3.8 fold. [^{14}C]palmitic and [^{14}C]sodium butyrate were also incorporated by *S. fiordicum* (the former also by *S. ekmani*); see Table 6.22.

Siboglinum atlanticum was dredged from a depth of 1800 m in the Bay of Biscay and shown to be permeable to mixed labelled amino acids and, in two experiments, to glucose at atmospheric pressure and 6 °C (Table 6.23). The tube of this animal is likely to be less of a barrier to the diffusion of small organic solutes than the tube of *S. ekmani* which Little and Gupta (1969) found to reduce the rate of amino acid uptake to a moderate extent.

Southward and Southward (1970) have shown that the post-annular region of the body of *S. ekmani*, which was absent in most of the experiments reported by Little and Gupta, accumulated labelled amino acids at a rate 60 per cent slower than the pre-annular portion and contained a larger ethanol-insoluble labelled fraction. Thus the post-annular body is likely to be more involved in the storage of nutrients. There is general agreement about the permeability of the pogonophores to nutrients in these two independent sets of experiments.

TABLE 6.23. *The permeability of the body wall of* Siboglinum atlanticum, *recovered from* 1800 *m depth, to labelled amino acids and glucose at atmospheric pressure and* 6 °*C*. (Southward & Southward, 1970)

A. Uptake of mixed [^{14}C]amino acids by *Siboglinum atlanticum*, at different concentrations. (Animals in tubes (*i*), animals out of tubes (*o*))

Concentration (M)	Time	No. of animals	Mean wet wt (mg)	Conc. factor, mean	Rate of uptake mean ± s.d. ($\mu g\,g^{-1}$ per hr)	% 'insoluble'
9×10^{-7}	15 min	3 *o*	2.46	1.53	0.5158 ± 0.1392 (both rows)	5.38
9×10^{-7}	30 min	2 *o*	2.16	2.32		
9×10^{-8}	30 min	1 *i*	11.28	(2.46)	(0.0544) —	—
9×10^{-8}	1 h	1 *i*	13.66	1.71	0.0189 —	8.4
9×10^{-8}	10 h	2 *i*	11.05	19.90	0.0219 ± 0.0086	7.6
9×10^{-8}	All the above combined			...	0.0303 ± 0.0325	

The results shown in parentheses are less the 'insoluble' material, which was not measured.

B. Uptake of [^{14}C]glucose by *Siboglinum atlanticum*, at different concentrations. (Animals out of tubes: period 2 hr)

Concentration (M)	*N*	Mean wet wt (mg)	Concentration factor, mean	Rate of uptake mean ± s.d. ($\mu g\,g^{-1}$ per hr)	% 'insoluble'
1.9×10^{-4}	2	1.74	0.68	12.61 ± 0.70	19.2
1.8×10^{-5}	2	2.30	1.96	3.39 ± 0.36	14.3

Additionally, a third approach to the study of food uptake has been used, namely autoradiography. Little and Gupta (1968) and Southward and Southward (1968) both demonstrated the presence of labelled amino acids in the epidermis and secretory cells. These labelled substances would probably correspond to the small fraction of ethanol-insoluble material previously mentioned. Southward and Southward used *S. atlanticum* from 1600 m depth and found [^{3}H]glycine to be particularly evident in the cuticle. Little and Gupta used *S. mergophorum* and found [^{3}H]phenylalanine in epidermal cells although the cuticle was relatively free of label. Ferritin, was also demonstrated to enter the epidermal cells after 18 hours of incubation, so pinocytosis may be an uptake mechanism for nutrients (Little & Gupta, 1970).

Measurements of the uptake of labelled amino acids or other nutrients alone provide no information as to *net* uptake. In fact autoradiography has indicated that some of the labelled compounds were incorporated into secretory cells. It is therefore difficult to assess the metabolic significance of the influxes until comparable effluxes are determined. In one case it will be noted a net transport of glucose was demonstrated by chemical analysis (Southward & Southward, 1970). There can be little doubt that the labelling experiments have revealed an important aspect of the animals' nutritional physiology, and the low K_m values for the uptake processes (Southward & Southward, 1970) are exceedingly interesting, suggesting the presence of an uptake system geared to low concentrations of solutes.

The problem of the source of nutrients in nature has not yet been investigated in any detail. The pogonophores grow in clusters in the mud where nutrients may be present in suitable concentrations for immediate uptake. The tentacle of *S. atlanticum* and *S. fiordicum* does not seem particularly proficient as an uptake organ so early ideas about its importance in this role are not substantiated. Does the animal simply rely on the external supply of suitable solutes or can it hydrolyse externally located protein to boost the influx of small molecules? Little and Gupta's suggestion (1970) that the hydrolysis of material occurs within the tube (but still external to the animal's body) is a strong possibility.

Some conclusions

Deep sea nutrition is interesting because it involves problems of food supply and economy of consumption.

Feeding involves dramatic contrasts, ranging from the carnivorous practices of angler fishes, gulper eels and squid, to the steady intake of nutrients by mud-eating benthic animals and the absorption of nutrients by the pogonophores.

A number of free swimming deep sea Crustacea have a low metabolic rate as judged by oxygen consumption.

The flux of organic material into sedimentary bacterial carbon is an important link between the benthos and the eutrophic zone. The benthos is relatively well provided with food in contrast to bathypelagic animals which appear dependent on a flux of material through food chains.

Deep sea nutrition

We have little or no information on the nutritional efficiency and normal energy expenditure of deep sea organisms.

Distance from primary production restricts metabolic activity and numbers of animals in the deep sea, but there is no evidence that other physical factors such as pressure and temperature exert any significant limitation.

7 DEEP SEA BIO-ENGINEERING

Introduction

Deep sea bio-engineering is dealt with in this chapter under four main headings. First, the technology and physiology of human diving is outlined because the diver is part of many underwater engineering operations and because he provides an interesting case of comparative physiology. Second, deep submersible vehicles (or submarines), which can work at depths far greater than divers, are described. They are already in use in deep sea biology and a general appraisal of their role is a natural sequel to diving physiology. Third, the role of surface ships in deep sea work is considered. They are an exceedingly expensive scientific facility, although much cheaper than manned submersible craft, which also require a surface support ship. Methods of simulating the high pressures characteristic of the deep sea are considered in the final section.

It is important to note that these four areas of bio-engineering are complementary and in certain projects may be used simultaneously. Looking at the entire field of deep sea bio-engineering it will be argued that many of the existing techniques are far from being fully exploited.

Deep sea man

The role of free swimming SCUBA* divers in shallow water marine biology is well established. Here we examine the physiological problems which arise when such divers attempt to penetrate the deep sea. For the reader who is unfamiliar with diving physiology a summary of the depth capability of humans will serve as a useful point of departure.

Commercial diving is carried out at depths to 200 m and at the time of writing one commercial diving firm operating in the North

* Self Contained Underwater Breathing Apparatus = SCUBA.

Sea oil field has announced its intention of carrying out commercial work to depths of 330 m, a depth first achieved by Keller in 1962. Man is therefore at the point of joining mesopelagic animals and working at depths exceeding that of the continental shelf.

In simulated 'dry dives' carried out in laboratories, men have experienced hydrostatic pressures of up to 60 atm without serious ill effects and there is every indication that an even greater pressure (i.e. depth) could be tolerated. Speculation on man's future usefulness as a technician and explorer in the mesopelagic depths is therefore timely. Much will depend on the ease and safety with which SCUBA divers can descend to these depths.

Other mammals dive to a significant depth in the sea but all involve breath holding which imposes a severe limit on the duration of the dive. Weddell seals descend to depths in excess of 400 m in dives lasting a maximum of 16 min, but a much longer period of time is spent in dives to lesser depths (Kooyman, 1972). The physiology of marine mammals is not dealt with in this book.

Diving stresses many facets of man's physiology, in particular respiration and thermal regulation (see Bennett & Elliott, 1969, for a survey of this subject). An important hazard in all but the most shallow dives is failure to achieve safe decompression. Deep diving, which is largely conducted in simulated form in laboratory pressure vessels, has brought to the fore three areas of physiology which are attracting intensive study. They are the process of gas exchange in the lungs, the phenomenon of inert gas narcosis and the direct effects of pressure on the human nervous system. Problems of thermal regulation, decompression, communication and many others are not particularly related to extreme depth and are not considered here, although of major interest to diving physiologists.

GAS EXCHANGE IN THE LUNGS OF MAN AND TEST MAMMALS

In regular deep dives to a depth of 200 m, special helium–oxygen (heliox) gas mixtures are breathed to maintain a normal partial pressure of oxygen in the alveoli. A P_{AO_2} of more than 1 ATA damages the lungs (Wood, 1969). Nitrogen, which impairs the mental performance of men at a P_{AN_2} of as little as 3.2 ATA, is normally excluded from the gas mixture (Bennett, 1969).

The range of pressures in which heliox mixtures can sustain adequate gas exchange has been the subject of theoretical and experimental studies. Compression of the gas mixture increases its density but not its viscosity. In general, an increase in gas density increases the work required to achieve an adequate tidal flow of gas in the lung. It is the turbulent flow component of tidal ventilation, and of expiration in particular, which is reduced by an increase in gas density; the laminar flow component is unaffected (Macklen & Mead, 1968).

Lanphier (1969) has predicted how the maximum voluntary ventilation (MVV) obtained breathing a heliox mixture will vary with pressure. Fig. 7.1 shows how the MVV obtained experimentally with human subjects at pressures up to 15 ATA is approximately proportional to the reciprocal of the square root of the density of the gas mixture breathed. Using this relationship, Lanphier predicts a MVV compatible with moderate exercise at pressures to 100 atm with a gas mixture containing 99 per cent helium and 1 per cent oxygen, or to 200 atm using a mixture of hydrogen and oxygen. The prediction is clear that tidal ventilation should not be a limiting factor in respiration at depths of less than 2000 m at the very least. The prediction has been partly confirmed in a number of simulated dives in which humans have breathed heliox mixtures satisfactorily at over 50 ATA pressure, although vigorous breathing during exercise at these pressures has not yet been studied.

Experiments with test mammals have produced conflicting results. Goats which were subjected to pressures of up to 56.2 ATA over a 10-day period (Chouteau, 1969 & 1971) showed signs of respiratory failure. At 39.7 ATA the animals were normal, but at 49.4 ATA they lay down, apparently in some discomfort, and made uncoordinated movements. When the P_{O_2} was increased from 150 mmHg to 190 mmHg the animals recovered a normal posture but they succumbed again when the ambient P_{O_2} was lowered. There is little doubt that the goats were experiencing a form of respiratory deficiency which was alleviated by the rise in P_{O_2}. The limiting factor was probably not tidal ventilation, which continues normally in other test mammals at much higher pressures, but it may have been the result of an unfavourable change in the $\dot{V}$:Q ratio due to a change in the diffusional mixing of the compressed gases in the airways (Wood, 1971).

The exchange of gases by diffusion and mixing processes in the alveolar airways even under normal conditions is incompletely

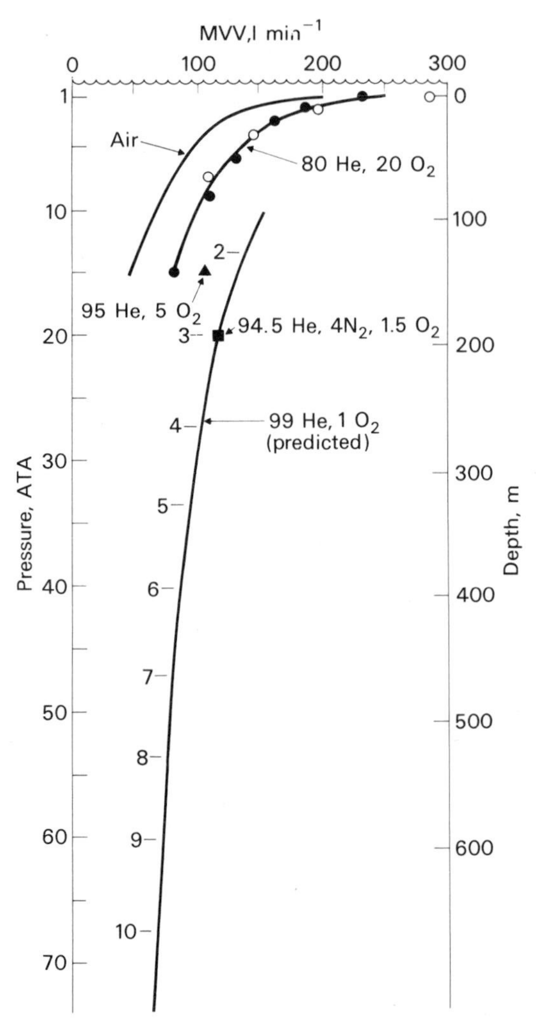

Fig. 7.1. Relationship between density of gas breathed by human subjects and the maximal voluntary ventilation (MVV). Figures indicate the percentage of each gas. Points for air are omitted. Numbers along the predicted curve refer to equivalent pressures of air in ATA. (After Lanphier, 1969)

understood and the phenomena at high pressure presents problems of widespread interest. On general grounds one might anticipate a reduction in gas exchange by diffusion in regions of high gas density. Lanphier (1972) has suggested how an intermediate zone of 'disturbance mixing' between regions of tidal flow and diffusion might create a situation in which an increase in density would actually assist in gas exchange.

Whatever the limiting stage proves to be in gas exchange it is likely to be encountered at a pressure well in excess of those being currently contemplated in human diving. Test mammals have survived 300 atm pressure in a heliox atmosphere, and human subjects have successfully breathed gas mixtures simulating the density of heliox at a pressure of 160 atm (Lambertsen, 1975). For the limit to man's pressure tolerance we have to turn to different facets of his physiology.

THE HIGH PRESSURE NERVOUS SYNDROME IN TEST MAMMALS

Hydrostatic pressure experiments have been carried out on mammals in the absence of a gas phase by means of the liquid breathing technique. This procedure has been pioneered by Kylstra and his colleagues (Kylstra *et al.*, 1967). Test mammals, mice in the following example, have their lungs filled with a fluorocarbon liquid in which oxygen is highly soluble. With the animals immersed in the fluorocarbon and with the body temperature reduced to between 17 °C and 25 °C, self-ventilation of the lungs achieves an exchange of dissolved gas adequate for several hours survival. Hydrostatic pressure has been applied to mice in these conditions at a rate of 10 atm min^{-1}. At pressures of 50–80 atm the limbs were seen to tremble, voluntary movements became uncoordinated and the paws clenched. At 100 atm the neck became bent forwards so the animal's chin pressed on the chest, and both hind and forelimbs extended caudally. Ventilation ceased. Note the similarity to the effects of pressure on Crustacea (Chapter 4) and to a limited extent to the effects seen in humans (p. 331.) Control experiments with mice at a similar body temperature but compressed in a heliox mixture showed a similar sequence of tremor, uncoordinated movement and finally spasms. Mice with transected spinal cords showed muscular contractions anterior to the severed cord but not posterior to it. Further, mice with a normal body temperature showed no spasms at pressures up to 100 atm when compressed at the same rate in a heliox mixture, although tremor and uncoordinated movements were apparent (Kylstra *et al.*, 1967).

MacInnis and colleagues (1967) have observed mice for 4 hours in heliox at pressures up to 122 atm following compression at a rate of 1 atm min^{-1}. Limb tremor was seen during compression at 75–80 atm. Ventilation changed at about 90 atm to a slower, deeper rhythm involving some gasping. At pressures of 90 atm or more, fairly normal behavioural activity was seen to alternate with periods when the mice lay down. The eyes would also remain closed for long periods, even when the animals were moving about. The mice were successfully decompressed with 80 per cent surviving four weeks after the experiments. The similarity between the periodic inactivity of the mice at pressure and the intermittent sleep experienced by some human subjects at pressures in the region of 30 atm, is striking (see below and Brauer, 1968).

The most detailed description of the high pressure effects on test mammals is provided by Brauer, Jordan and Way (1972) who used *Saimiri sciureus*, the squirrel monkey. When compressed at a rate of 24 atm hour^{-1} the monkeys underwent generalised motor seizures at 65 atm and showed marked changes in EEG. At lower pressures mild tremors were seen and these were not associated with EEG changes. Brauer *et al.* suggested that there are two distinct components in this animals' response to compression. Such a view is substantiated by their finding that the threshold pressure for tremors was not influenced by the rate of compression, whereas the threshold pressure for convulsions was raised by a reduced rate of compression.

The study of Brauer *et al.* with the squirrel monkey is notable for the attempt made to control or monitor the experimental variables. Blood P_{O_2} (peripheral and internal carotid), P_{CO_2}, and glucose levels were monitored. The chamber temperature was controlled to give a satisfactory rectal temperature in the animals. The EEG and the ventilatory rhythm were observed, even during motor seizures, and all the evidence which was accumulated leads to the conclusion that high pressure was responsible for both tremor and convulsions. A generalised deterioration in the condition of the animals due to impaired respiration or obscure stress effects, and which might have elicited the symptoms, was not detected.

In summary, these and other experiments show effects of high hydrostatic pressure on excitable tissue in intact animals in widely different conditions, from hypothermic liquid-breathing mice to animals in 'comfortable' conditions in heliox mixtures. Comparative studies of convulsion pressures in warm-blooded animals (Brauer *et al.*, 1971) and invertebrates (Macdonald, 1975) show that, with the exception of some deep sea animals and shallow water Protozoa, most animals undergo spasms or convulsions in the range of 50 to 100 atm (Fig. 7.2).

THE HIGH PRESSURE NERVOUS SYNDROME IN HUMANS

During the past decade simulated dives have extended the pressure range known to be safely tolerated by humans from 22 atm (Hamilton, MacInnis, Noble & Schreiner, 1966) to the 60 atm pressure which was experienced by French divers in 1972. This latter work has been briefly reported at the Fifth Symposium on Underwater Physiology, 1972.

A particularly significant 'dive' was that of Brauer and his fellow diver (Brauer, 1968; Brauer, Dimov, Fructus, Gosset & Nagnet, 1969) who experienced the relatively rapid mean compression rate of 0.30 atm min^{-1}, to approximately 36 atm. In addition to experiencing muscle tremor (referred to as helium tremor) the subjects were prone to drift in and out of a sleep-like condition and exhibited changes in EEG, notably an increase in theta-wave activity. These symptoms were apparent at 30 atm; the compression was stopped at 36.5 atm and the subjects were eventually safely decompressed.

Subsequent deep simulated dives have avoided serious disturbance of the central nervous system by employing slower rates of compression. Although the 60 atm simulated dive of the French is the world record at the time of writing, an earlier dive carried out at the Royal Navy Physiological laboratory (UK) provides a detailed and instructive picture of human physiology at high pressure (Bennett, 1970; Morrison 1975). The dive profile is given in Fig. 7.3. Compression was fairly rapid but interrupted for 24 hours three times to allow accommodation to take place.

Changes in the EEG were noted in the two subjects, one of whom showed a marked increase in theta-wave activity which faded over 12 hours at constant pressure and which reverted to normal after decompression. No sleep-like state was experienced. Muscle tremor was measured by a tremor transducer attached to the subjects' middle fingers. Fig. 7.4 shows the difference in the

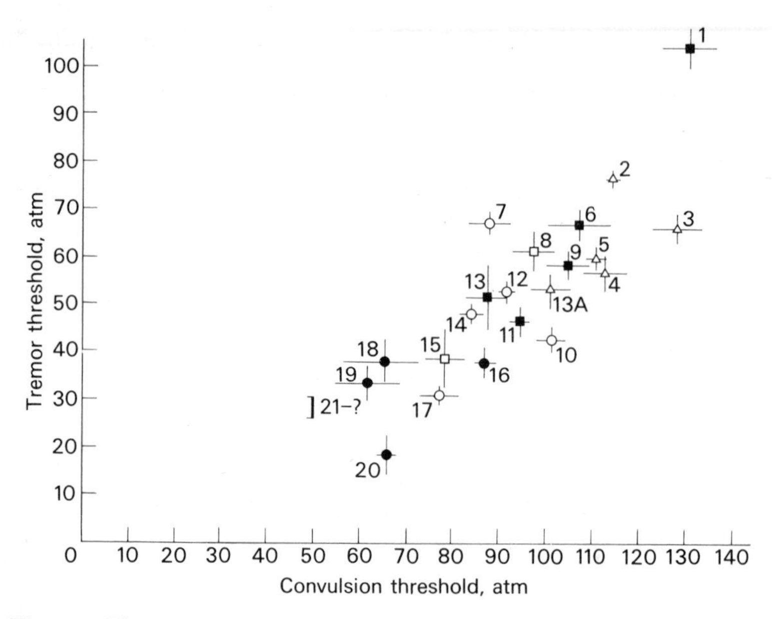

Fig. 7.2. The pressures which cause tremor (vertical axis) and convulsions (horizontal axis) in selected vertebrates subjected to a compression rate of 24 atm hour^{-1}. The numbers correspond to the species listed in the table below; each point refers to the mean ± standard deviation. (Brauer, 1972)

Class	Order	Species	No. in figure
Pisces		*Achirus fasciatus*	1
		Paralichtys dentatus	6
		Anguilla rostrata	9
		Symphurus palguisa	13
Reptilia		*Thamnoptulis sistalis*	3
		Anolis carolinensis	4
Aves		*Serinus canarius*	2
		Ectopistes carolinensis	5
		Gallus domesticus	13A
Mammalia	Insectivora	*Erinaceus europaeus*	23*
	Marsupialia	*Didelphys virginiana*	15
	Edentata	*Dasypus novemcinctus*	8
	Rodentia	*Epimys rattus* S. D.	7
		Mus musculus CD-1	12
		Mus musculus AJ	14
		Cavia porcellus	17
	Lagomorpha	*Oryctolagus cuniculus*	10
	Carnivora	*Procyon lotor*	11
	Primates	*Tupaia tupaia*	20
		Saimiri sciureus	19
		Macaca mulatta	16
		Papio papio	18
		Homo sapiens	21

* No. 23 – location overlaps No. 13A.

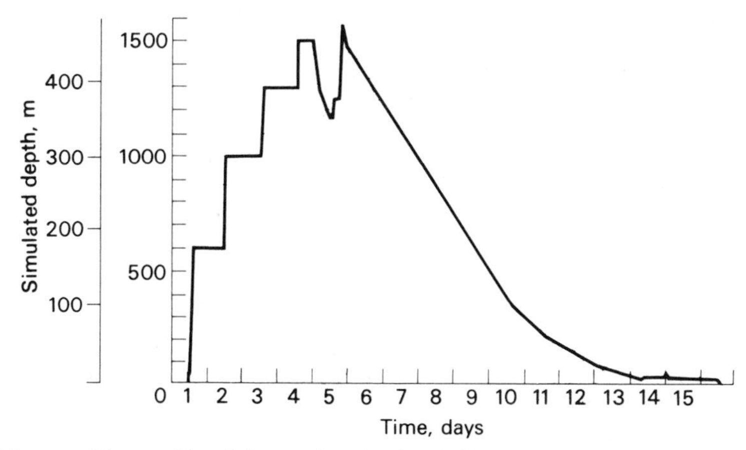

Fig. 7.3. The profile of the 15-day simulated dive to 500 m depth (approximately). The decompression was interrupted by a recompression due to a vestibular bend and much later due to a leg bend. The 'dive' lasted from 3 March to 18 March, 1970. (Bennett, 1970)

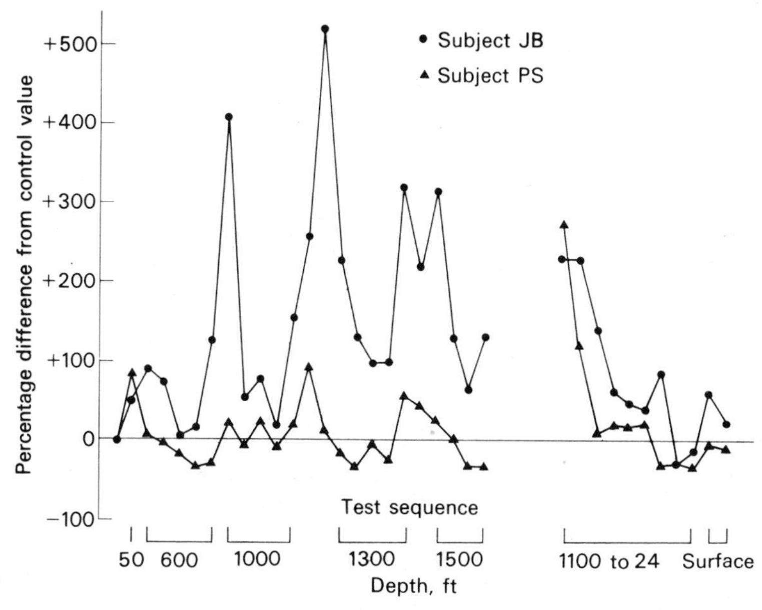

Fig. 7.4. The percentage change in the amount of tremor measured from the tremor transducer attached to the middle finger of each of two subjects during the dive (see Fig. 7.3). Note that JB shows evidence of a marked increase in tremor, especially after compression. (Bennett, 1970)

susceptibility of the two subjects and the extent of the tremors which, like the theta-wave activity, faded after a compression stage. Tests of manual dexterity showed a significant decrement in performance due to tremors. For example, the task of picking up ball bearings with tweezers was completed at rates as low as 70 per cent below normal after compression, but the subjects' ability in this task improved with time to a level of 35 per cent below control levels. Mental performance was normal.

The subjects breathed a P_{O_2} of 0.45 ATA in heliox. Light physical exercise (300 watts min^{-1}) yielded no particular respiratory difficulty although the divers easily overheated in the warm temperature of the chamber which is required to offset the high thermal conductivity of helium. Other tests on heart rate, kidney function and blood chemistry revealed minor changes which may become important in longer dives. Fig. 7.3 shows the decompression schedule which proved too fast for one subject although the slower rate which was used after re-compressing to approximately 50 ATA was satisfactory.

The conclusion from this work is that humans can tolerate up to 50 atm pressure for a short time under precisely controlled conditions experiencing only a few minor physical ailments chief of which is muscle tremor in the hands. It is clearly important to understand the cause of muscle tremor and the other disorders of excitable tissue.

INERT GAS NARCOSIS AND CONTROL OF THE HIGH PRESSURE NERVOUS SYNDROME

Human divers breathing compressed air (80 per cent nitrogen, 20 per cent oxygen) may experience serious narcotic effects at pressures of 5–10 atmospheres due to the high partial pressure of nitrogen. The narcotic* or anaesthetic effect of nitrogen is essentially similar to that produced by clinical anaesthetics at much lower partial pressures. The potency of these chemically inert substances was first related to their lipid solubility by Meyer and Overton. Although there has been much discussion of their other

* In this context, 'narcosis' means a readily reversible depression of function caused by chemically inert substances which typically are gaseous or volatile, and which do not form covalent bonds. Ethanol (p. 336) behaves as an inert narcotic although it is chemically reactive. Anaesthesia is a reversible loss of consciousness caused by a narcotic agent.

TABLE 7.1. *Narcotic potency of selected general anaesthetics and inert gases as judged by the depression of the righting reflex of mice.* (Data from several authors, modified from Smith, 1969)

	atm
He[a]	[190]
Ne	110.0
H_2	85.0
N_2	35.0
Ar	24.0
CF_4	19.0
SF_6	6.9
CH_4	5.9
Kr	3.9
N_2O	1.5
Xe	1.1
C_2H_4	1.1
C_2H_2	0.85
CF_2Cl_2	0.4
cC_3H_6	0.11
Ether	0.032
Halothane	0.017
$CHCl_3$	0.008

[a] The value for helium is derived by extrapolation and ignores the antagonistic effect of hydrostatic pressure. In the righting reflex of mice the potency of helium is probably zero.

physical properties there is now general agreement that the solubility of the chemically inert narcotic gases in hydrophobic solvents, such as olive oil or benzene, is an excellent guide to their potency. Helium is one of the series of narcotics listed in Table 7.1 from Smith (1969), and in view of the use of helium in breathing mixtures at pressures of more than 50 atm the question of its narcotic potency is of considerable interest.

Measurements of the potency of helium are complicated by the appearance of the high pressure nervous syndrome which, superficially, is the opposite of a narcotic effect. Brauer and Way (1970) used a method of estimating the potency of helium and hydrogen without applying the full narcotic partial pressure of either gas which would have caused concomittant hydrostatic pressure effects. It was argued that the narcotic pressure of hydrogen, using the righting reflex in mice as a criterion, is 25 per cent of that of nitrogen. Helium yielded a negative potency, that is, it antagonised other narcotics present.

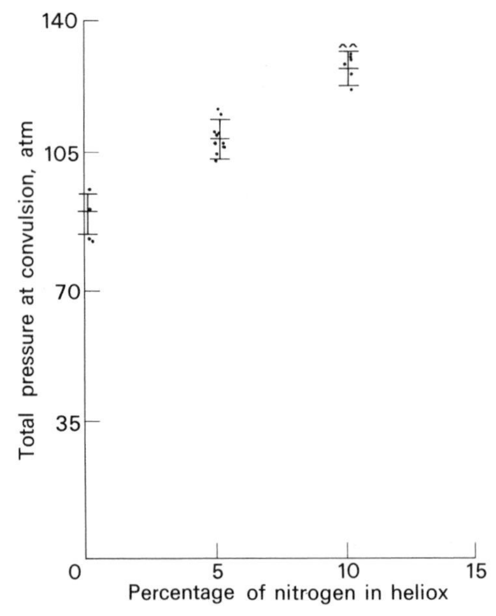

Fig. 7.5. The effect of nitrogen added to a helium oxygen mixture on the convulsion pressure in mice. (Brauer, Way, Jordan & Parrish, 1971)

In practical diving, helium in heliox mixtures appears to behave as an ideal physiologically inert fluid which transmits pressure through the airways of a diver and dilutes the oxygen to a safe level.

If free swimming men are to descend to depths greater than 200 m in the sea then the high pressure nervous syndrome has to be understood and avoided. Slow compression is obviously one way of controlling the symptoms (Fig. 7.3) but pharmacological possibilities also exist.

Brauer and his colleagues (1971) found that a small quantity of nitrogen in the heliox breathed by mice considerably raised the pressure at which the mice convulsed (Fig. 7.5). Since pressure stimulates locomotor activity and anaesthetics diminish it, this is perhaps not too surprising. The antagonism between hydrostatic pressure and a narcotic was first demonstrated in animals by Johnson and Flagler (1950) who showed that 150–200 atm hydrostatic pressure revived tadpoles which were anaesthetised with ethanol. However, the level of pressure used in this experiment was claimed to exert little effect on the un-narcotised animals so the result indicated that pressure was able to reverse the narcotising reaction of ethanol. The reversal of narcosis by hydrostatic pressure has been studied in a variety of biological systems in an attempt to understand its mechanism.

A quantitative model which accounts for the antagonism between narcotics and pressure has been developed by Lever, Miller, Paton and Smith (1971) and Miller (1972) principally using the interactions between inert narcotics and hydrostatic pressure in newts. The model provides a rational, although simplified, approach to the use of inert narcotics to control the high pressure nervous syndrome. Newts were used in the study as they possess a convenient rolling reflex and may be compressed in gas or water. The rolling response of the animals in heliox is abolished by 260 atm when their limbs undergo a paralytic spasm. In the presence of a P_{N_2} of 34 atm the rolling response is only 20 per cent below the control level (Fig. 7.6). Thus at high hydrostatic pressures the presence of the inert narcotic nitrogen offsets the pressure spasms (Fig. 7.6*c*) which may be regarded as the final stage in the high pressure nervous syndrome.

The model put forward to account for these phenomena assumes that narcotic molecules dissolve in a hydrophobic region of excitable tissue, probably a synaptic region. Their presence causes an expansion in the volume of the hydrophobic region, which in turn causes the diminution of excitability characteristic of narcosis or anaesthesia. Pressure compresses hydrophobic regions, and the antagonism between pressure and anaesthetics is seen as a balance of compressive and expansive forces in the hypothetical hydrophobic region.

According to this model, a narcotic will increase the volume of a hydrophobic region according to its partial molar volume ($\overline{V}_2$), its mole fraction solubility (x_2), and its partial pressure (P_a). Thus the per cent expansion of volume in a hydrophobic region is given by:

$$\%\ \text{expansion} = \frac{\overline{V}_2 x_2 P_a}{V_m} \times 100 \qquad (1)$$

V_m is the molar volume of the membrane or the target molecules therein.

Hydrostatic pressure, according to the model, will compress the same hydrophobic region according to its bulk compressibility β,

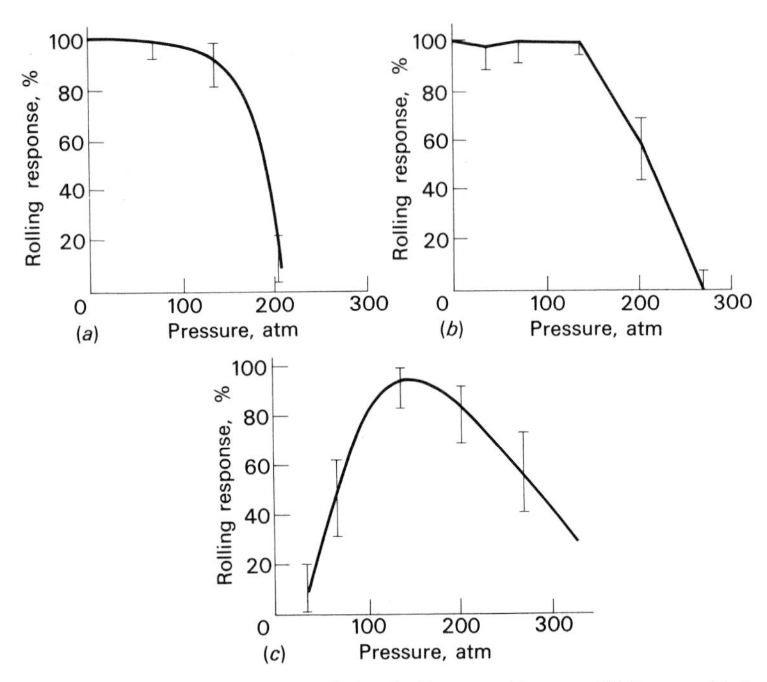

Fig. 7.6. The rolling response of the Italian crested newt (*Triturus cristatus carnifex*) as a function of hydrostatic pressure. Ten animals were used (*a*) in water and (*b*) in helium oxygen mixture. In (*c*) 34 atm of nitrogen were present before the pressure was raised with helium. (Lever, Miller, Paton & Smith, 1971)

and the pressures involved P_T, and a change in function will ensue.

$$\% \text{ compression} = \beta P_T \times 100 \qquad (2)$$

Thus a hydrophobic region of excitable tissue is thought to be expanded by the solution of narcotic in it. At high pressure the normal volume of the hydrophobic region may be restored by the presence of narcotic molecules and normal function returns. 100 atm pressure would decrease the volume of a plausible hydrophobic region by about 0.5 per cent, assuming β is equal to 0.5×10^{-4} atm^{-1}, the value for the compressibility of olive oil or benzene. Narcotising partial pressures of any inert narcotic will cause a similar expansion in the volume of the hydrophobic region. When 50 per cent of a group of animals (newts in the case of the experimental data in Fig. 7.6) are able to perform their

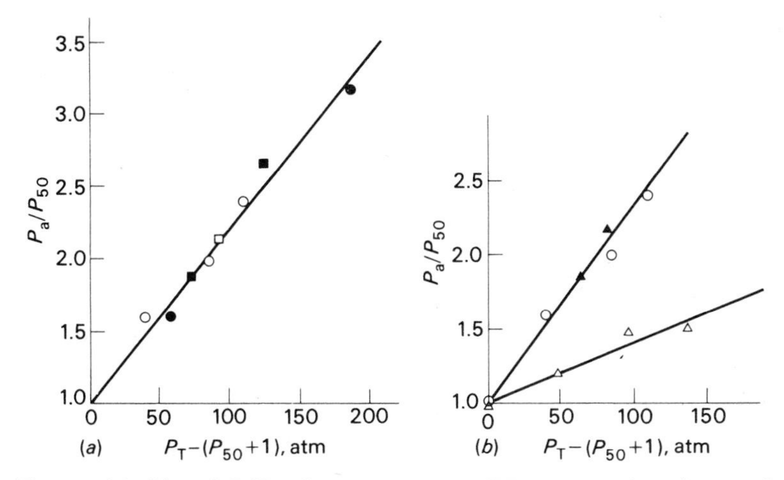

Fig. 7.7 (*a*). Plot of P_a/P_{50} (inert gas pressure/ED_{50} pressure) against total pressure P_T less $(P_{50}+1)$ required to restore response to 50% of animals in groups of about ten newts at 20 °C. ●, N_2; ○, N_2O; □, CF_4; ■, SF_6. See text. (Miller, 1972) (*b*) Same plot as in (*a*). ○, N_2O for newts at 20 °C; ▲, N_2O for newts at 30 °C; △, N_2O for mice. See text. (Halsey & Eger, 1971)

righting reflex at high pressure and in the presence of anaesthetic, then

$$\beta P_T = \frac{\bar{V}.x_2}{V_m}(P_a - P_{50}) \tag{3}$$

P_a is a selected pressure of anaesthetic (narcotic) and P_{50} its partial pressure which on its own anaesthetizes 50 per cent of the animals (alternatively known as the experimental dose, ED_{50}). As hydrostatic pressure increases, more anaesthetic is required to offset its compressive effect. A plot of hydrostatic pressure against P_a/P_{50} gives a straight line for a variety of anaesthetics which is consistent with the idea that each anaesthetic contributes an increase in volume according to equation (1) and is otherwise indifferent in its effect (Fig. 7.7). Note that the partial pressure of nitrogen is subtracted from the total hydrostatic pressure and the curve passes through the origin. Halsey and Eger (1971) have also made measurements of the antagonism between hydrostatic pressure and nitrous oxide in mice and their data are plotted in Fig. 7.7, also from Miller (1972).

It has to be pointed out that although the experimental data are neatly accounted for, the bulk concepts involved in the model

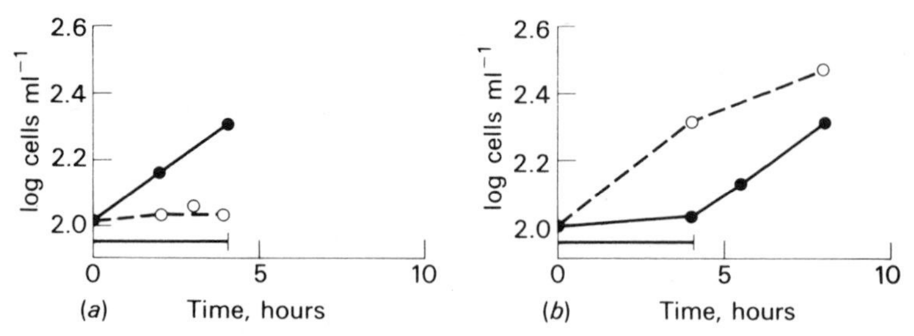

Fig. 7.8. The interaction between the anaesthetic halothane and hydrostatic pressure in dividing *Tetrahymena pyriformis*. Vertical axis, log cells ml^{-1}, standardised to 100% at the start of the experiment. The horizontal bar shows the duration of experimental treatment. (*a*) A low dose of halothane. 0.001 ATA, on its own (●) is without effect. Similarly 100 atm on its own is without effect, but with 0.001 ATA an inhibition of division is seen (○). (*b*) An inhibitory dose of halothane. The effect of 0.01 ATA (●) is reversed by 100 atm pressure (○). Similar results are seen with $C_3 H_8$ and N_2. (Kirkness & Macdonald, 1972, also Kirkness & Macdonald, in preparation)

($\bar{V}_2$, β) are doubtless oversimplifications; in Chapter 2 a variety of molecular changes which are induced by pressure are discussed. However it provides a general explanation for the way inert gases might be used to offset the high pressure nervous syndrome by contributing their partial molar volume in some critical molecular site and opposing the more drastic effects of pressure.

The antagonism between anaesthetics and hydrostatic pressure has also been demonstrated in the squid axon (Spyropoulos, 1957*a*), liposomes (Johnson & Miller, 1970), bacterial luminescence (Johnson *et al.*, 1945) and cell division (Kirkness & Macdonald, 1972). In the latter case a more complicated interaction between pressure and anaesthetics was discovered. The inhibitory action of clinical anaesthetics and nitrogen on dividing *Tetrahymena* cells was found to be relieved by the application of 100 atm hydrostatic pressure which on its own is without effect (Fig. 7.8). At a dose of anaesthetic or nitrogen which is too low to exert an effect on cell division on its own 100 atm has the surprising effect of causing division to be inhibited. Thus 100 atm raises the potency of both clinical anaesthetics and nitrogen in low doses in these dividing cells. The previous arguments show there is good reason to believe that a mild dose of an inert narcotic gas will diminish the symptoms of tremor in a diver who is subjected to rapid compression, but the same treatment may inhibit growing cells. If the effect on the cells is no worse than that experienced during normal clinical anaesthesia then it would probably be acceptable. However, this clearly introduces a complication in the possible control of the high pressure nervous syndrome by means of inert gases and other pharmacological methods may prove superior.

The point has been made elsewhere (Macdonald, 1975) that the control of the high pressure nervous syndrome has presumably been accomplished by nature in both diving mammals and in pressure-tolerant deep sea animals which experience severe pressure changes without functional disorder (Chapter 4). In addition to having technological significance the direct effects of pressure on excitable tissue is clearly of fundamental biophysical and evolutionary interest.

TECHNIQUES USED IN REAL AND SIMULATED DIVES

Commercial deep diving usually involves a pressure vessel in which to transport the divers to and from the work site. Undersea habitats are also under development to allow prolonged 'saturation' diving. It seems likely that if diving is to play a role in deep sea biology it will follow procedures developed in commercial diving which currently favours methods in which the resting divers are held in a safe hyperbaric environment on a ship or oil drilling platform, rather than in an underwater habitat (see Table 7.2, p. 344; Fig. 7.11, p. 347).

Simulated dives in gaseous environments with either humans or experimental mammals require suitably capacious pressure vessels (normally called hyperbaric chambers). Small mammals can be accommodated in chambers no larger than those often used for experiments with rather higher hydraulic pressures (p. 388). Pipes, valves and other components may be the same as for hydraulic work, but hyperbaric vessels normally possess a number of distinctive features. Mammals in helium based gas mixtures require an environmental temperature quite close to their body temperature as helium conducts heat very effectively. Accordingly, provision must be made to control the temperature inside the vessel. In the case of experimental animals it is also usual to mount a fan driven by a brushless motor inside the vessel to circulate the gas over carbon-dioxide absorbing material. Gas samples may be readily drawn off from the vessel for analysis by conventional

Fig. 7.9. The hyperbaric chamber at the Royal Naval Physiological Laboratory, Alverstoke, UK, which is capable of simulating 750 m depth shown during the record dive to approximately 500 m in March 1970. (Bennett, 1970)

techniques. The compressibility of gas is such that sampling the hyperbaric vessel causes only a negligible pressure fluctuation.

Hyperbaric vessels for experiments with human subjects are major installations. Fig. 7.9 shows a chamber in the Royal Naval Physiological laboratory, UK. It may accommodate two men to a pressure of 68 atm. It has a personnel lock in/out chamber as well as a smaller lock to allow food or physiological samples to be exchanged between the occupants and the scientific party outside. Another chamber filled with water provides realistic exercise facilities.

Human divers usually participate in the physiological investigation, making psychometric testing and blood sampling relatively easy. Brauer *et al.* (1972), however, have used a procedure for obtaining blood samples from monkeys inside hyperbaric chambers.

CONCLUSIONS

Men have successfully tolerated 50 atm and 60 atm pressure in laboratory experiments and 33 atm in an open water dive. Physiological limits to deep human diving probably lie in the range of

100–150 atm. Time may be an important factor; the depth limit for a short exposure may prove much greater than for a prolonged exposure.

Marine biology is at present a minor field of activity for the deep water commercial diver and will probably remain so. Divers can collect, observe, and operate equipment but many of these tasks may be accomplished from manned vehicles. The extent of the SCUBA divers' contribution to the biology of the mesopelagic zone and upper part of the continental slope is likely to be influenced by the development of deep sea vehicles which are also able to operate in depths far greater than we can imagine men safely reaching. A similar argument applies to commercial underwater work where divers and submarines work together. It is by no means clear at what depth free swimming men will be completely supplanted by vehicles for practical purposes.

Deep sea vehicles

A variety of vehicles has been designed to penetrate the deep sea; they are classified in Table 7.2.

MANNED VEHICLES

The bathysphere of William Beebe

The bathysphere is a steel pressure-resistant sphere which is lowered into the sea from a ship at the surface. The one designed by Barton, Butler and Barret in 1930, and used by Beebe, is illustrated in Fig. 7.10. The internal diameter of the sphere was 1.38 m (54 in), sufficient to accommodate two cramped observers who viewed the exterior through three 15 cm diameter quartz windows. A supply of oxygen for respiration was supplied from a cylinder and carbon dioxide was removed by a carbon dioxide-absorbent material.

The sphere weighed nearly three tons in air and was lowered into the sea by a steel cable of the non-spinning type, 22 mm in diameter and of ultimate strength 29 tons. The maximum dynamic loading on the cable would have been in the region of 8 tons. While the weight of the sphere is reduced in water the weight of the steel cable paid out to a depth of 1000 m would have amounted to nearly 2 tons.

TABLE 7.2. *Classification of deep sea vehicles*

Manned
- Suspended on a wire from a ship at the surface, e.g.
  - Bathysphere of Barton, Butler and Barrett (Beebe, 1935)
  - Transfer capsules for diving work
- Floating and self powered, e.g.
  - Bathyscaph (Piccard, 1956)
  - Deep research submarine such as *Alvin* and *Deep Quest* (Wenzel, 1969)
  - See also *Undersea Technology Handbook* (Anon, 1968)
- Pressure-resistant, articulated diving suit, under development in Britain

Unmanned
- Torpedo type (Anon, 1968)
- Tethered and powered vehicle *CURV* (Talkington, 1969)

Communication between the surface mother ship and the occupants of the sphere was achieved by a telephone cable passing through the wall of the pressure sphere by way of a stuffing box packed with flax! The sphere was negatively buoyant and although its calculated collapse pressure is not reported, the lives of the occupants would certainly have been lost if the cable had parted in deep water. The chances of this happening could only be minimised by good design and careful seamanship, and never entirely eliminated. The motion of a ship on the surface of the sea can produce great stress in a long suspended cable and can translate motion to the end of it. The more important limitations in the concept of the bathysphere are its dependence on the mother ship and the inability of occupants to manoeuvre their vessel to explore the sea floor. Designers of modern deep sea vehicles have opted for buoyancy and manoeuvrability, and the bathysphere principle is only retained for vehicles whose purpose is the transfer of divers between a work station and shipboard decompression chambers. Nevertheless, the bathysphere pioneered both types of modern vehicle.

Legend for Fig. 7.10

Fig. 7.10. Bathysphere of 1934. From left to right: Carbon dioxide absorber with its blower, four trays and pan; oxygen tank and valve; telephone coil and battery box – the telephones are plugged into this box and it is connected by the wire shown on the two hooks above the oxygen valve with the telephone wires in the communication hose; air-temperature recorder, and below it the left-hand sealed window; barometer; switch-box at top of sphere; central observation window immediately below switch-box; oxygen tank and valve; searchlight. The communication hose is shown as it enters the bathysphere through the stuffing-box. (Beebe, 1935)

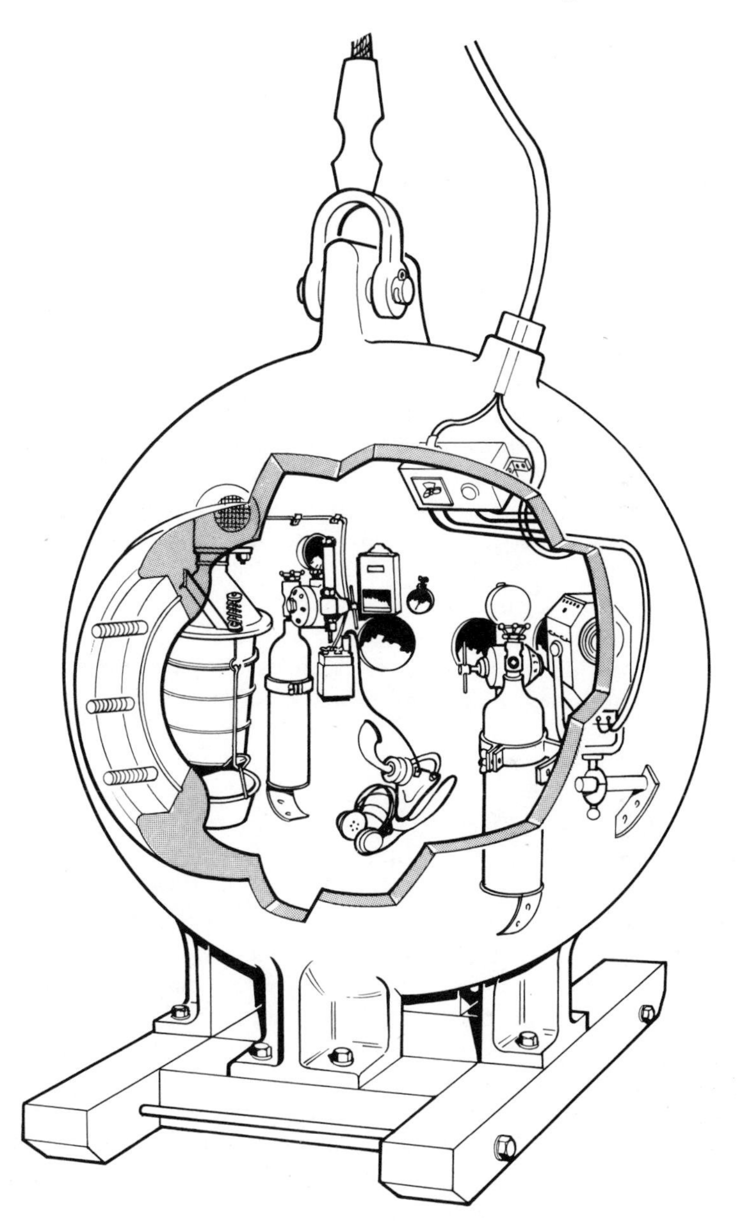

Fig. 7.10. For legend see opposite page

Diver-transfer vehicles

One of the most advanced ship-board diving systems at the time of writing is carried by the French ship *Triton*. Fig. 7.11 shows in diagrammatic form the interconnecting hyperbaric chambers and diver-transfer vehicle. The latter is pressurised with gas and is lowered on a cable to a maximum depth of 250 m and divers may then enter or leave. The transfer vehicle mates to a hyperbaric chamber in the ship which contains a matching pressure thereby allowing divers to return to a heated, comfortable chamber in which they may rest or be decompressed. Simultaneously a fresh team of four divers can enter the transfer vehicle and be lowered to the work site without experiencing a pressure change. The interconnecting pressure chambers on the ship (Fig. 7.11) permits a further group of divers to be decompressed from (or compressed to) the working pressure, so this remarkable system provides for saturation diving with the maximum security and comfort for the divers.

Triton also has the accommodation for a full-sized deep submersible vehicle of the type to be described shortly.

A contrasting diver-transfer vehicle is the small self-powered submersible Johnson-Sea-Link which can transport a diver to and from a work site (Fig. 7.12).

Manned and self-powered deep sea vehicles

It is interesting that the first manned descent into the greatest depths of the oceans was achieved before the development of the commercial deep sea vehicles which have a restricted depth range. The historic descent took place on 23 January, 1960. J. Piccard and D. Walsh (Piccard & Dietz, 1961) in the bathyscaph *Trieste* descended into the Challenger Deep and briefly moored their vehicle on the sea floor at an ambient pressure of 1140 atm. A flat fish is reported to have swum slowly past the observation port as if to make the point that engineering may have taken man thus far; now it was up to the biologist. The point is even more forcefully made in view of Wolff's opinion (Wolff, 1961) that the flat fish may have actually been an echinoderm!

Since then a variety of more manoeuvrable craft have been built, generally only capable of descending to a limited depth but

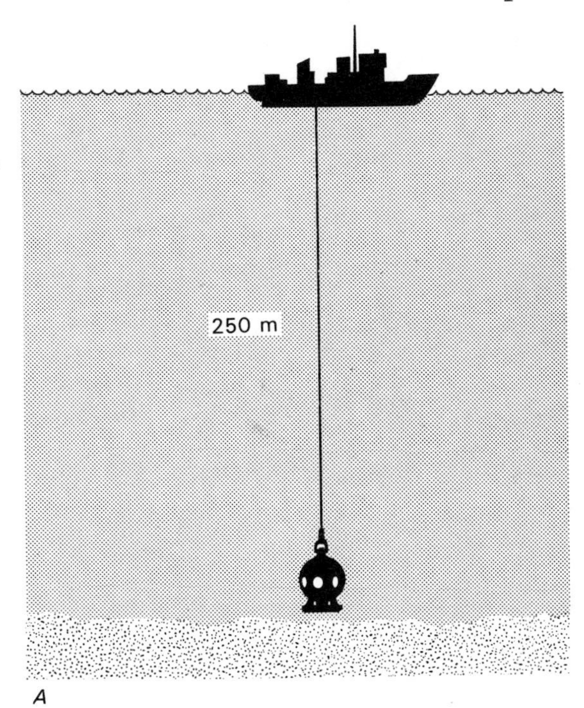

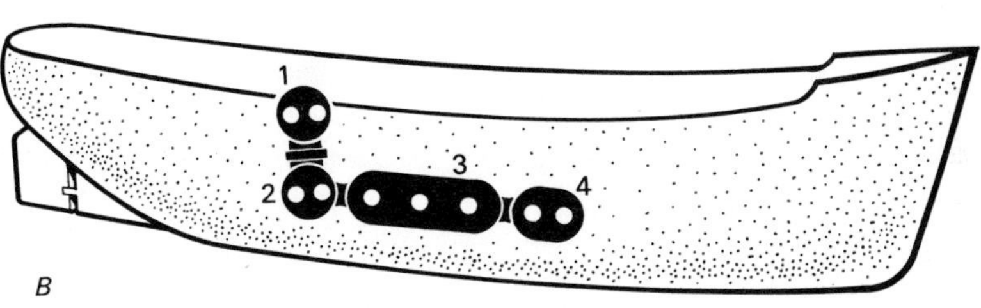

Fig. 7.11*A*. Diagram illustrating the diver-transfer facilities on the French ship *Triton*. Divers may be lowered to and recovered from a depth of 250 m.
B. The *Triton*'s interconnecting hyperbaric chambers (not to scale). The transfer capsule 1 (illustrated in *A*) is shown locked on to the decompression chambers 2 and 3; chamber 4 allows personnel to lock in or out of the system without affecting the pressure in the main chamber 3. See text.

equipped to tackle specific functions in geology, biology, physical oceanography or marine salvage. A notable recent development is the armoured suit which is capable of withstanding an ambient pressure of 30 atm yet allows the occupant, who is at atmospheric pressure, to walk about on the sea floor and undertake tasks

Fig. 7.12. Johnson-Sea-Link research submarine. The two-man acrylic plastic sphere is seen forward of the separate diver compartment. Batteries supply 28 V DC to a number of electric motors, of which five propel the craft forward at 2.5 kt and three give lift and steerage. Gas for breathing and for blowing the ballast tanks is stored in cylinders.

requiring manual dexterity (Barton, 1973). For the experimental deep sea biologist these craft offer the opportunity of procuring material or carrying out experiments at depth. They are undoubtedly one of the most decisive (and expensive) pieces of apparatus we can hope to use.

The bathyscaph of Auguste Piccard

The vessel *Trieste* was designed by Piccard (1956) and followed two smaller bathyscaphs. *Trieste* comprises a pressure resistant steel sphere which holds two men and which is buoyed by a large float which equilibriates with the ambient pressure (Fig. 7.13). The first pressure sphere made for *Trieste* had an outside diameter of 2.18 m. The wall was 9 cm thick with local thickening around portholes. The sphere was forged and machined as two hemispheres which sealed face-to-face. The initial sealing compression for the two metal faces was originally provided by, incredibly, an

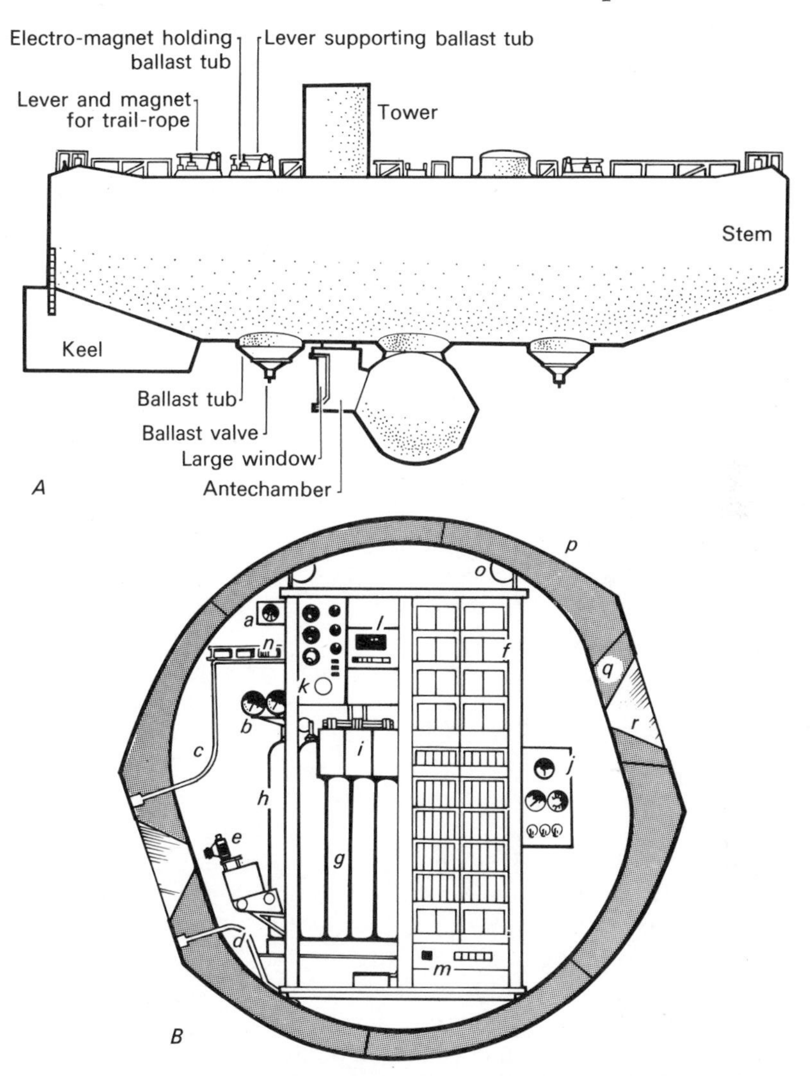

Fig. 7.13. The bathyscaph *Trieste*, *A*. Side view. *B*. Section of sphere which accommodates two men. Details: *a*, chronometer; *b*, pressure gauges; *c*, high pressure tube for connecting the pressure gauges with the outside water; *d*, lead-through for electric cables entering the cabin from the outside; *e*, still camera; *f*, battery rack and silver–zinc batteries; *g*, bottles of compressed air for blowing the entrance chamber; *h*, oxygen for breathing; *i*, alkali for carbon dioxide absorption; *j*, electric control panel; *k*, acoustic telephone (wireless); *l*, ultrasonic echo sounder; *m*, various electronic oceanographic equipment; *n*, tachometer for measuring vertical speed for ascent or descent; *o*, indirect interior lighting, *p*, shell of the sphere; *q*, entrance door; *r*, windows. (After Piccard, 1956)

external rubber band; subsequently a more permanent bond was used. The sphere was capable of safely sustaining 1200 atm ambient pressure. The float which is some 19 m long and approximately 4 m wide contains petrol (gasoline) sufficient to buoy the steel sphere, whose weight in water is 5 tons, and additional structures, instruments and ballast. The volume of petrol normally used is approximately 120 cubic metres.

The bathyscaph descends by flooding compartments containing air fore and aft of the petrol tanks. Fine adjustment of negative buoyancy is achieved by releasing petrol. Lift is obtained by dropping fine iron pellets. Precise control of the craft's buoyancy is essential to avoid bumping on to the sea floor. The buoyancy provided by the petrol diminishes with decrease in temperature and increase in pressure to the extent that a ton of ballast has to be released each 1000 m in order to keep the craft in slight negative buoyancy. Rapid descents cause the petrol to heat and thus delay the need to jettison ballast. Although the bathyscaph was designed mainly for ascending and descending, some horizontal progression is made possible through electrically powered propellors.

The main observation port is a conical Plexiglass window of 10 cm viewing diameter. It is situated forward and is inclined downward at 18°, giving a field of view illuminated by external lamps. Two men can occupy the pressure sphere for up to 18 hours without recourse to emergency oxygen supplies. This is plenty of time for most purposes; the dive to the bottom of the Challenger Deep lasted only 9 hours. A large portion of the sphere is occupied by instruments which include an echo sounder, an acoustic telephone, and batteries (Fig. 7.13).

The bathyscaph was a major advance on the bathysphere. In particular, it may safely descend to the greatest depths with fine control over its buoyancy and with some control of forward motion. It can also provide a stable and quiet observation platform at depth. Its disadvantages are its high cost of sea time, its inability to work in anything but the calmest weather and its lack of manoeuvrability.

Deep sea research submarines

Deep research submarines are generally smaller and much more manoeuvrable than bathyscaphs. As an example we may consider *Alvin*, operated by the Woods Hole Oceanographic Institution (Fig. 7.14). *Alvin* can normally carry three men, one pilot and two observers for 8 hours, and can safely descend to 2000 m. A new pressure-sphere will soon double that depth range. Four windows are fitted; three give forward views and one a vertical view upwards. An external mechanical arm can be operated by the pilot to collect specimens or operate equipment. External lights and a closed circuit TV system allow the occupants to observe and search around them. An echo sounder and horizontal scanning sonar system give longer range 'visibility'. An acoustic telephone provides communication between the mother ship and the submersible. Batteries exposed to the ambient pressure provide power for hydraulic motors which drive propellors and hydraulic pumps.

The vessel's positive buoyancy which is required to keep the craft at the surface is provided by air tanks which are flooded to start a dive. Once submerged, *Alvin* can dive and climb by means of its three manoeuvrable propellers at a speed of 1 knot over a range of 15 miles (see Table 7.4). The fine control over the craft's buoyancy is determined by the distribution of oil between rubber bags and pressure tanks; to decrease buoyancy, oil is pumped into the pressure tanks; to increase buoyancy, oil is allowed to expand into the rubber bags, thereby displacing more water. The vessels trim is controlled by pumps which distribute mercury between several tanks (Fig. 7.14), and by this means any fore and aft angle of up to 20° from horizontal may be adapted without relying on hydrodynamic effects. The same system can offset uneven loads, while the buoyancy control pumps can offset reasonable payloads and variations in the density of seawater. *Alvin* is designed to explore the sea floor and is therefore exposed to the hazards of cable and crevices which might prevent it from surfacing in addition to the hazards of a power failure whilst in mid-water. Positive buoyancy under these conditions can be produced by successively shedding the mechanical arm, batteries, mercury, and finally, by releasing the buoyant pressure sphere from the 'hull' and conning tower.

More recently built submarines such as *Deep Quest* are bigger,

Fig. 7.14. The deep submersible research vehicle, *Alvin*.
A. Alvin being lowered into the water by a deck crane. (Photograph by courtesy of Woods Hole Oceanographic Institution)

go deeper and for longer periods than *Alvin*, but are fundamentally similar (Table 7.3). The instrumentation fitted to *Deep Quest* is similar to that in *Alvin*, comprising three television cameras, two echo sounders (surface and bottom) side scanning sonar, optical periscopes and two mechanical arms.

The biggest and most advanced submarine is currently the US Navy's deep submergence rescue vehicle (*DSRV*) which is designed to rescue the occupants of foundered submarines (Fig. 7.15).

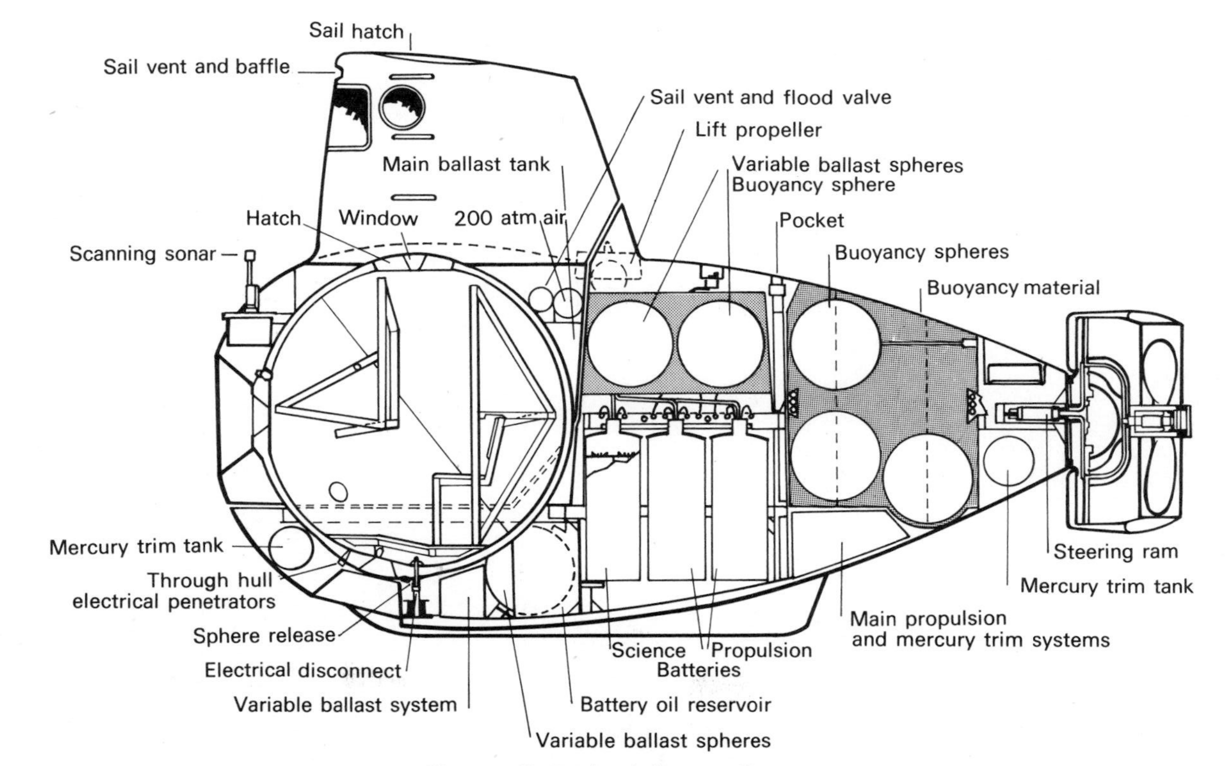

Fig. 7.14B. Sectional diagram. See text.

TABLE 7.3. *Features of two deep submersible research vehicles.* (From Wenzel, 1969, and various other sources)

	Deep Quest	*Alvin*
Length	12.1 m	7 m
Beam, maximum	5.7	—
Draft, surfaced	2.3	2.1
Displacement	50 long tons	15 tons
Maximum depth	2600 m	2000 m (to be extended)
Maximum speed, submerged	4.5 knots	3 knots
Personnel		
crew	2 } 4	1 } 3
observers	2	2
Trim angle variation	±30°	±20°
List angle variation	±10°	—
Endurance		
power	12 hr at 4 knots 24 hr at 2 knots	Range 10–15 miles
life support	192 man-hr 12 man-hr emergency	72 man-hr
Main power plant	230 kWh, 115 VDC	—
Emergency power	3.6 kWh, 28 VDC	—
Jettisonable weight	3400 kg	See text

All deep submergence craft, including bathyscaphs, require a mother ship. *Alvin*'s mother is *Lulu*, a catamaran fitted with an elevator platform between the hulls. Once *Alvin* is lifted clear of the water the catamaran may transport *Alvin* between stations. *Deep Quest* is tended by *Transquest*, a 100 ft long tug-like ship with a catamaran stern and elevator. The advanced DSRV may be transported by special aircraft and supported at sea either by a surface ship or another submarine.

The tendency in the design of recent submersibles has been for increased flexibility of function and manoeuvrability, in direct contrast to the bathyscaph or bathysphere. An elegant newcomer is the Johnson-Sea-Link vehicle previously mentioned on p. 348 (Fig. 7.11). This vessel is made up of two personnel chambers. One is a pressure-resistant plastic (acrylic) sphere which houses a pilot and observer in a 'shirt-sleeved' environment. The other chamber is positioned aft and consists of a diver lock-out chamber. The vessel provides great manoeuvrability and visibility for observers but is restricted to 300 m depth. Work in deep research submarines is not free from danger. *Alvin* sank after capsizing

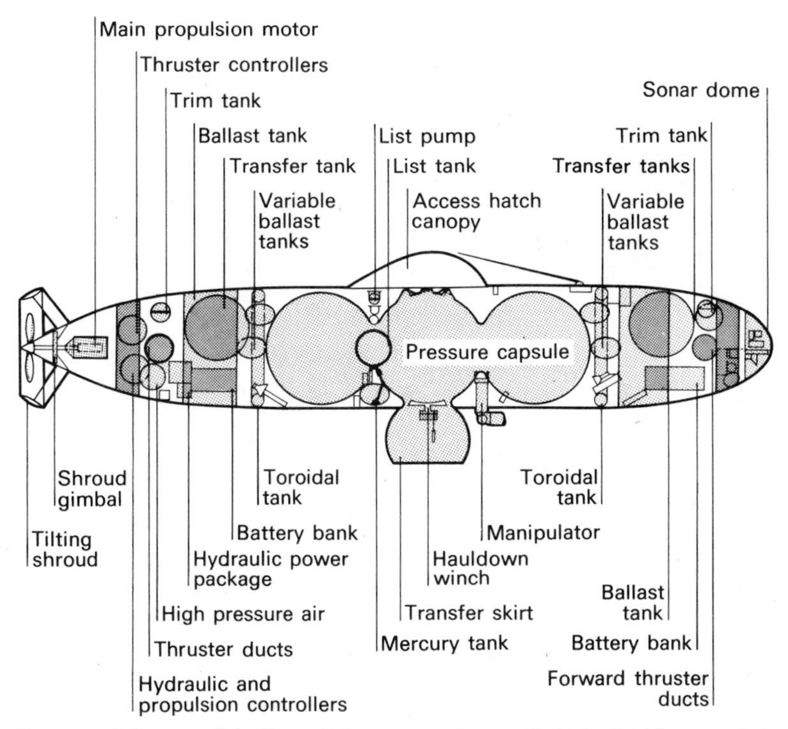

Fig. 7.15. Diagram of the Deep Submergence Rescue Vehicle. (Feldman, 1969)

when being launched from its mother ship and has subsequently been salvaged. The Johnson-Sea-Link became ensnared with a sunken ship with the result that two divers died in the lock-out chamber.

UNMANNED VEHICLES

A torpedo sized vehicle approximately 6 m long and 50 cm in diameter has been built at the Applied Physics Laboratory, University of Washington (Fig. 7.16). The vehicle carries a 45 kg pay load at 6 knots to a depth of 4000 m. Its underwater course is determined before each dive but an acoustic signal (p. 386) can override the preset course and bring the vehicle to the surface. In the event of a breakdown the craft's slight positive buoyancy will also bring it to the surface.

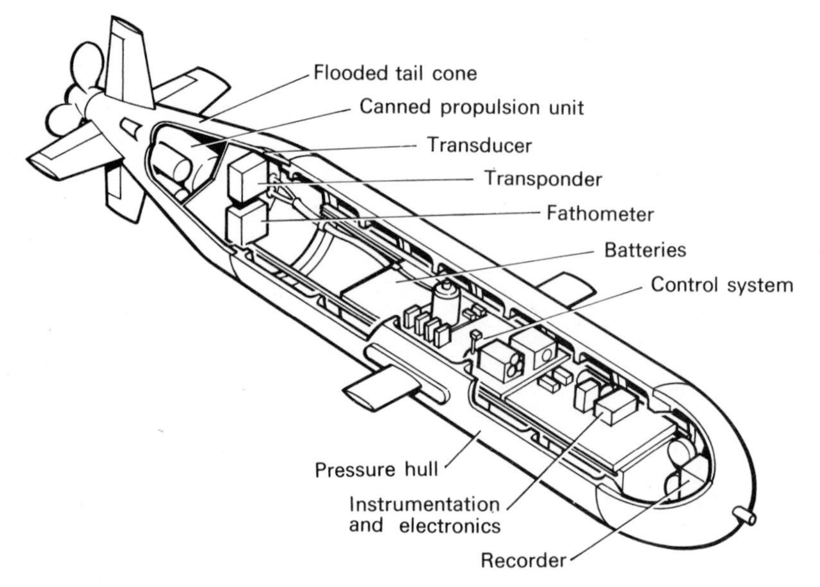

Fig. 7.16. Unmanned, torpedo-style research vehicle built at the Applied Physics Laboratory, University of Washington. (Aron, 1962)

In view of the great cost of manned submersible craft, vehicles of this level of complexity and comparative cheapness might well be used for purposes such as photography and plankton collecting, especially from isolumes or other contours of interest.

An unmanned underwater vehicle which is controlled and powered through a cable paid out from a surface ship is shown in Fig. 7.17. The so-called Cable Controlled Underwater Recovery Vehicle, *CURV*, has been used to retrieve a lost hydrogen bomb from the sea floor (the bomb was located by the manned submersible *Alvin*), and to collect biological material. A later version is designed to dive to a depth of 2300 m (Talkington, 1969). The advantages of unmanned vehicles of this type as compared to manned vehicles are, in general, fairly obvious, and so are their limitations.

SOME CONCLUSIONS

This brief survey of deep sea vehicles shows that engineering science has made the deep sea accessible for exploration and

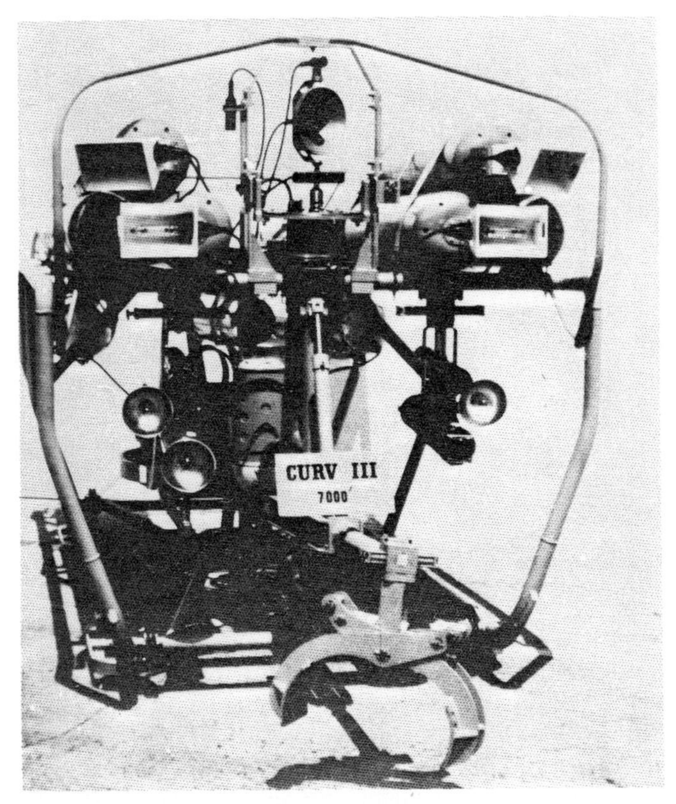

Fig. 7.17. *CURV*, an unmanned deep sea vehicle which is controlled by a cable from the surface. (Talkington, 1969)

scrutiny by man. For the experimental biologist deep submersible vehicles open up opportunities for making accurate measurements *in situ*. We have already noted the observations which Beebe (1935) and Barham (1966) have made on the behaviour of deep sea fish (Chapter 5). An interesting example of a deep sea experiment carried out *in situ* is the current attempt to measure the flux of oxygen into undisturbed deep sea sediments at depths of 600 m and greater (see Chapter 6). The submersible *Alvin* is being used for the purpose and by means of its mechanical arm a bell jar containing an oxygen electrode and acoustic transmitter is placed over a carefully selected patch of mud. Information about the P_{O_2} in the bell jar is received at the surface. Similar measurements have been made at a depth of 180 m by Pamatmat and Banse (1968) working from the surface, but it would appear that use of a submersible is an enormous advantage in deeper water. The 600 m working depth which is being attempted by workers at the Woods Hole Oceanographic Institute is still a long way short of the ocean floor, whose average depth it will be recalled, is nearly 4000 m.

Other interesting observations have been made at depth by Milliman and Manheim (1968) (see p. 381), by Bernard (1958) and by Peres (1965) aboard the bathyscaph *Archimède* in the Puerto Rico Trench. Peres observed how the suspension of plankton which had been noted during the descent diminished abruptly at a depth of 6600 m. It is difficult to believe this observation could have been made by other methods although an investigation into this type of discontinuity may well utilise unmanned vehicles and probes from surface ships.

The role of surface ships

The role of surface ships in experimental deep sea biology is to provide a platform for (i) the collection of organisms, (ii) the operation of deep sea probes, and (iii) service as a field laboratory. Most workers prefer a ship sufficiently large to be comfortable, but with a low freeboard for handling equipment over the side. Ocean going research ships are normally designed for a wide range of oceanographic work, which traditionally includes the collection of organisms, but few have laboratories particularly suited for experimental biology. Portable laboratories have been used on some ships but this logical solution to the problem of making a ship versatile is not yet popular.

In the following section we will consider how deep sea organisms are collected by surface ships and how probes, manipulated from the surface, can provide information about deep sea life. At this point it will be worth recalling the enormous volume of the deep sea and the fact that the deep sea floor covers an area of 300×10^6 km^2 (Fig. 1.1, p. 2).

COLLECTING ORGANISMS

Micro-organisms

Bacteria from bulk seawater can be collected in samplers of the type used by chemical oceanographers or by the apparatus shown in Fig. 7.18. This sterile water sampler was first described by ZoBell in 1941 in a review paper. Several samplers can be lowered vertically on a hydrographic wire, and filled by the action of the weight (messenger) which is sent down the wire. Some of the problems of enumerating the bacteria in water samples are outlined on p. 272.

Micro-organisms from deep sea sediments have been obtained by incubating isolates from the interior of solid cores obtained from conventional geological cores.

Mid-water animals

Carrying out mid-water trawling and plankton hauls, like many other biological techniques, is easy to do badly and difficult to do accurately. A plankton haul is likely to be made either with the object of collecting animals in good condition for experimental work or to quantitatively sample the natural population. Methods of plankton collection involve concentrating the animals from a dispersed condition, usually by filtration and normally by means of a fine mesh net towed by a ship. Collecting animals from deep water involves a considerable length of time and has to overcome the problems of the sparse distribution of the animals and operating equipment several miles distant and out of sight. Here we will consider some basic principles.

On towing objects through water. Before considering the behaviour of towed nets it is worth noting the phenomena which are observed on towing a compact body through water. The object and the wire are subjected to hydrodynamic drag and gravity. The tension in the wire is the product of complicated interacting forces and the resultant curved towline is a complex shape. Hydrodynamic resistance increases with the square of the speed of the towed body so tension in the wire also increases with speed. However, as the increased resistance causes lift, the main effect of an increase in speed is to elevate the towed body in the water and cause only a

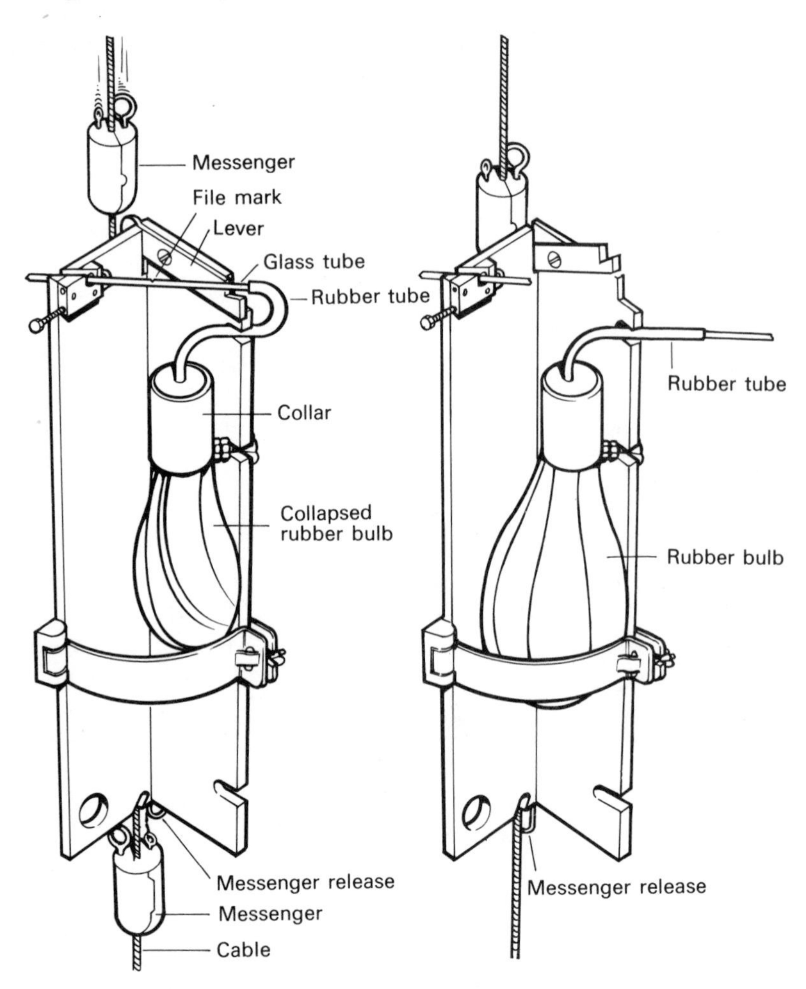

Fig. 7.18. J–Z bacteriological sampler fitted with a pressure-resistant collapsible rubber bottle, before and after collecting a sample of water. In the second diagram the messenger has caused the lever to swing up and break the glass inlet tube. (ZoBell, 1941)

small rise in tension in the towing wire. Table 7.4 illustrates the point using the results of Rather and co-workers (1965) working from the research ship R.V. *Chain*. The table shows how trebling the towing speed caused an increase in tension in the wire of approximately 50 per cent, much less than the increase in resistance

TABLE 7.4. *Towing speed and wire tension*

Length of wire 600 m, diameter 12 mm, faired, towing a 980 kg 'fish'

Speed of ship (knots)	Wire tension relative to 100 at zero speed
4	127
8	141
10.25	156
12	180

to the towed 'fish'. The depth of the towed body also varies in a complicated way when towing speed and wire length are altered. With a heavy streamlined 'fish' whose weight to drag ratio is high a curved tow line is found. The curvature of the wire increases as it approaches the 'fish'. At a certain depth, dependent on the towing speed and the weight of the fish, the point of proportional depth is reached, when further wire paid out causes the fish to descend a similar vertical distance. With less compact towed bodies the curve of the towline is determined by the accumulative weight of the wire. Fig. 7.19 shows the results of tests at sea in which the depth of a body towed 200 m behind a ship with a faired cable changes from 440 to 190 m when the towing speed was increased from 4 to 8 knots. Fairing is often used to decrease the towline drag and to increase the depth at which the towed body equilibrates, but this is not normally the practice when towing nets. For this purpose a multi-stranded galvanised steel cable or an armoured and electrically insulated cable is normally used (Aron *et al.*, 1964).

Fig. 7.19 shows how the use of fairing can double the depth at which a towline of given length tows a body; alternatively, fairing clearly reduces towline tension and hence the risk of losing valuable equipment. For a thorough treatment of the problem of towing objects in water the reader is referred to Pode (1951).

Nets and associated devices. Plankton nets and trawls are not compact and have a low weight: drag ratio. Their towing characteristics are more kite-like, indeed experience with plankton nets at the end of long towing wires suggests they may never reach their point of proportional depth, even at low speeds. Special depressor plates are often mounted on the front of nets to overcome this difficulty.

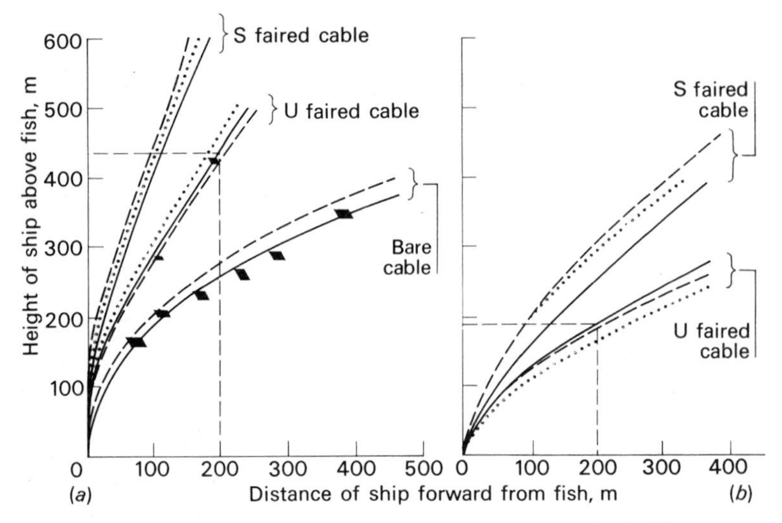

Fig. 7.19. The relationship between the depth of a compact body ('fish') towed with faired or unfaired cable and its horizontal distance from the ship. S and U fairing refer to different designs of fairing. In (*a*) the ship speed is 4 kt, in (*b*) 8 kt. ------, configuration constructed using angle information and length paid out; ·····, computed configuration; ■, data plotted from sonic measurements. The computed configuration and the experimentally observed configurations relate to a body 900–1120 kg in weight. (Rather, Goerland, Hersey, Vine & Dakin, 1965)

If the object of a tow netting or mid-water trawling operation is to sample a population of animals then the following features of the net are particularly important. It should sample different species in an unselective manner, with constant efficiency (see p. 365), and the volume of water filtered and the depth of sampling must be known. It is not possible for any one net to sample all grades of animals equally; indeed it is difficult to demonstrate that any one net can sample even a restricted group of animals properly.

For collecting small planktonic organisms a net such as the NF 70V net which derives from a design used by Nansen (1915) is used. It is normally hauled vertically but may be towed horizontally. The size of the apertures between the meshes during filtration is about 170 μm (Fig. 7.20).

For collecting larger animals, such as fish and large crustaceans, a mid-water trawl is often used. Fig. 7.21 shows the Isaacs–Kidd mid-water trawl which is towed permanently open (Devereux &

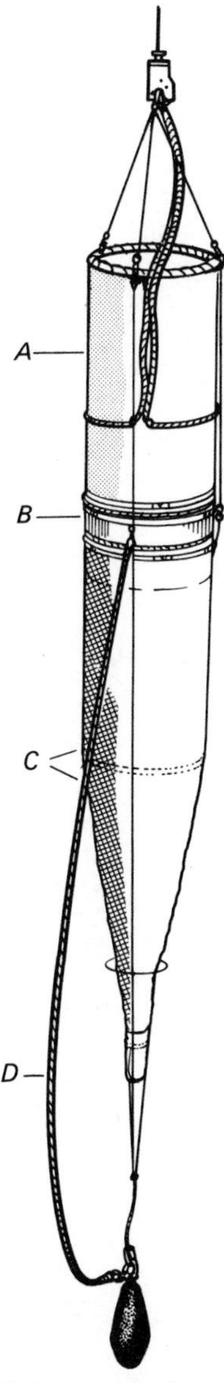

Fig. 7.20. The NF 70V plankton net *A*, canvas tube, 70 cm in diameter; *B* brass tube housing a mechanical flow-meter and depth recorder; *C*, filter, typically 30 meshes per cm for collecting small animals. The mechanism at the top of the towing bridle is actuated by a weight which slides down the main wire and releases the bridle. The slack rope around the canvas section *A* throttles the net for hauling in. *D*, hand line attached to a sinker. (Currie & Foxton, 1957).

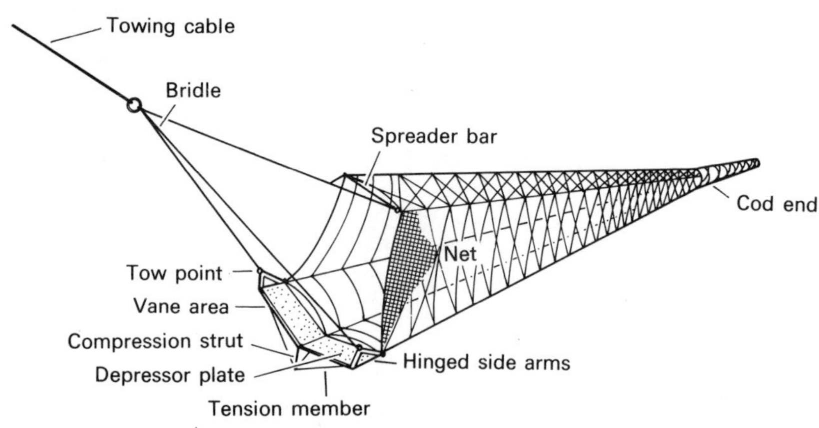

Fig. 7.21. The Isaacs–Kidd mid-water trawl. A trawl with a 5 m mouth would measure 24 m long, terminating in a cod end of approximately 40 cm in diameter. (Devereux & Winsett, 1953)

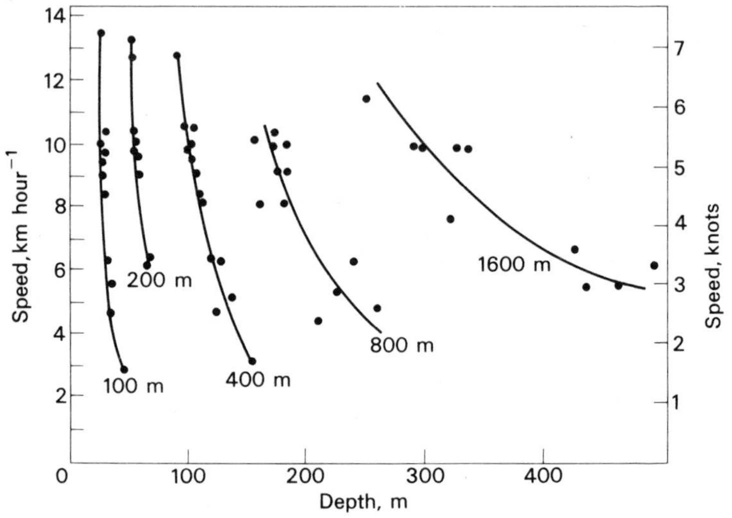

Fig. 7.22. Towing characteristics of the Isaacs–Kidd mid-water trawl. Towing speed (vertical axis) is plotted against depth for five different wire lengths given in m and a 2 m-mouth trawl. (Aron, Ahlstrom, Bary, Be & Clarke, 1965)

Winsett, 1953); note the depressor plate. Fig. 7.22 shows how doubling the speed of towing from 3 to 6 knots causes the net to rise from a depth of approximately 500 m with 1600 m of wire paid out, to a depth of approximately 250 m.

Analysis of the filtration process in plankton nets has been prompted by a need to improve the catching performance of quantitative samplers. Nets tend to disturb the water ahead of them, clog, and damage the small animals they collect (Tranter & Smith, 1968). Forward disturbance can elicit escape responses in most animals and must therefore be minimised. The disturbance often arises from the presence of bulky towing bridles and from the pattern of water flow through the net. According to Clutter and Anraker (1968) a 1 m diameter net may set up such a disturbance 1.5 m ahead of itself. Animals detect either the flow of water or the slight increase in pressure and take avoiding action. Nets may be seen by animals, even at great depth where bioluminescence can be reflected by the net material. The escape velocities of animals normally caught by means of a plankton net or mid-water trawl vary widely. An example from the middle of the range would be *Calanus helgolandicus* which has been filmed swimming at speeds of 67 cm sec^{-1} over a 7 cm distance. Generally, escape speeds can only be sustained over distances which are short compared to a large plankton net or trawl. Perhaps the euphausiids are the fastest of small invertebrates with cruising speeds of 17 cm sec^{-1}. Clutter and Anraker (1968) conclude after a mathematical discourse on the problem of the avoidance of nets, that good collection is favoured if as large a net as possible is towed as fast as possible and at constant speed. Its filtration efficiency should not fall below 85 per cent and it should be dark in colour. In order to collect undamaged animals for experimental work the net gauze should have a low pressure differential across it and a large volume of stationary water should be provided at the collecting cod end. Separation of a catch by filters can be worthwhile; a large jellyfish can smother many smaller and useful animals, and squid can bite, if not eat, all the occupants of a cod end.

The filtration pressure in a net is largely determined by the velocity at which water approaches the net gauze, which is not simply the towing velocity. So far, little attention has been given to the task of designing a large deep water plankton net suited to filtering the sparsely distributed animals in horizontal tows, with minimal damage to the organisms. With extra large nets it may also be necessary to smooth the ships motion which is transmitted along a tow wire. Apart from damaging the animals in the net and warning others off, the dynamic loading on a large net can become

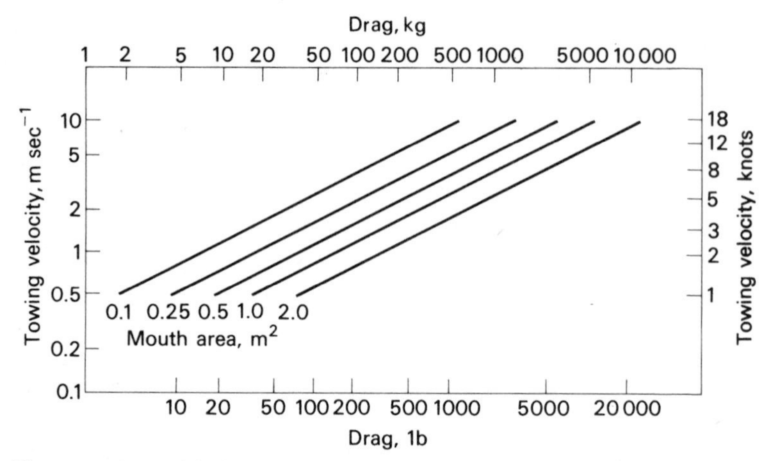

Fig. 7.23. A graphical method for estimating the drag of nets with mouth areas of 0.1–2.0 m^2, towed at velocities of 0.5 m sec^{-1} (1 knot) to 9.5 m sec^{-1} (18 knots), assuming a drag coefficient of 1.33. (Data from Motoda, see Tranter & Smith, 1968)

severe in a heavy sea. A 35 mm circumference wire paid out to 5000 m will, by its own weight in water, generate a static tension equal to a third of its own breaking strain.

Fig. 7.23 shows the drag–velocity relationship estimated for a variety of small nets by Motoda and discussed by Tranter and Smith (1968). The figure shows how drag increases steeply with towing speed, and, as already mentioned, causes lift. If the depth–velocity relations of a net are known it may be towed at constant depth but it is much better to monitor the depth of the net continually. To filter animals from specified depths it is also necessary to open and close the net at depth. This may be done with devices attached at the mouth of the net and at the cod end.

Grice and Hulseman (1968) found that a plankton net of the NF 70V type became contaminated with surface organisms when it was lowered vertically in the open position. To prevent this they used a pressure-powered net opening device. Foxton (1963) has described a device for use with larger nets which separates the catch according to the depth of filtration (Fig. 7.24). A more complicated cod end catch-separating device is shown in Fig. 7.25.

The most widely used net for macroplankton and nekton collecting in deep water is the Isaacs–Kidd mid-water trawl (Fig. 7.21).

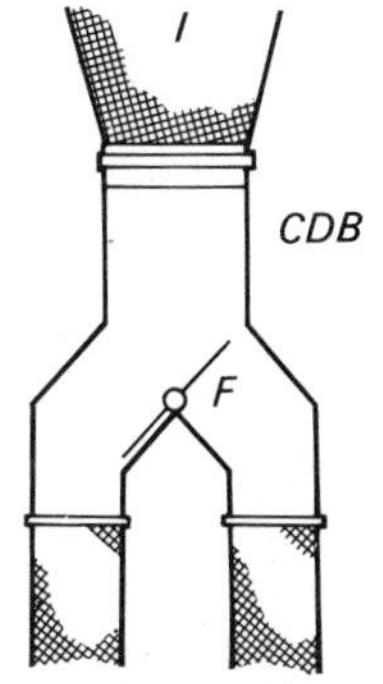

2m nets of 31-mesh nylon

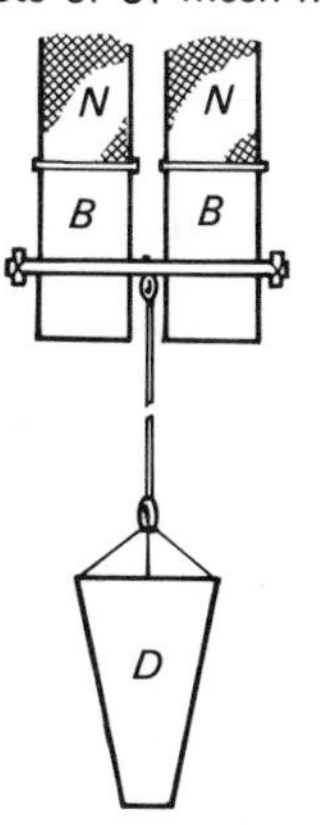

Fig. 7.24. The catch-dividing bucket (*CDB*). Schematic drawing showing *B*, polyethylene collecting buckets; *D*, canvas drogue; *F*, flap, *I*, cod end of an Isaacs–Kidd mid-water trawl; *N*, collecting nets. The tube leading to the Y-piece is 20 cm bore. The position of the flap valve *F* is determined by a piston mechanism which responds to ambient pressure. As the net is paid out, animals are first collected in one bucket and then, at a predetermined depth, the flap valve is actuated and the catch is retained in the second bucket. On hauling in the mechanism reverses, although due to hysteresis the depth which separates the two catches is subject to an error of about 10% at 1000 m. (Foxton, 1963)

The National Institute of Oceanography, UK, has developed a worthy successor which may be opened and closed at selected depths. The practical model (Clarke, 1969) has an effective mouth area of 8 m^2 and a total filtering surface of 24 m^2. The mouth is rectangular and opens and closes in the ingenious way illustrated in Fig. 7.26. The towing wire connects to the top of the net so no obstacles precede the net mouth through the water. Attention has been given to the design of the cod end bucket which is large and free of turbulence. This trawl appears to be capable of both quantitative sampling and recovering healthy animals from considerable depths. Advanced instrumentation enables the net to be towed at controlled depths in the following way. The trawl is paid out at 0.5 m sec^{-1} relative to the ship whilst the ship makes 2 knots through the water. An acoustic pinger attached to the top of the

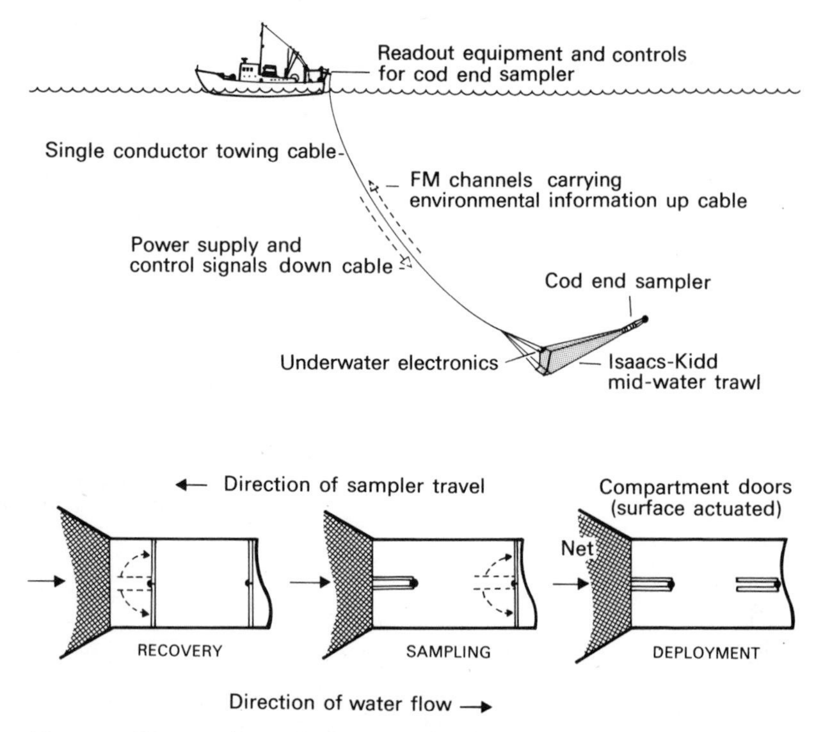

Fig. 7.25. Discrete depth plankton sampler. The cod end collecting vessel is subdivided. The towing cable is electrically conducting, triggering solenoids in the cod end vessel which successively closes partitions, thereby confining animals caught at selected depths. Depth and temperature data are relayed up the conducting cable to the towing ship. (Aron, Baxter, Noel & Andrews, 1964)

net emits two pulses the interval between which is proportional to depth. The net is allowed time to equilibriate at the selected depth ± 10 m before an acoustic signal (517.5 Hz) is emitted from the ship. The signal starts an electric motor which is housed in a pressure casing on the net bridle, and a reduction gear box rotates a shaft which passes through the pressure housing into the ambient seawater. The rotation of the shaft releases the net opening mechanism, and at this point the drag of the net increases and it tends to rise slightly in the water. This is a welcome sign to those on the ship who then know that the net is almost certainly open and fishing. A second acoustic signal (502.5 Hz) can be sent from the ship at the end of a tow to cause the net to close. Clarke

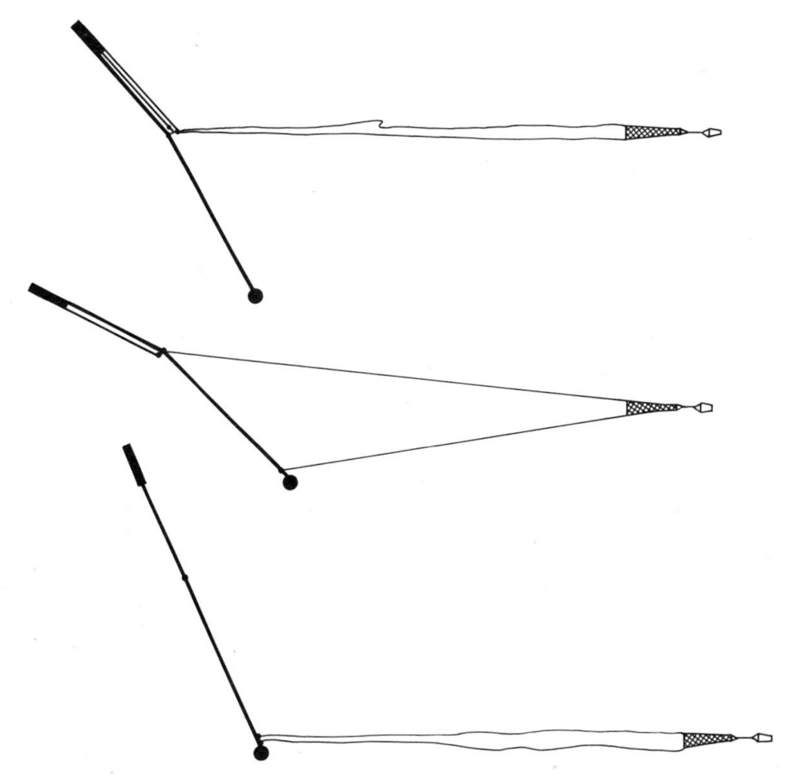

Fig. 7.26. A large opening and closing mid-water trawl. Wires taking the strain are drawn thicker. See text. (Clarke, 1969)

reports that some large contaminating animals may enter a fold of the closed net during its descent but cannot enter the cod end. Small animals capable of passing through the 4.5 mm mesh which is normally used have to be discounted from the quantitative sample.

Large free swimming deep sea animals such as sharks, bony fish and squid are not caught very readily by nets. Long line fishing has been used to catch deep water sharks for studies of their retinae and buoyancy mechanisms (see Forster, 1964, Harrison, 1967).

Phleger and Soutar (1971) used a 'pop-up' method for catching the fish *Coryphaenoides* on baited hooks (see Chapter 3). The apparatus consists of a sinker weight, fish hooks, a release mechanism and a buoyant element (Fig. 7.27). A slowly dissolving

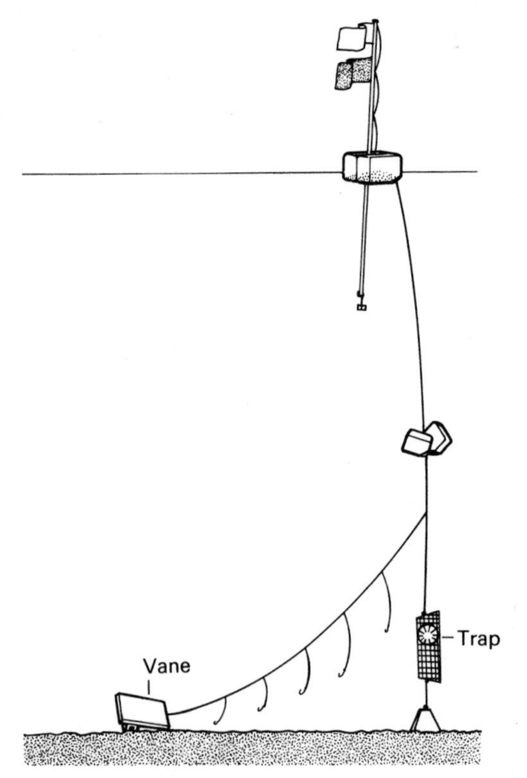

Fig. 7.27. Pop-up trap and baited hooks. Top, buoyant package with flag and radio. Two more buoyant vessels are positioned above the junction between the line of hooks which terminates in a vane, and the trap which is attached by a soluble release mechanism to a sinker. (Phleger & Soutar, 1971)

magnesium wire, or similar device, releases the buoyant section of the apparatus after a delay of several hours. A radio transmitter and flag help the surface ship to locate the catch.

Animals from the deep sea floor

Organisms which live on top of and buried in the sea floor sediments may be recovered by a surface ship operating a corer, a dredge or a grab. Corers sample a small cross section but penetrate deep; a grab samples a larger surface area but penetrates only 0.3 m or so. A dredge or trawl skims the surface and stands a good chance of collecting some active animals which live in the

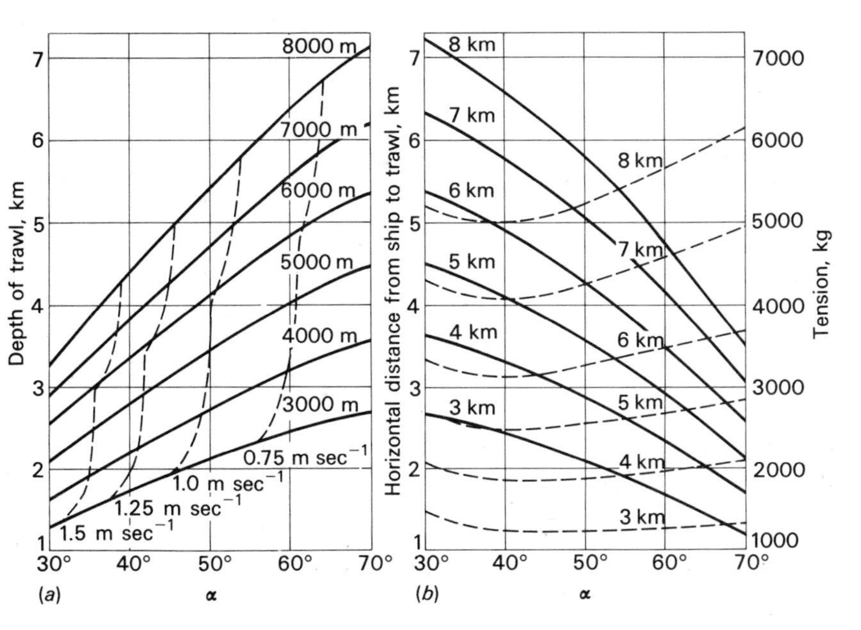

Fig. 7.28*A*. Diagram showing the dimensions and stresses in the trawl cable used by the *Albatross* when trawling on the bottom with a 10 m trawl. The depth of the trawl is shown as a function of the angle α between the cable as it leaves the ship and the horizontal for various values of the length of cable paid out (full curves). The broken curves indicate the velocity of the trawl.

B. Diagram for the cable of the *Albatross* when trawling with a 10 m trawl showing the horizontal distance from the ship to the trawl as a function of the angle α (definition as in *A*) for various values of the length of cable paid out (full curves). The broken curves indicate the tension in the cable for these lengths. (*A* and *B* Kullenberg, 1957)

water above the bottom. Benthic collecting is a much more hazardous task than mid-water trawling, even when carried out on a sea floor whose contours have been mapped by echo sounder. Gear on the sea floor can become snagged and short of continuously monitoring the operation with television or a submersible vessel, the operator has to trust to luck. The length of wire required to trawl or dredge on the deep sea floor is very considerable.

The deep trawling technique used during the Swedish Deep Sea Expedition of 1947–8 is dealt with in a comprehensive manner by Kullenberg (1957). Fig. 7.28 from his paper illustrates the interrelationship between depth of trawl, wire paid out, tension generated in the wire and other factors. In Fig. 7.28*A* we see that to lower the 10 m wide trawl used on the Expedition to the sea floor at a depth of 5000 m, 8000 m of wire is required if the angle of the wire at the surface is 45° and the trawl velocity 1.25 m sec^{-1}. Fig. 7.28*B* shows that for a given amount of wire paid out, tension increases as the angle at the surface increases and the horizontal distance between ship and trawl decreases. The increase in tension is small and is due to the increased drag on the near-vertical wire.

Rowe and Menzies (1967) have described a method of deep bottom trawling which involves the minimum of wire and a steep wire angle. Acoustic telemetry and a record of wire strain indicate when the trawl is lifted clear of the sea floor and when a reduction in speed or an extension of the towing wire restores it to the floor to trawl again. Bottom sampling equipment is amply illustrated and discussed in a review by Holme (1964).

The retrieval of deep sea animals without pressure change

For experiments on intact deep sea animals it seems particularly important to use material which has suffered neither decompression nor serious temperature change. Enzymic and other chemical studies may also require animals retrieved with similar care.

The problem of the retrieval of deep sea animals at constant temperature is relatively easily solved, whereas constant pressure retrieval involves serious engineering.

Animals which are to be recovered without change in pressure have first to be located within a pressure vessel, which is most likely to be a cylinder with end closures. We will return to the ways and means of persuading deep sea animals to enter a pressure vessel shortly, but first, consider the fate of a cylindrical stainless steel pressure vessel which is lowered into deep water, sealed at depth in an ideally rigid manner and raised to the surface (Macdonald & Gilchrist, 1972). The vessel will expand as the external pressure falls, and consequently the internal pressure will decrease. Seawater is relatively incompressible and steel is fairly elastic, so the fall in internal pressure can be appreciable. It is shown in Fig. 7.29 that an elastic expansion in the volume of the idealised pressure cylinder causes a significant fall-off in internal pressure (ΔP_E). Even when the vessel wall is infinitely thick ΔP_E is close to 5 per cent of the trapped pressure, and when a practical wall thickness is used ΔP_E is approximately 7 per cent. ΔP_E is

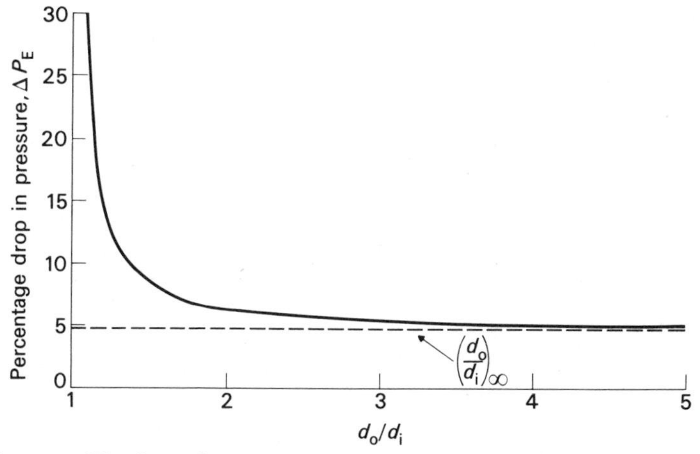

Fig. 7.29. The loss of pressure trapped in a hypothetical cylindrical vessel which is sealed in an ideally rigid manner at depth in the sea and raised to the surface. Vertical axis, pressure loss as per cent of pressure at the time of sealing. Horizontal axis, ratio of outside diameter (d_o) to inside diameter (d_i). (Macdonald & Gilchrist, 1972)

mainly caused by the elasticity of steel but a small contribution comes from the compressibility of steel which is about one sixtieth of that of seawater.

When the pressure vessel is sealed by a practical method there will be further pressure losses during recovery due to the compression of seals, ΔP_S. If the vessel experiences a rise in temperature then the pressure fall-off will be offset to some extent, but constancy of temperature is required on physiological grounds. To counter the total loss of pressure ($\Delta P_S + \Delta P_E$) it is necessary to introduce a highly compressible element in hydraulic connection with the interior of the pressure recovery vessel.

At this point we can no longer deal with generalities as the practical details in the design of a pressure-recovery vessel have considerable influence on its performance. We have now to consider the problem of trapping or collecting animals for constant pressure retrieval.

The collection of benthic animals is probably best carried out by a deep submersible vehicle. In the absence of such vehicles in the UK Macdonald and Gilchrist (1972) have attempted to design and build equipment suited to the recovery of small planktonic animals without pressure loss. The aim has been to build a pressure

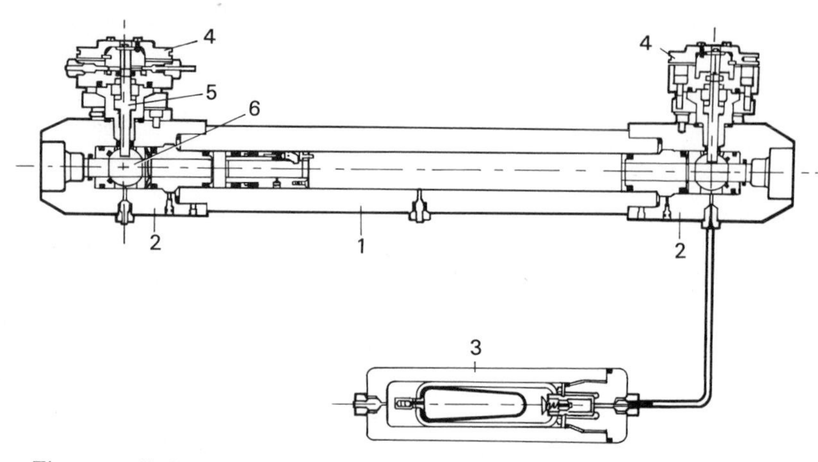

Fig. 7.30. Cod end pressure vessel for recovering deep sea plankton without pressure loss. 1, body of vessel; 2, globe valves (shown open); 3, hydraulic accumulator; 4, valve closing mechanism; 5, rotating shaft which actuates 6, steel sphere. When the latter is rotated through 90° it closes the valve. (Macdonald & Gilchrist, 1972)

vessel which attaches to the cod end of a net and which can be operated from an ordinary research ship. The plankton pressure-recovery vessel was therefore conceived as a small cylindrical pressure vessel which is lowered open at the end of a plankton net, closed at depth and brought to the surface. For reasons which may be set aside at present, it was decided that the pressure recovery vessel should be separate from the vessel in which observations of the animals were to be carried out. Thus, following the retrieval of the pressure vessel, the suspension of plankton is transferred to a separate experimental pressure vessel without change in pressure or temperature.

Fig. 7.30 shows the cod end vessel. The body of the vessel is 56 cm long, its internal diameter 5.1 cm and its external diameter 10 cm. At each end are fitted large globe valves which provide a 2.5 cm bore entry to or exit from the vessel and are closed by rotating the central steel sphere 90°. There are a number of seals which deform under pressure causing a significant pressure fall-off, ΔP_S, which has been measured by simulating the deep sea recovery operation.

For this purpose a globe valve was fitted to a blank-ended recovery vessel body and placed in a large pressure vessel. The globe valve was closed at selected pressures and the pressure in the large vessel was slowly decreased to simulate the hauling in of the recovery vessel. The resultant trapped pressure in the recovery vessel body was monitored. Fig. 7.31 shows the results of these experiments which were carried out by Gilchrist. At high pressures the total pressure fall-off approaches that to be expected in an ideal vessel, but at low pressures it attains unacceptable large proportions. At 100 atm for instance, $\Delta P_E + \Delta P_S$ is about 35 per cent. The inclusion of a hydraulic accumulator, which is a separate vessel containing gas at a selected high pressure (Fig. 7.30), alters the performance of the vessel dramatically. It now retains pressure to within 7.5 per cent of the closing pressure, and if due allowance is made for the relatively small pressure vessel which was used in the tests it may be shown that the practical cod end device will recover seawater from 10000 m depth or 1000 atm with a pressure loss of less than 3.5 per cent.

In order to test the pressure recovery vessel at sea it was mounted in a handling frame and a pressure transducer and pressure recorder were attached (Fig. 7.32). Seawater from 1400 and 2000 m depth was recovered with the vessel fitted with a globe valve at one end and a blank plug at the other. Table 7.6 shows that the pressure change in the trapped seawater following recovery from 1400 m was zero, in accordance with data in Fig. 7.31. The increase of 6 per cent following recovery from 2000 m was almost certainly due to the greater temperature change in the latter case.

For collecting plankton the pressure vessel is suspended at the end of a 2 m diameter, vertically-hauled ring trawl. During tests it has recovered a suspension of plankton from 580 m depth with negligible change in pressure (Macdonald, 1970). Fig. 7.33 shows the method of collapsing the net; the globe valves are closed by springs triggered by a large messenger device.

The plankton suspension inside the pressure recovery vessel is transferred to a second vessel for experimental study. The hydraulic circuit for the purpose is shown in Fig. 7.34. Coupling the pressure recovery vessel to the transfer pipe (Fig. 7.34) is a delicate task requiring a few minutes, but thereafter the transfer operation is semi-automatic. Pressure in the receiving vessel is generated by an air powered hydraulic pump and matched to that in the pressure recovery vessel. Pressure is held constant by a method which is described on p. 406. The globe valve connecting

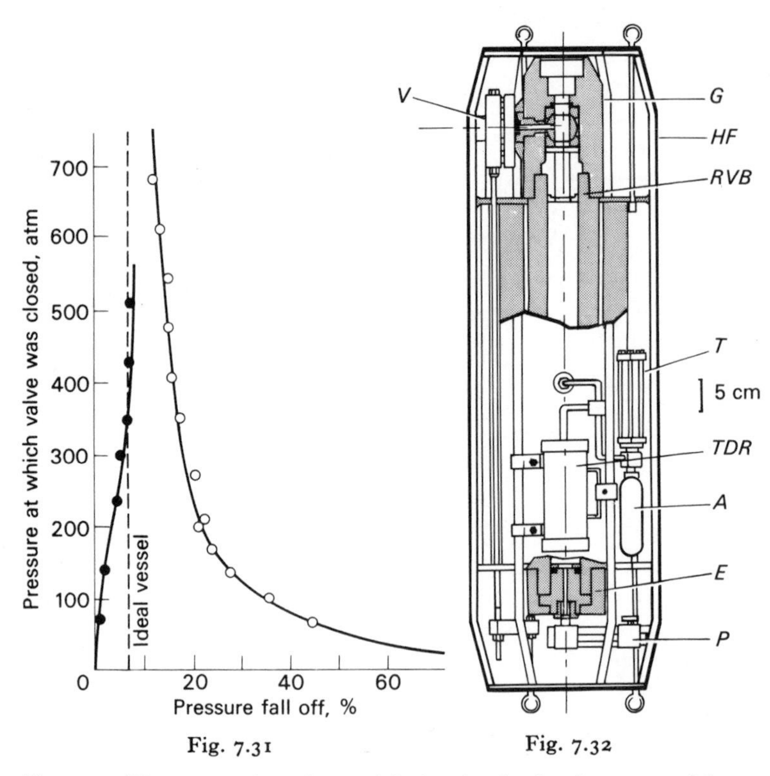

Fig. 7.31

Fig. 7.32

Fig. 3.31. The pressure loss observed during the simulated recovery of deep sea plankton. Vertical axis, pressure at which valve was closed; horizontal axis, decrease in pressure within the recovery vessel as a percentage of the initial trapped pressure. See text. Circles, vessel without hydraulic accumulator; dots, vessel with hydraulic accumulator. Vertical dashed line, ideal vessel, explained in text. (Macdonald & Gilchrist, 1972)

Fig. 7.32. Recovery pressure vessel, mounted in its handling frame, *HF*. *V*, valve closing mechanism; *G*, body of globe valve; *RVB*, recovery vessel body; *T*, pressure transducer housing; *TDR*, time–depth recorder; *A*, hydraulic accumulator; *E*, end-cap; *P*, connection to hydraulic circuit. (Macdonald & Gilchrist, 1969)

the recovery vessel to the hydraulic circuit can only be opened when there is no pressure differential across it so the fluctuations in pressure during the transfer operation are necessarily very small and not demonstrable on the recording shown in Fig. 7.35.

This equipment has not yet had sufficient sea-time to yield biological results but it has demonstrated a practical solution to

TABLE 7.5. The recovery of deep sea water without pressure loss (Macdonald & Gilchrist, 1969)

Cast no.	Depth of recovery apparatus from wire paid out, and calculated pressure		Recorded pressures at which globe value closed	Recovery pressure (atm)		Pressure change during recovery	Pressure fluctuation during transfer
	(m)	(atm)	(atm)	Depth recorder	Transducer	(%)	(%)
3	1400	140	144	144	—	0.0	—
4	1400	140	144	144	144	0.0	±2.3
5	2000	200	200.6	211	—	+5.5	—
6	2000	200	200.6	211	215	+6.5	±2.0

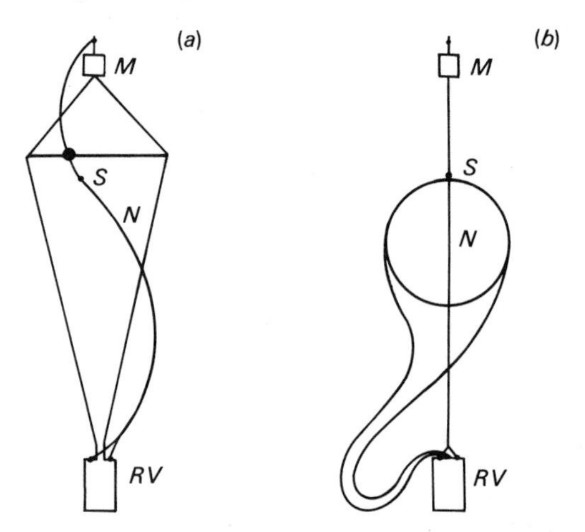

Fig. 7.33. The recovery pressure vessel, *RV*, suspended at the cod end of a 2 m diameter ring trawl by wires, not shown. A messenger-activated device releases the towing bridle. The net cants and the weight of the recovery vessel is taken on a second wire which triggers the valve closing device. *M*, messengers; *S*, stop; *N*, net. (Macdonald & Gilchrist, 1972)

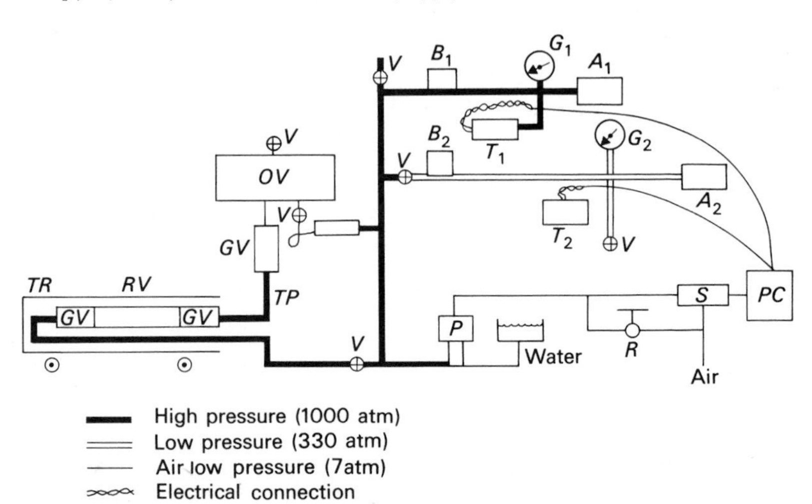

Fig. 7.34. Hydraulic circuit used to transfer a suspension of deep sea plankton at high pressure from the recovery vessel (*RV*) to the observation pressure vessel (*OV*). *A*, hydraulic accumulator; *G*, pressure gauge; *T*, pressure transducer; *B*, bursting disc; *V*, valve; *GV*, globe valve; *TP*, transfer pipe; *TR*, trolley and handling frame; *P*, air-powered high pressure pump; *S*, solenoid valve on air supply to pump; *R*, air regulator (by-passes solenoid valve if required); *PC*, potentiometric controller. (Macdonald & Gilchrist, 1972)

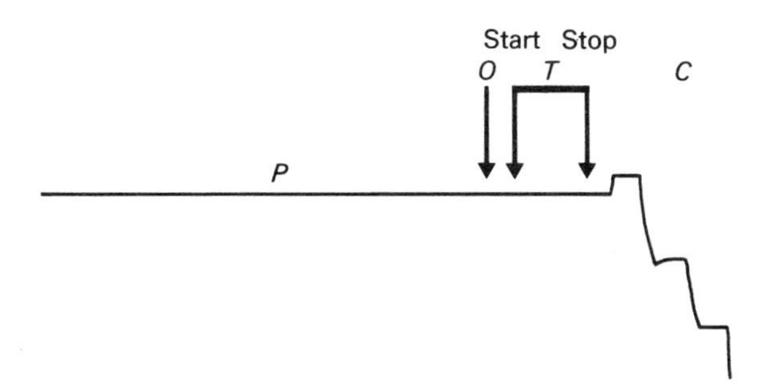

Fig. 7.35. Recording of the pressure inside the recovery pressure vessel during the transfer of deep sea water which was recovered without pressure loss. The recording refers to cast 4, Table 7.6. The steady pressure at *P* is 144 atm. At *O* the globe valve connecting the recovery vessel to the receiving observation vessel is opened. *T* shows the $4\frac{1}{2}$ minute period when water flowed through the recovery vessel to the observation vessel. *C* denotes calibration steps to zero pressure at the end of the operation. (Macdonald & Gilchrist, 1972)

the mechanical problem of retrieving deep sea animals without pressure loss. A pressure vessel which can be operated by the mechanical arm of a deep submersible to retrieve large benthic or trapped mid-water animals will obviously differ from the plankton vessel, but the principles underlying the retention of pressure will be the same. The plankton vessel may be readily modified for use with a deep submersible to recover small benthic animals.

PROBES FROM THE SURFACE OF THE SEA

Detecting animals in the deep sea

Animals may be detected in the deep sea by workers on surface ships using acoustic techniques (Hersey, Backus & Hellwig, 1962), photography (Hersey, 1967) and television (Barnes, 1963).

Animals may be listened to by lowering a hydrophone, or their presence may be detected by the reflection from an acoustic pulse. The acoustic impedance of animals has to be sufficiently different from that of the background if they are to be detected by acoustic methods. The target strength of an animal varies with the profile it presents to the oncoming acoustic pulse which is emitted from a transmitting transducer (Fig. 7.36). In fish it is the gas-filled swimbladder which is the major acoustic reflector. The detection of fish in shallow water by acoustic means was practised on a small

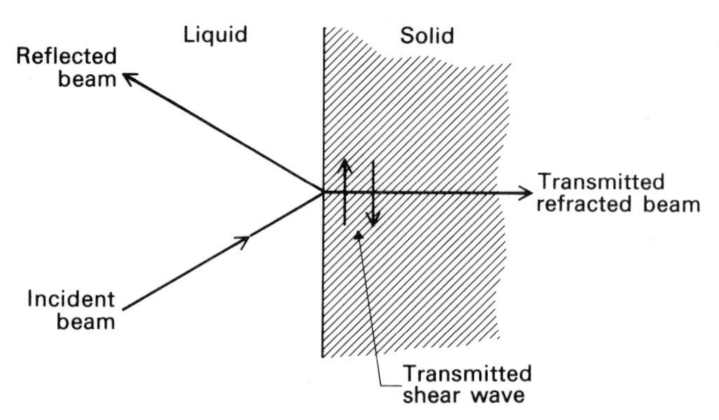

Fig. 7.36. Reflection of sound by an object underwater. (Jones & Miles, 1969)

scale in the 1930s. During World War II submarine acoustic techniques were developed and the widespread occurrence of deep scattering layers (DSLs) was established. A DSL is a region of acoustic reflectance comprising many targets whose depth varies, often rising by night to near the surface and descending by day to depths of several hundred metres. A number of methods have been used to identify the animals which are the cause of the sound scattering but none can compare with the acute sensitivity of the acoustic method. According to Hersey and Backus (1962), one hundred suitably reflecting gas-filled swimbladders at a depth of 300 m and at a concentration of 1 in 10^4 cubic metres will yield a substantial scattering or echo. 10^4 cubic metres is a volume equivalent to a public auditorium and the single fish occupying such a volume need only be about 10 cm long to be detected as a scattering object. One may readily imagine the chances of locating such a fish with a suspended camera or a blindly towed mid-water trawl.

The principal factors which limit the range at which animals can be detected may be set out in the following elementary fashion. The target strength is important; it is defined as the ratio of the intensity of the echo 1 m from the target to the incident intensity i.e. $TS = 10 \log_{10} \left(\frac{i_2}{i_1}\right)$ (Tucker, 1967). Quite obviously, big targets have larger target strengths. At the receiver the returning echo has to exceed the background noise. As a target is moved further away from the receiver it emits a much reduced echo, principally due to geometrical spread. The echo is reduced even

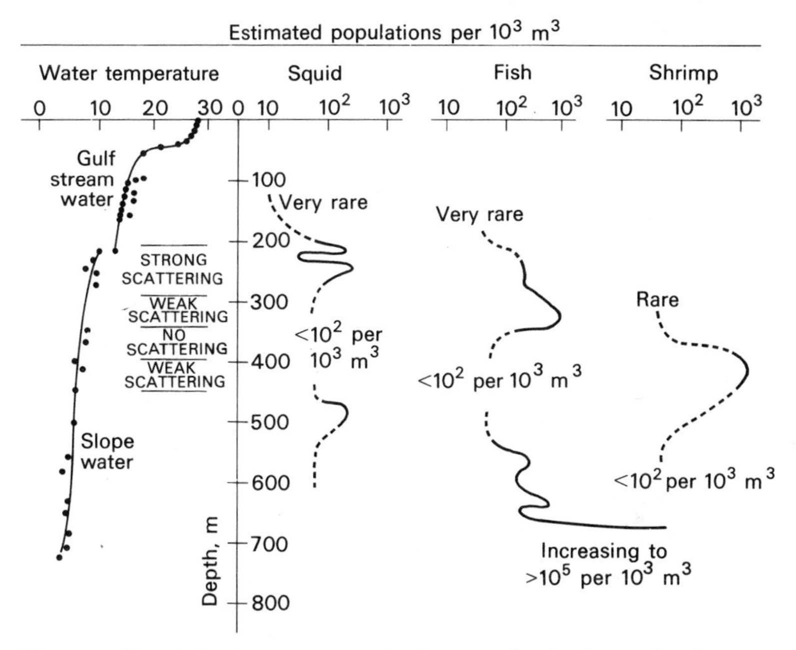

Fig. 7.37. Correlation between scattering layers and animals seen by observers from the deep submersible *Alvin*. Temperature, on left, was measured during the dive with a thermistor mounted on the outer hull; measurements are accurate within 0.2 °C. *Alvin* dived at 1050 hours, reaching a depth of 240 m at 1120 hours; when informed by the surface vessel that the submersible was being carried eastward by the Gulf Stream, *Alvin* drove westward into the colder water, and resumed descent, reaching the bottom (707 m) at 1226 hours. Worsening weather forced termination of the dive at 1320 hours. (Milliman & Manheim, 1968)

further relative to the background noise because of reverberation, i.e. the scattering of the transmitted pulse by other objects.

Acoustic methods are highly developed in fishery biology and a good account is provided by Tucker (1967). In practice it appears that a population of animals below 1000 m are, at present, out of range of acoustic echo ranging equipment mounted on surface ships, largely because of noise and reverberation effects.

In densely populated deep scattering layers mid-water trawls have recovered myctophids, siphonophores, which also have gas-filled floats, and crustaceans, which do not. Observers in deep submersibles have seen these animals in the DSLs and, in addition, squid (Fig. 7.37, Milliman & Manheim, 1968). Backus and Barnes

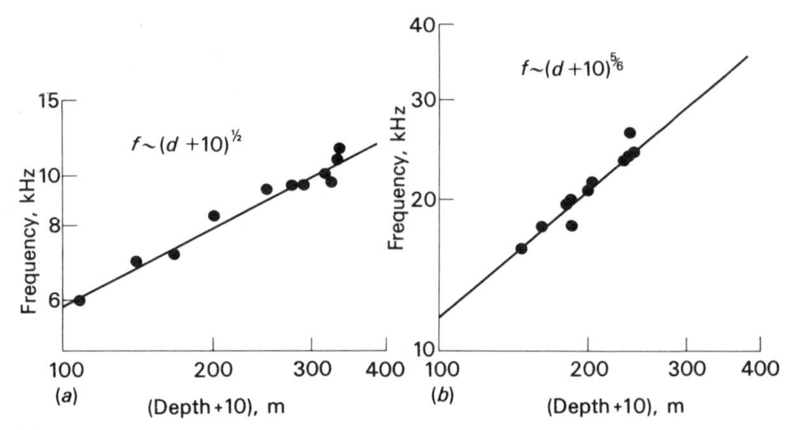

Fig. 7.38. Examples of relationships between peak scattering frequency and pressure in a deep scattering layer. (*a*) peak scattering frequency varies as $P^{\frac{1}{2}}$, (*b*) as $P^{\frac{5}{6}}$. (Hersey, Backus & Hellwig, 1962)

(1956–7) have used a television camera aligned with an acoustic transducer to correlate sound reflections with the presence of individual fishes and, in a number of cases, established unequivocably the cause of certain echoes. Hersey (1967) has similarly coupled acoustic transducer and photographic camera.

The reflection of sound by a bubble of gas is most marked at the frequency which induces resonance in the bubble. The resonance frequency giving peak scattering from a bubble of constant volume will vary as $P^{\frac{1}{2}}$ where P is the hydrostatic pressure. The swimbladder of small fish approximates to a gas bubble and in fish undergoing vertical migration with a constant swimbladder volume a peak scattering frequency should be obtained which varies as $P^{\frac{1}{2}}$. In a remarkable series of observations, Hersey, Backus and Hellwig (1962) detected vertically migrating reflecting bodies in the sea whose peak scattering frequency varied in just this manner. Presumably, the fish whose swimbladders were providing the sound scattering were maintaining their swimbladders at constant volume and thus close to neutral buoyancy during their migration. Other reflectors were observed whose peak scattering frequency varied as $P^{\frac{5}{6}}$, and these were interpreted as fish whose swimbladders were compressed on descent and therefore contained a constant mass of gas; such fish would lack buoyancy (Fig. 7.38).

The resonance of fish swimbladders and the extent to which they deviate from ideal gas bubbles is important to know. Capen (1967) has argued that in response to frequencies of 12 kHz and 3 kHz mesopelagic fish such as *Lampancytus leucopsarus* will not resonate at any point in their natural depth range. Others, such as *Lampanyctus omostigma* which has a large swimbladder, will resonate at 12 kHz over the depth range of 556–1655 m and would therefore be susceptible to acoustic observations. This complex subject is dealt with by Weston (1967).

Photography is a useful tool in quantitative deep sea ecology and has a role in experimental studies. Deep sea cameras differ from shallow water cameras in requiring a pressure housing, a pressure window and a special lens. The subject of deep sea photography is fully dealt with in a monograph edited by Hersey (1967).

Deep sea feeding experiments may be monitored with cameras, for example the device of de Baker (1957) (Fig. 7.39) but no observations beyond the exploratory stage have yet been reported.

The results obtained with the luminescence camera of Breslau, Clarke and Edgerton (1967) have already been mentioned. The camera is triggered by light acting on a photomultiplier tube sensitive to 5×10^{-6} μW cm^{-2}. 800 frames can be obtained from a single lowering and the maximum working depth is 6000 m (Fig. 7.40*A*). The interruption camera, triggered when a light path is interrupted, should be able to photograph large, fast swimming, mid-water animals at depth, although it has had limited use to date (Fig. 7.40*B*).

Television can give observers in surface ships or deep submersibles a view 'in real time', while cameras are essentially recording devices. The development of a long range acoustic link between a submarine television camera and shipboard receiver will open up the scrutiny of the deep sea floor in a way which is impossible at present (Borrows, 1969).

Acoustic and visual probes can contribute to field-experiments, particularly in the areas of sensory and nutritional physiology. An interesting example of such an experiment in moderately deep water is the attempt of Blaxter and Currie (1967) to influence a diurnally migrating scattering layer by lowering lights into it. A 1.5 or 1.0 kW light source was used and the reflections from the

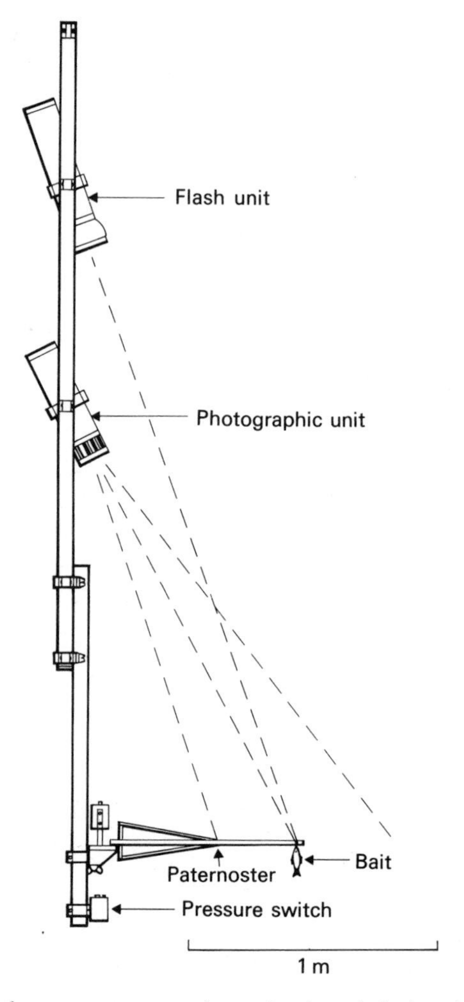

Fig. 7.39. Underwater camera triggered when bait is taken. (de Baker, 1957)

animals sonified with 10 kHz and 67 kHz from an echo sounder were recorded. In a number of cases it appeared that the presence of artificial light caused the sound scattering animals near it to descend, or move away from the light source. Preliminary estimates of the migration speeds of the animals, both in the light and when it was switched off, were made.

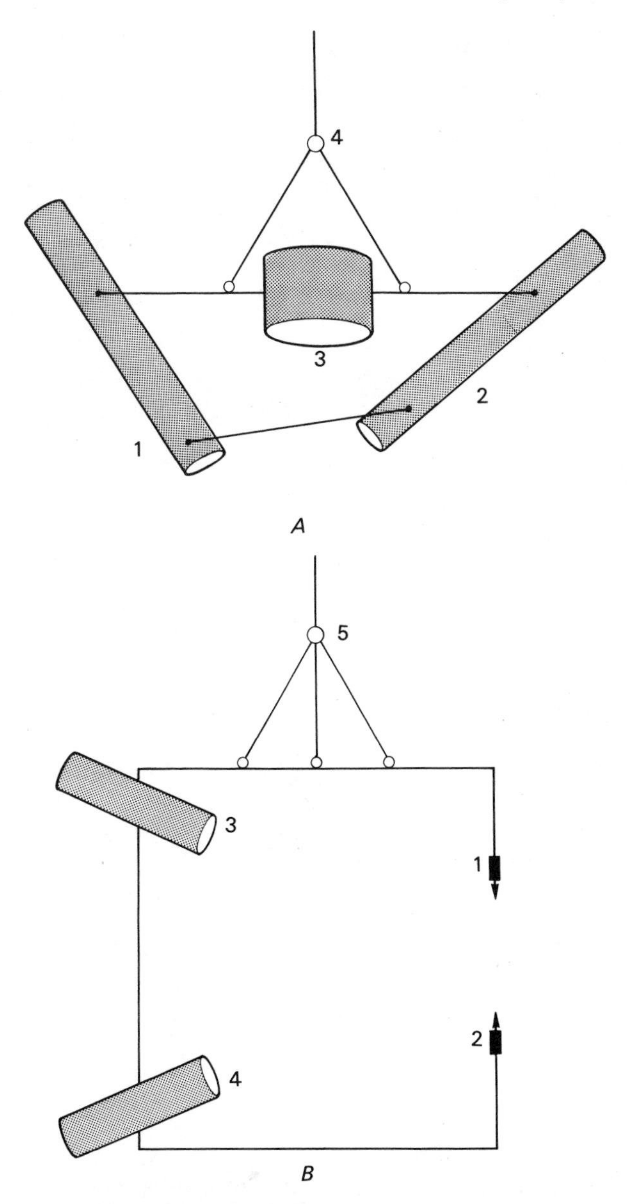

Fig. 7.40. *A*. Luminescence camera layout. 1, camera; 2, electronic flash; 3, sonar pinger (depth sensor); 4, suspending cable.

B. Diagram of the interruption camera. 1, infra-red light sensor; 2, light source and filter; 3, camera; 4, electronic flash; 5, suspending cable.

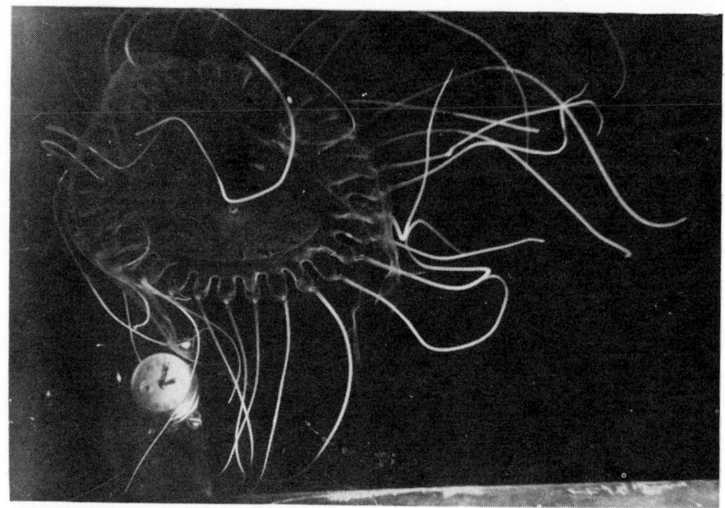

Fig. 7.40 *C* Medusa photographed by a luminescence camera at 1000 m depth. (*A*, *B* and *C*, after Hersey, 1967)

Deep sea telemetry

Acoustic energy may be used to switch instruments remotely or to transmit information. The acoustic command apparatus developed at the National Institute of Oceanography, UK is a good example because it is employed in mid-water trawling (p. 367). The apparatus consists of two parts, a ship board transmitter and an acoustic receiver (Harris, 1969, see Fig. 7.41).

Acoustic transmitters may be used to relay data over a considerable distance from instruments, tagged divers or animals. Their use in the deep sea presents few problems additional to those met with in long range work in shallow water (Skutt, Fell & Hagstrom, 1972).

Simulating the deep sea in the laboratory

INTRODUCTION

The environmental features of the deep sea which are of particular interest to the experimental biologist include a low constant temperature, high pressure, a low noise level, absence of sunlight and diurnal light-rhythms (except in the twilight zone where a

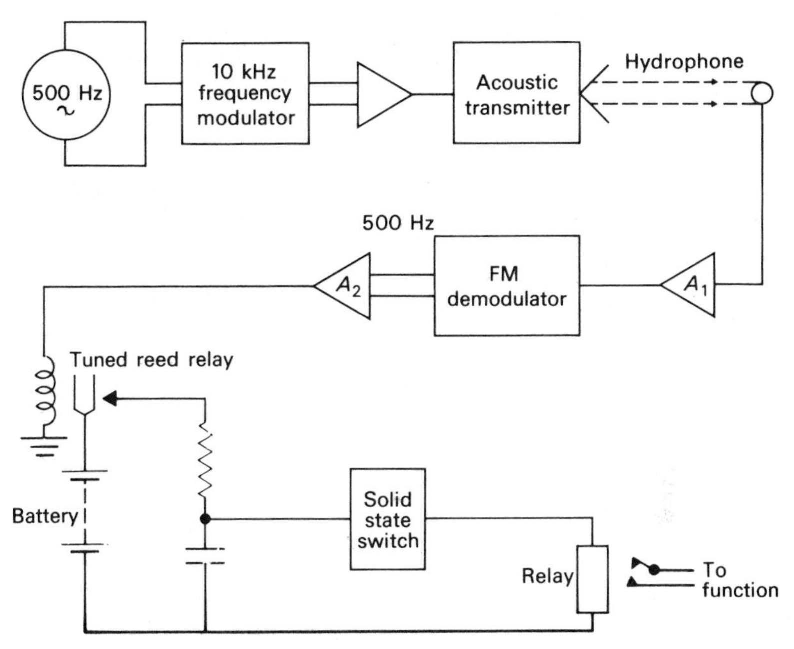

Fig. 7.41. An acoustic command system. 10 W of electrical energy is converted into 5 W of sound at a frequency of 10 kHz. Dotted lines indicate path-length through the sea, up to 5000 m. The hydrophone is made of a barium titanate sphere with a flat frequency response near 10 kHz and which is unaffected by the ambient pressure. The amplifier A_1 feeds into the frequency demodulator from which emerges an amplitude-modulated signal, further amplified in A_2. The reed serves as a switch. On receiving the appropriate frequency (499–501 Hz) it charges a capacitor. This causes a thyristor to conduct at a characteristic voltage, discharging the capacitor and actuating a relay. The instrument has to discriminate against ambient noise triggering the relay but cannot be too selective as sound from a rolling ship is subject to a Doppler effect, and the signal arriving at the hydrophone is not of the precise band width emitted. Background noise is excluded in three ways. The command signal is frequency-modulated whereas ambient noise is predominantly amplitude-modulated. Amplification of the signal and operation of a tuned reed to charge the capacitor is a further selective process. The leakage of the capacitor requires that a 5 sec period of continuous charge is required to actuate the thyristor. (Harris, 1969)

characteristic spectrum prevails) and low levels of food and sensory stimuli. Light, or the absence of it, and low temperature are readily available as experimental variables and techniques concerned with them need not be pursued here. Similarly, food and sensory stimuli are factors which can be controlled experimentally by well known procedures; the provision of controlled noise levels, however, is far from easy and work is often carried out on a field scale in convenient lakes or lochs. High pressure techniques are less well known and high pressure equipment is not available in many biological laboratories. Because of the emphasis given to high pressure in this book some comments are called for on high pressure techniques. Examples will be drawn from experiments already discussed in previous chapters.

MEANS OF GENERATING AND CONTAINING DEEP SEA PRESSURES

Deep sea pressures, up to 1100 atm, are not high by the standards of high pressure engineering. However, such a statement has little meaning unless the scale and function of the vessel which contains the pressure is indicated. Consider two pressure vessels, each capable of simulating the greatest deep sea pressure. One may be designed for testing the pressure resistant spheres of deep submersibles, the other for containing a solution for spectrophotometry. The former vessel, which may be cylindrical and 2 m or more in diameter, will exert a maximum thrust of the order of 40000 tons on its end closure. Its design and construction will have posed serious high pressure engineering problems, and in fact such vessels have only recently been constructed. A small pressure vessel mounted in a spectrophotometer may have a closure only 1 cm in diameter which will restrain a force of less than one ton at maximum pressure. Most ordinary bolts have a shear strength in excess of this, and the design of such a closure is a modest problem. However, the windows of the spectrophotometer vessel also have to confine the pressure, and these in turn present a different design problem.

A typical laboratory high pressure vessel is cylindrical with machined end closures. The safe maximum working pressure is determined by the strength of the material and the ratio of outside diameter to inside diameter. For a high strength material, such as a suitable steel or aluminium alloy, a ratio of approximately 1.5 is commonly used for a 1000 atm vessel, but a variety of details can affect the final ratio selected. A typical vessel and end closure is shown in Fig. 7.42 which also illustrates a number of high pressure components. High pressure is normally generated by a hand operated hydraulic pump or by an air powered pump, both of

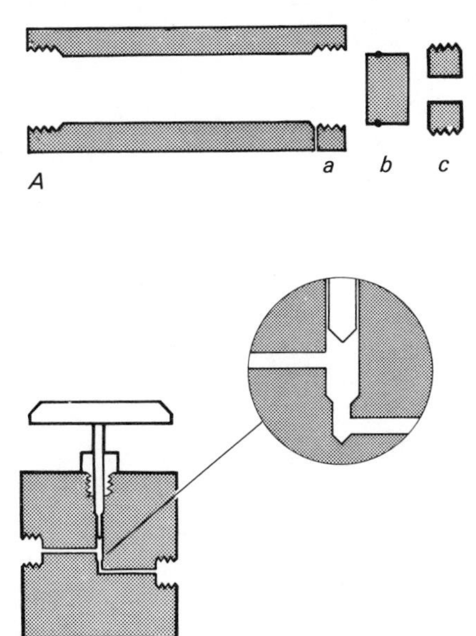

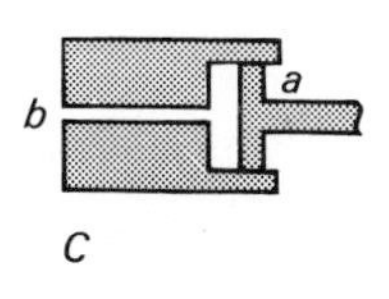

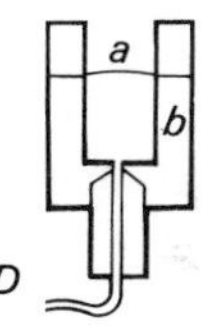

Fig. 7.42. High pressure components, not to scale.
A. *a*, cylindrical pressure vessel (note the safety bleed hole) with *b*, end-plug and O-ring seal and *c*, threaded ring.
B. Needle valve, showing valve seat (inset).
C. Hydraulic intensifier principle used in hydraulic pumps, low pressure acts at *a* to generate high pressure at *b*.
D. Rupture disc *a* in holder *b*.

which serve a variety of industrial purposes and may therefore be purchased. In either case the same principle of operation is used; a small pressure exerted over a large cross-sectional area acts through a small cross sectional area to give a larger pressure (Fig. 7.42*C*). High pressure pipe, couplings, valves, branching pieces, filters and capillary are available because of their commercial use, chiefly in the chemical industry. Direct reading gauges are also commercially available and work by magnifying and displaying the tendency of a coiled tube (a Bourdon tube) to straighten when its interior is pressurised. Absolute pressure measurements are made with a dead weight tester, an instrument which uses weights acting on a piston to give a known force per unit area, or pressure, when due allowance has been made for friction and the viscosity of the hydraulic fluid. Convenient recordings of hydrostatic pressure can be obtained from potentiometers driven by Bourdon tubes or by strain gauges which usually form part of a bridge circuit.

The hydraulic fluids which are commonly used in biological work are water, liquid paraffin or some similar oil, and helium gas. The energy content of a laboratory vessel containing fluid at high pressure is determined by the compressibility of the fluid, its volume and the pressure. Gas-filled vessels contain a great deal of energy at pressure and should be regarded as potentially more dangerous than hydraulic liquid-filled vessels. Here we are concerned with the difference between the energy which may be released from a small bomb and that released from, say, a cartridge. The pressure-fuse or rupture disc (Fig. 7.42) is worth noting in this context. Thin discs of steel, or other materials, can be purchased which yield to within 2 per cent of stated values, preventing excessive pressures building up in equipment with catastrophic results. Spring loaded relief valves may also be used.

AN ELECTRICAL FEEDTHROUGH, A HIGH PRESSURE WINDOW, AND A ROTATING SHAFT

Recent developments in underwater technology have encouraged the manufacture of a wide range of electrical feedthroughs for high pressure equipment. These are frequently rather bulky for laboratory instruments but small electrical feedthroughs can be made by the experimenter and should be carefully tested (Fig. 7.43). Note the stuffing box packed with flax used in Beebe's bathysphere (Fig. 7.10); it worked well.

The design of high pressure windows is not a subject for amateurs. Plastic, glass, quartz and sapphire have all been used at high pressure. Acrylic (perspex) is a particularly convenient material from which windows can be machined for use at physiological temperatures and at pressures over the deep sea range. A useful article on the use of acrylic materials at high pressure is to be found in Stachiw (1970). A practical high pressure window designed by Gilchrist (1972) is illustrated in Fig. 7.43; note the absence of any sealing element. Miniature windows for medium power light microscopy can be made in similar proportion.

A rotating shaft is a particularly versatile feature of a pressure vessel and the example shown in Fig. 7.43 may be easily rotated by a small motor at approximately 100 rpm, whilst confining several hundred atmospheres pressure. The rotating seal requires a good surface finish and careful assembly.

HIGH PRESSURE APPARATUS USED IN CELL BIOLOGY

Large cells may be viewed inside a pressure vessel with an ordinary light microscope provided a suitably long range objective is used in conjunction with a thin window. For observing the growth and division of cells as large as sea urchin eggs various workers have used inverted microscopes or hanging drop preparations (Fig. 7.44).

For histological purposes cells and tissues have been fixed at high pressure by a device which mixes fixative and test material together rapidly. Fig. 7.45 shows a method used by Landau and Thibodeau (1962). Pressure itself does not seem to have had any obvious effect on fixation.

The pressure-induced shift in the sol–gel equilibrium in cytoplasm was measured with a high pressure-centrifuge device which has been described in Chapter 4.

The long term culture of both eukaryote and prokaryote cells at high pressure awaits serious study. Some techniques which are likely to be involved in high pressure continuous culture vessels have, to a limited extent, been tried out in high pressure sea water aquaria. The process of monitoring cell growth at pressure is particularly important because cells, especially eukaryote cells, are unlikely to survive the shear forces experienced in decompression through an aperture. Microscopic observations and optical density measurements are relatively simple methods of observing growth at high pressure. An adaptation of the Coulter counting technique has been used to monitor cells in suspension at high pressure (Fig. 7.46).

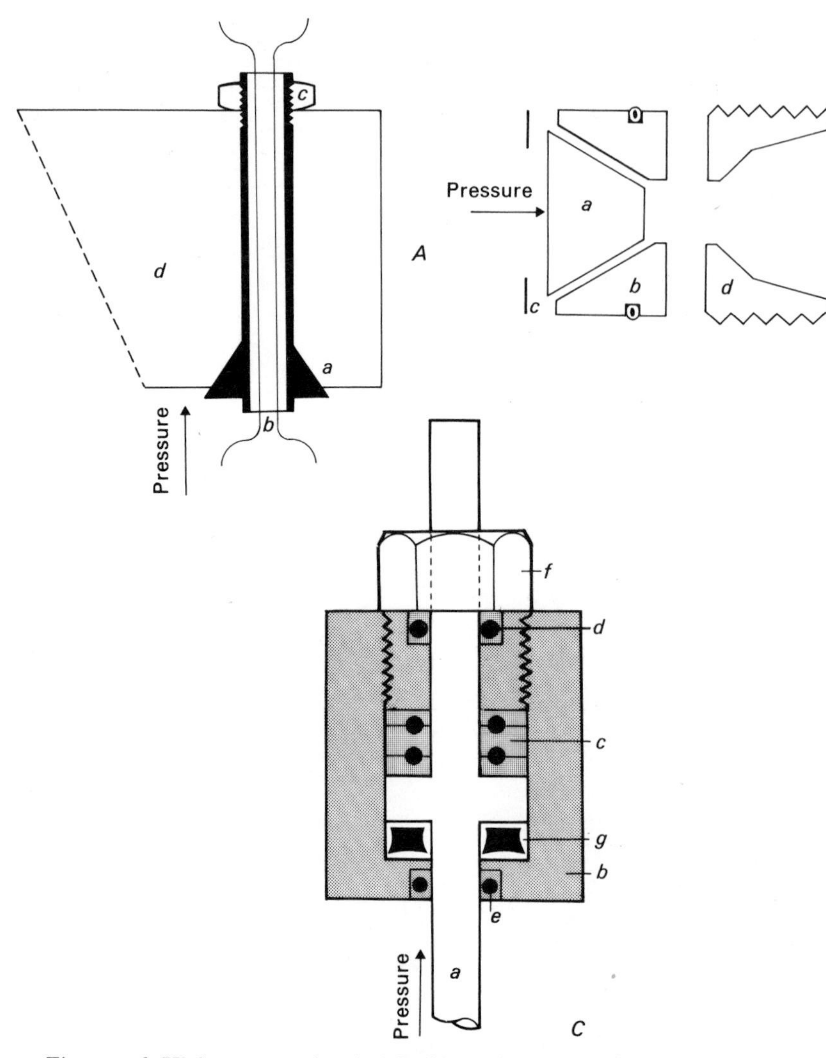

Fig. 7.43*A*. High pressure electrical feedthrough. *a*, cone of brass brazed onto metal sheathed, mineral insulated cable; *b*, conductors exposed, mineral insulation impregnated with epoxy resin under high pressure to seal against moisture; *c*, ring to tighten conical seat; *d*, end-closure as in Fig. 7.42.

B. Acrylic window, included angle 60°, ratio of thickness to viewing diameter = 2. Note absence of special seal for the window although an O-ring is used for the holder *b*. *c* is a retaining ring and *d* an end-ring.

C. Rotating shaft passing into a high pressure vessel. *a*, shaft inside the high pressure vessel; *b*, vessel wall or end-closure; *c*, thrust bearing; *d*, *e*, centre bearings; *f*, a honed finish is required on the shaft at the sealing ring, *g*.

Fig. 7.44. Inverted microscope, *A*, and vessel, *B*, used for observing cells at high pressure. (Marsland, 1950)

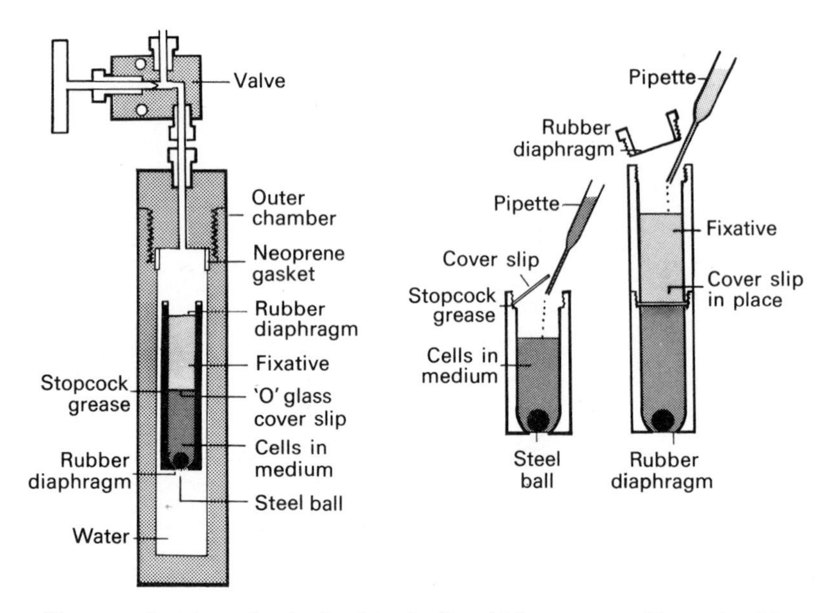

Fig. 7.45. Apparatus for the fixation of cells at high pressure, with, on the right, its method of assembly. (Landau & Thibodeau, 1962)

The principle of the Coulter counter is simple enough; electrodes immersed in saline on either side of a small orifice in a glass plate detect a resistance change when a particle (a cell) moves through the orifice. By means of a complex electronic circuit the number of particles and a close approximation of each particle volume is obtained. Counts of cells in suspension are normally carried out using a manometric apparatus which causes a known volume of liquid to move through the orifice. The compact hydraulic device shown in Fig. 7.46 works well at high pressure. Rotation of the shaft causes the piston to move up and down, forcing the cell suspension to pass to and fro through the orifice. The electrodes detect the transient resistance changes in the normal way and pass the signal to the electronic circuitry outside the pressure vessel via electrical feedthroughs. The glass plunger is caused to jerk up and down, stirring the cell suspension. If very great changes in cell size were encountered during long experiments the Coulter counting technique might run into difficulties, but results with batch cultures are promising, and are mentioned in Chapter 4.

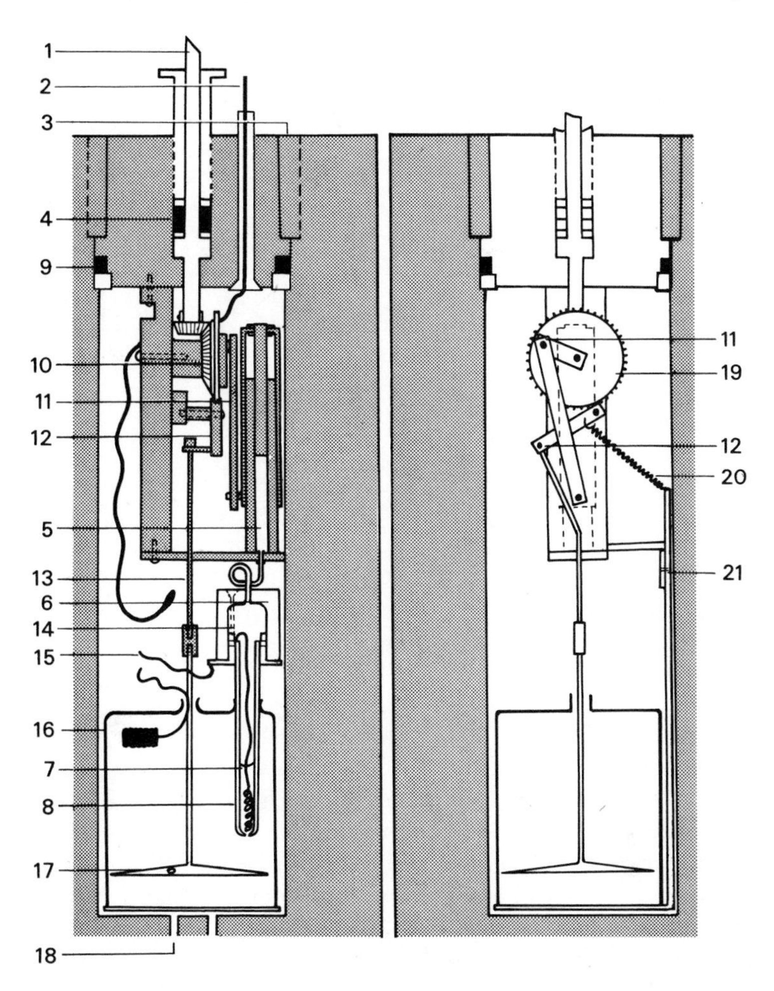

Fig. 7.46. Front- and side-view of an apparatus for counting and sizing growing cells under high pressure (modified Coulter counter). 1, drive shaft; 2, thermocouple wires; 3, retaining collar; 4, rotary seal; 5, piston connecting to the perspex cap, 6, which fits over the orifice tube, 8; 7, oil–water interface inside the orifice tube; 9, packing ring; 10, bevel gear; 11, pushrod; 12, lever which actuates the plunger-shaft, 13; 14, a clamping screw which holds the electrode against the brass collar fitted to the rim of the perspex cap, 6; 15, wires from electrodes; 16, glass vessel containing the cell suspension; 17, plunger; 18, inlet and outlet pipes connecting to the high pressure pump and drain cock respectively; 19, wheel bearing cams; 20, spring; 21, support for the platform which holds the growth vessel, 16. (Macdonald, 1967)

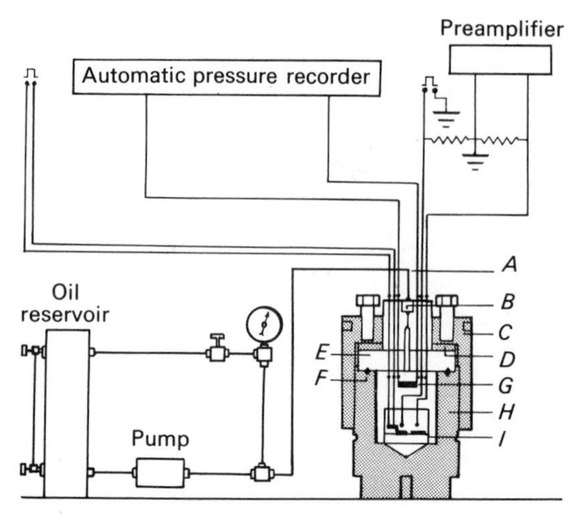

Fig. 7.47. Arrangement for applying high hydrostatic pressure to single fibre preparations. *A*, superpressure tubing connecting pump to interior of reaction vessel; *B*, pressure inlet; *C*, outside cap screwed onto body of vessel, cap was equipped with 16 thrust bolts; *D*, steel thrust ring; *E*, inner pressure head; *F*, 'Delta' type steel gasket; *G*, self-sealing electrical connector; *H*, body of reaction vessel; *I*, bridge-insulator. (Spyropoulos, 1957*b*)

ELECTROPHYSIOLOGICAL HIGH PRESSURE APPARATUS

In a study of the electrical properties of single nerve fibres at pressures up to 1000 atm, Spyropoulos (1957*b*) used a stainless steel vessel with 12.8 cm bore and an outside compression cap end-closure containing 6 screened electrical feedthroughs (Fig. 7.47). Pressure was recorded from an internal sensor. A pair of platinum stimulating electrodes were used in conjunction with two silver–silver chloride–Ringer-agar recording electrodes which seem unaffected by high pressure. The vessel shown in Fig. 7.47 has particularly good access which is an advantage when mounting apparatus in a pressure vessel. Mineral oil was used to transmit pressure and to insulate the wires, but as large a volume of aqueous solution as possible was included in the system to minimise heat of compression effects. Electrophysiological recordings made during human dives in gaseous environments similarly only require conventional electrodes.

One of the most interesting pieces of high pressure physiological equipment known to the author is that devised by Cattell and

Edwards (1928) to measure the heat and tension produced in muscular contraction. Fig. 7.48 shows a 6 cm bore pressure vessel with an end closure containing a small window, two pairs of electrical feedthroughs and a shaft whose rotation drove a spur gear. The test muscle was mounted on a torsion wire isometric transducer whose indicating pointer could be observed, or recorded on a photokymograph by means of a mirror. Pressures of up to 70 atm were applied by gas from a cylinder through an oil/gas separator. The muscle was either immersed directly in oil or in a small volume of saline surrounded by oil, and care was taken to avoid temperature changes by altering the pressure slowly. The thermopile which measured the heat emitted by the muscle was probably unaffected by pressure although its electrical insulation proved troublesome.

Measurements of the temperature inside an experimental pressure vessel are commonly carried out with either a resistance thermometer, a thermocouple, or a thermistor. Tavernier and Prache (1952) have studied the effect of deep sea pressures on thermistors in some detail. It is the author's experience, in agreement with ZoBell (1959), that the small pressure effect which is usually found is reproducible for a given thermistor. Thermocouples are not affected by pressures of interest here.

ELECTROCHEMICAL MEASUREMENTS AT HIGH PRESSURE

Electrodes sensitive to hydrogen and other cations have been used at high pressure by Distèche (1962), Culberson, Lester and Pytkowicz (1967) and Whitfield (1969). Distèche and Dubuisson (1960) have used the pH electrode assembly shown in Fig. 7.49 mounted on a bathyscaph to obtain a pH profile of the sea to 2350 m depth. No physiological measurements using these electrodes at high pressure seem to have been reported.

HIGH PRESSURE RESPIROMETERS AND RELATED EQUIPMENT

The results of a number of measurements of the oxygen consumption of cells and organisms at high pressure have been referred to in Chapter 4. Methods of obtaining a continuous measurement of the consumption of dissolved oxygen at high pressure either use an oxygen electrode inside a pressure vessel or measure the

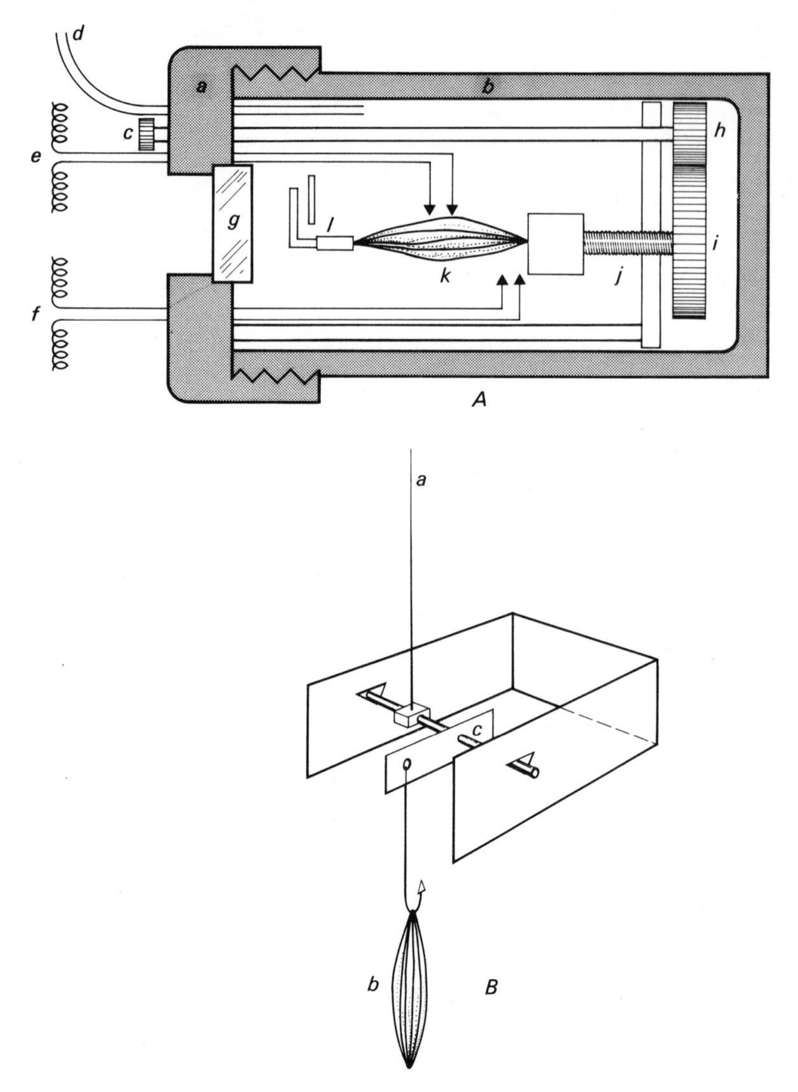

Fig. 7.48. Apparatus for studying muscular contraction under high hydrostatic pressure.

A. *a*, end closure of pressure vessel; *b*, body of vessel; *c*, rotating shaft (turned by hand); *d*, pressure connection (oil-filled); *e*, *f*, electrical feedthroughs; *g*, window; *h*, *i*, pair of spur gears driven by *c*; *j*, threaded shaft adjusting tension on the muscle, *k*; *l*, isometric torsion indicator (see *B*) and mirror.

B. The isometric torsion indicator. *a*, pointer which is reflected through the windows onto a photokymograph (see *C* opposite); *b*, muscle; *c*, torsion bar.

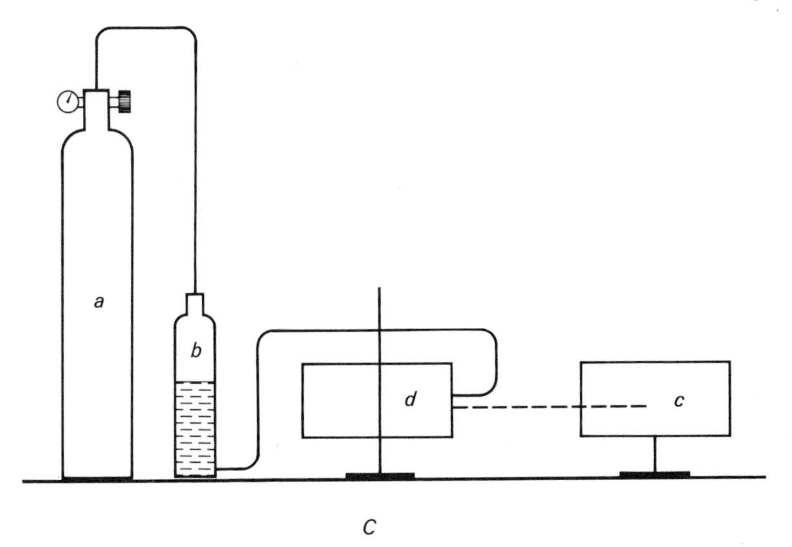

Fig. 7.48. *C*. The entire apparatus. *a*, gas cylinder which pressurised *b*, an oil reservoir; *c*, photokymograph and *d*, the pressure vessel. (*A* and *C* after Cattell & Edwards, 1928; *B* Fulton, 1925)

oxygen content of water flowing into and out of one. The latter arrangement is used in the high pressure aquarium of Naroska and co-workers (1968) and is described on p. 406.

Oxygen polarography at deep sea pressures involves few difficulties for those familiar with P_{O_2} electrodes. Both naked electrodes and electrodes covered with an oxygen-permeable membrane have been used. A naked platinum electrode with a silver–silver chloride reference anode has been used independently by Kono (1958) and Macdonald (1965). The 'oxigraph' of Kono (Fig. 7.50) appears to have yielded an increase in current with no change in half-wave potential at pressures up to 1500 atm; no stirring was used with the apparatus. Macdonald (1975) noted the absence of any effect of 500 atm on a naked electrode in sea-water. Fig. 7.51 shows a platinum electrode which is vibrated to achieve a stirring effect and which has been used at pressures to 1000 atm. The oil surrounding the solenoid and moving parts tended to dampen the motion of the electrode at high pressure so the calibration of the electrode was altered only indirectly by pressure. Despite this, reliable measurements of oxygen concentration were obtained. Naked electrodes are not particularly suited

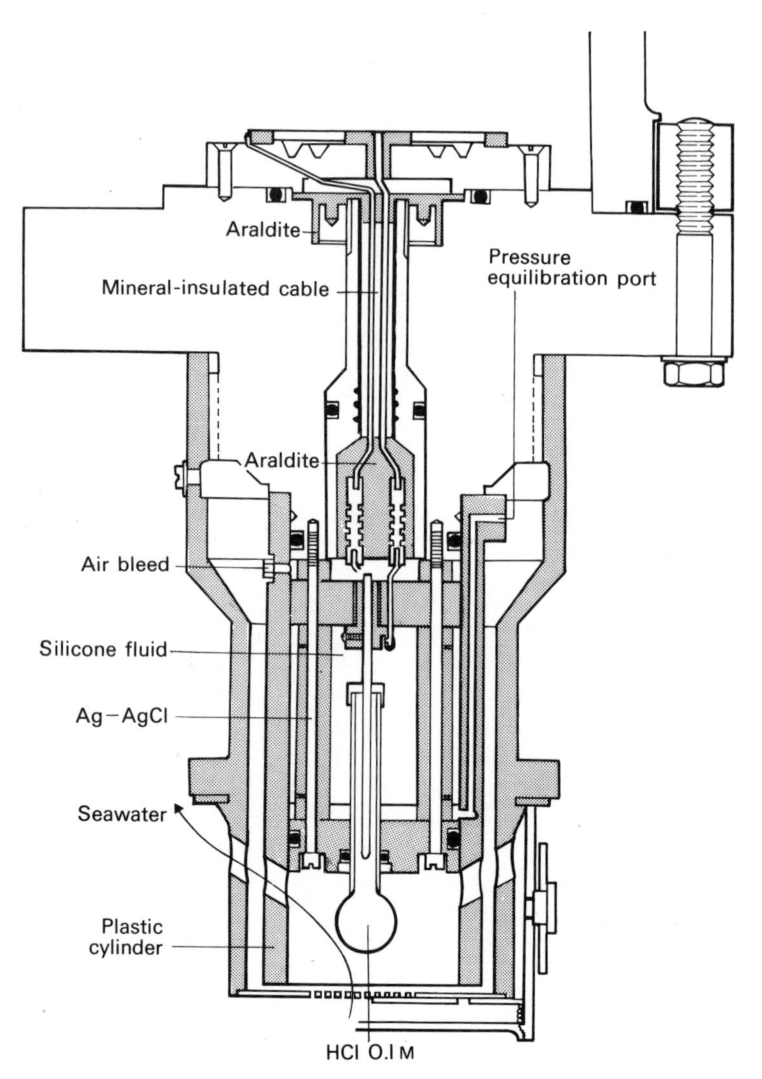

Fig. 7.49. pH electrode mounted on the outside of a bathyscaph (Distèche & Dubuisson, 1960)

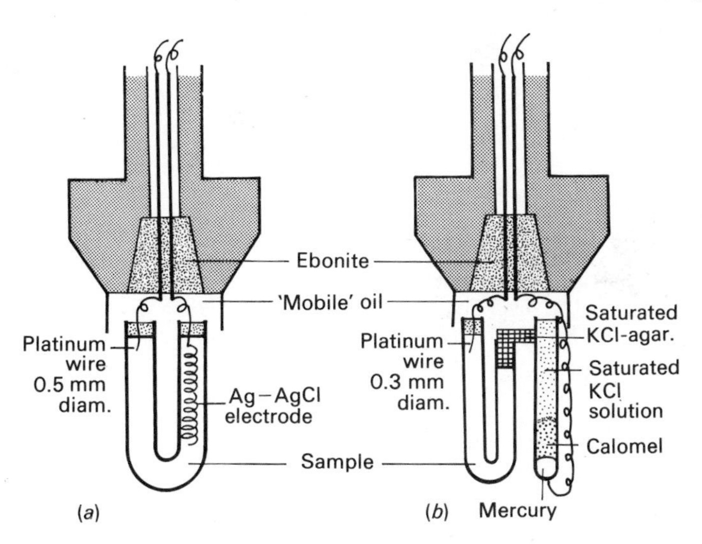

Fig. 7.50. Naked platinum electrodes used for oxygen determinations at high hydrostatic pressure. (*a*) Silver–silver chloride electrode. (*b*) Calomel electrode. (Kono, 1958)

for long experiments in seawater but those covered with an oxygen-permeable plastic membrane are. They give a steady current which is diminished at high pressure, presumably because of a decrease in the permeability of the plastic. Sometimes the pressure effect on the electrode output is a step change, but slow and irregular changes have been observed with certain membranes. High current levels in the absence of oxygen, the so-called nitrogen current, may be caused by pressure differentially compressing the anode or cathode in its epoxy or glass holder thereby exposing impure electrode surfaces. The first P_{O_2} depth sounding in the sea appears to have been made by Kanwisher (1962).

Two respirometers employing membrane-covered oxygen electrodes are shown in Figs. 7.52 and 7.53. One uses magnetic stirring and a non-magnetic pressure vessel; in the other, stirring is achieved by means of a rotating shaft which passes through the end closure. Both may have the water in the respirometer chambers flushed through either continuously or, more conveniently, discontinuously. The respirometers in Figs. 7.51 and 7.53 have the disadvantage of using oil as a hydraulic fluid which evolves some heat on compression.

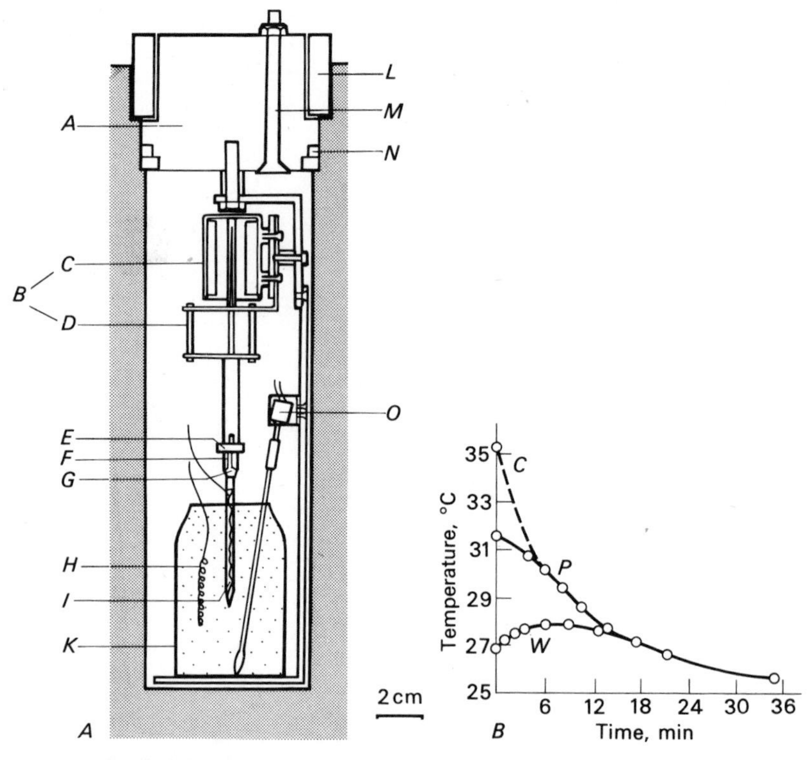

Fig. 7.51*A*. Vibrating naked platinum electrode used to determine the oxygen in a respiring cell suspension. *A*, pressure vessel end plug; *B*, vibrator, comprising *C*, solenoid and *D*, springs; *E*, ring which tightens on the chuck, *F*; *G*, plastic tube on *I*, glass tube supporting the platinum cathode, *H*, silver–silver chloride anode; *K*, glass vessel containing cells; *L*, retaining ring; *M*, electrical feedthroughs; *N*, neoprene seal: *O*, stirring motor with paddle. Liquid paraffin fills the vessel and floats on top of the cell suspension.

B. *W*, graph showing the heat change in the cell suspension on compressing the apparatus to 544 atm. *P* shows the heat change in the surrounding liquid paraffin and *C* shows the heat rise calculated for the paraffin. (*A* and *B*, Macdonald 1965)

Chandler and Vidaver (1971) have described an oxygen electrode mounted in a pressure vessel which responds very rapidly to an increase in P_{O_2} due to photosynthesis or a decline due to the respiration of plant tissues.

An apparatus in which experimental tissues were exposed to high partial pressures of gases is described by D'Aoust (1968) and shown in Fig. 7.54. The apparatus is of particular relevance here

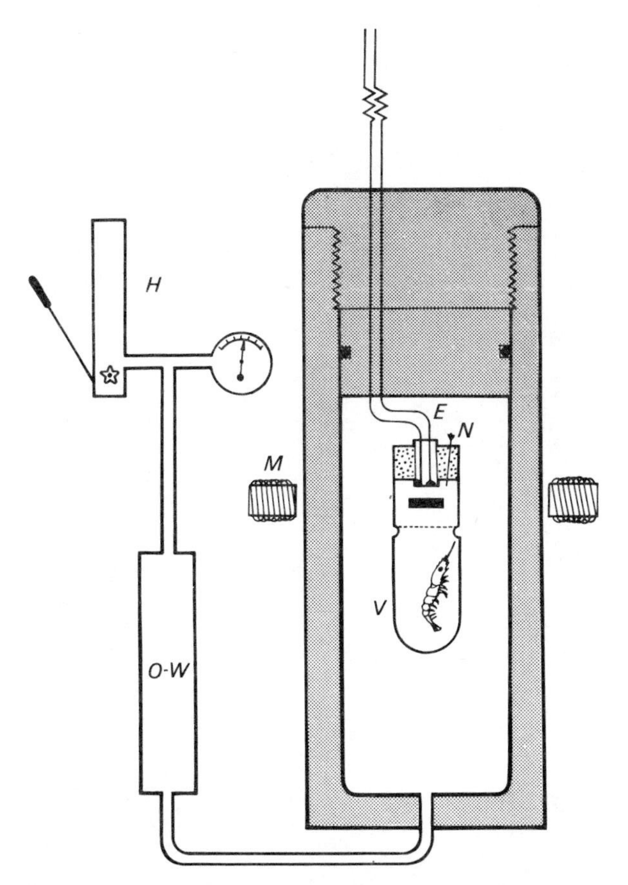

Fig. 7.52. High pressure respirometer of Teal and Carey (1967). The hydraulic pump, *H*, transmits pressure through an oil–distilled water separator *O–W*. The test animals are confined in seawater in a small glass vessel which is stirred by an array of electromagnets *M* switched sequentially. Underwater connectors insulated with moulded neoprene connect the oxygen electrode *E* mounted at the top of the glass vessel to the external circuit; a hypodermic needle *N* transmits pressure to the interior of the glass vessel *V*. (Teal & Carey, 1967)

because it was used to incubate gas gland tissue at a P_{O_2} of up to 130 atm. It has two notable features: a solenoid mechanism which allows liquid to be added to the experimental tissue, and the provision of valves to draw off fluid from the tissue at high pressure.

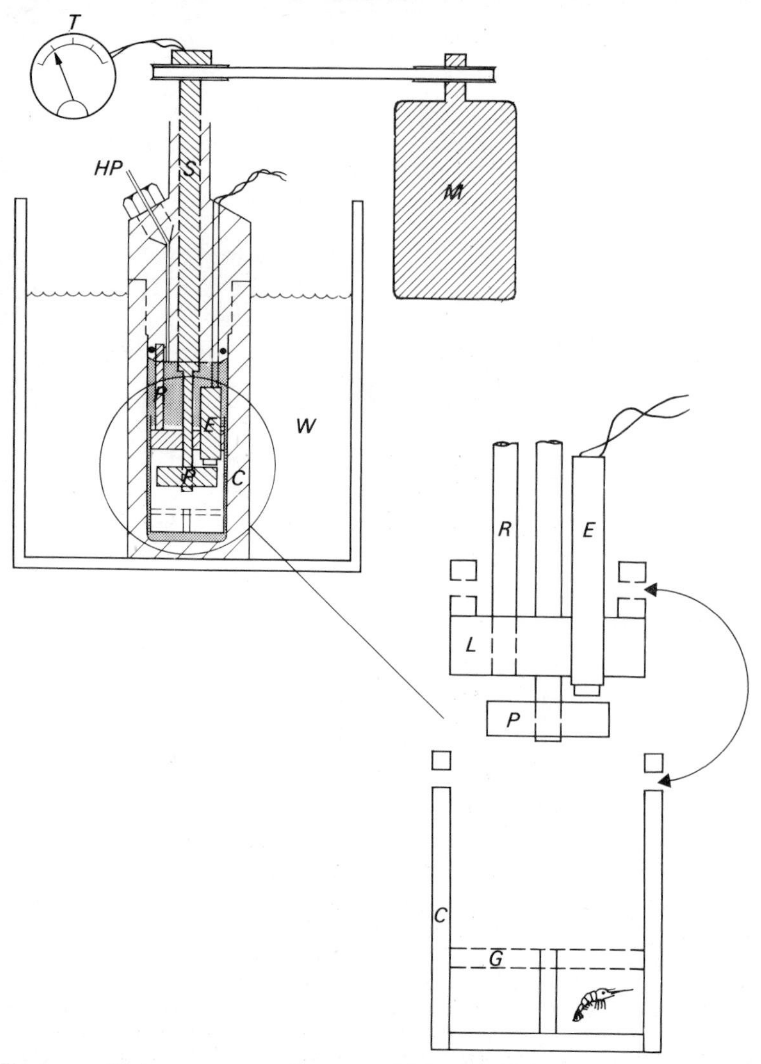

Fig. 7.53. High pressure respirometer of Macdonald (1975). *M*, motor which drives a rotating shaft *S* and paddle *P*; *T*, tachometer. The test animals are confined by a gauze *G* to a perspex chamber *C*. (inset) held to the end plug by a rod *R*; *E*, oxygen electrode; *L*, lid of the respirometer; *W*, water bath; *HP*, hydraulic connection. Liquid paraffin fills the pressure vessel. To flush the respirometer with seawater at high pressure the high pressure connection carries seawater which is ducted by an internal tube to the respirometer. Seawater overflows from the respirometer, falls through the oil to the bottom of the pressure vessel where it may flow out through a high pressure connection, not shown.

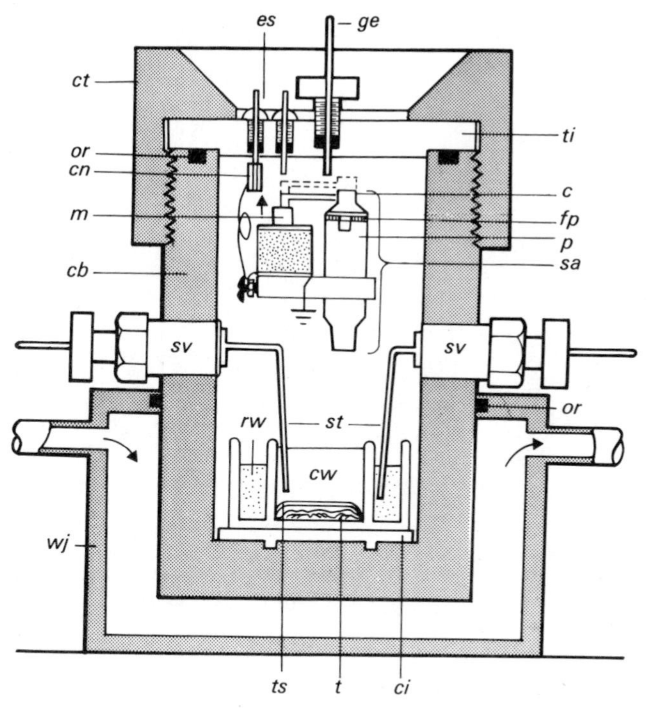

Fig. 7.54. Diagram of a gas-filled pressure vessel used to incubate tissue. *cb*, chamber body; *ct*, chamber top; *cn*, connector; *ci*, chamber insert (base); *c*, cap; *cw*, centre well of glass tissue vessel; *es*, electrodes; *fp*, filter paper seal of pipette; *ge*, gas exhaust; *m*, magnetic core of solenoid; *or*, O-ring seal; *p*, pipette; *rw*, ring well; *sa*, solenoid assembly; *st*, sample tubes; *sv*, sample valves; *t*, tissue; *ti*, top insert; *ts*, Teflon screen; *wj*, water jacket. (D'Aoust, 1968)

HIGH PRESSURE AQUARIA

The term aquarium can be applied to any vessel which can maintain aquatic animals in a healthy condition for a prolonged period. The high pressure respirometers through which water can flow are, therefore, aquaria. Because some marine animals, particularly invertebrates, can remain healthy for weeks without feeding, high pressure aquaria can be fairly simple items of equipment. One of the first, and certainly not the simplest of high pressure aquaria is that of Naroska (1968) built by the Aminco Company to operate at several hundred atmospheres pressure. Fig. 7.55 shows the hydraulic circuit employed. A number of features are

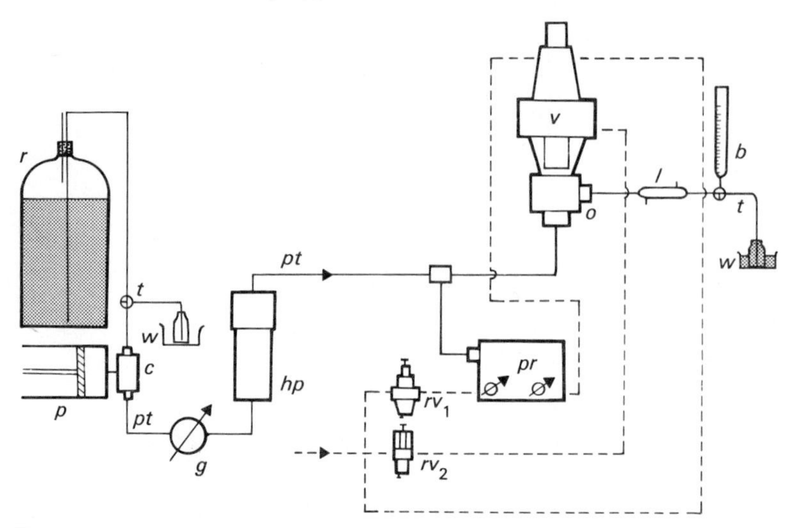

Fig. 7.55. High pressure aquarium and respirometer of Naroska (1968). *o*, outlet; *b*, burette; *t*, tap; *hp*, high pressure vessel; *c*, check-valve; *g*, gauge; *p*, pump; *pr*, pressure regulator; *v*, pressure regulating valve; *w*, flask for Winkler O_2 titration; *pt*, pressure tube; rv_1, rv_2, air pressure reducing valves; *r*, reservoir.

not altogether clear from the published description of the apparatus, but it appears that the pump is pre-set to work at any appropriate rate and that pressure is controlled by varying the rate at which water is bled through a needle valve to atmospheric pressure. The smoothness with which pressure and flow are controlled is not reported. The needle valve (*o*) which regulates the outflow of water from the high pressure circuit is actuated by a pneumatic device (*v*) in which two opposing air pressures balance the position of the needle valve to give the desired aperture. The relationship between the two air pressures is determined by the pressure regulating device (*pr*), although exactly how this is done is not described. Water samples are collected at intervals and dissolved oxygen subsequently determined.

The high pressure aquarium of Macdonald and Gilchrist (Fig. 7.34) employs a different control system. An air powered pump is used to generate pressure which is independently controlled by making use of the output from a transducer. This feeds a potentiometric controller which in turn operates solenoid valves on the air line supplying the high pressure pump. Thus, within limits, pressure tends to hold a pre-set level regardless of the rate at which water is drawn off from the high pressure circuit. The presence of an accumulator minimises pressure fluctuations which, in steady flow conditions, are of the order of ± 2 per cent. The way in which water is drawn off from the circuit may vary according to needs. Careful manual control of a metering valve (i.e. a needle valve with the minimum of backlash on the valve stem) is sufficient for short periods of flow, as when transferring water from the recovery pressure vessel to the experimental vessel, as described on p. 375. Means of providing an automatically controlled continuous flow of water through the vessel have not been devised. Instead a pulsed flow has been used in which a pair of pneumatically operated needle valves in series are actuated alternatively to decompress or 'lock-out' a small volume of water. The rate at which the pneumatically operated valves are driven determines the flow rate, and the presence of the accumulator in the high pressure circuit smooths the pressure changes as before. The system operates well at pressures up to 500 atm and with the modification of a number of components is capable of working at 1000 atm. A high pressure aquarium through which seawater is flushed under manual control at constant high pressure once every 24 hours has been used to maintain small groups of Crustacea at 100 atm for periods of 18–20 days (Macdonald, unpublished).

Aquaria for maintaining marine animals at pressures up to 20 atm have been described by Sundnes (1962), and Blaxter and Tytler (1972). In the latter example, a continuously operating rotary pump generates pressure which is controlled by the flow of water through both spring-loaded and manually adjustable relief valves (Fig. 7.56). The authors have also investigated the noise level in their vessel which was primarily used for conditioning experiments and not as a normal aquarium.

Conclusion

It is no longer difficult to simulate selected deep sea conditions in the laboratory, nor is it particularly novel for observers to descend into the depths in research vehicles. The deep sea is therefore accessible for a variety of studies involving both field and laboratory experiments.

The depth capability of free swimming human divers has yet to

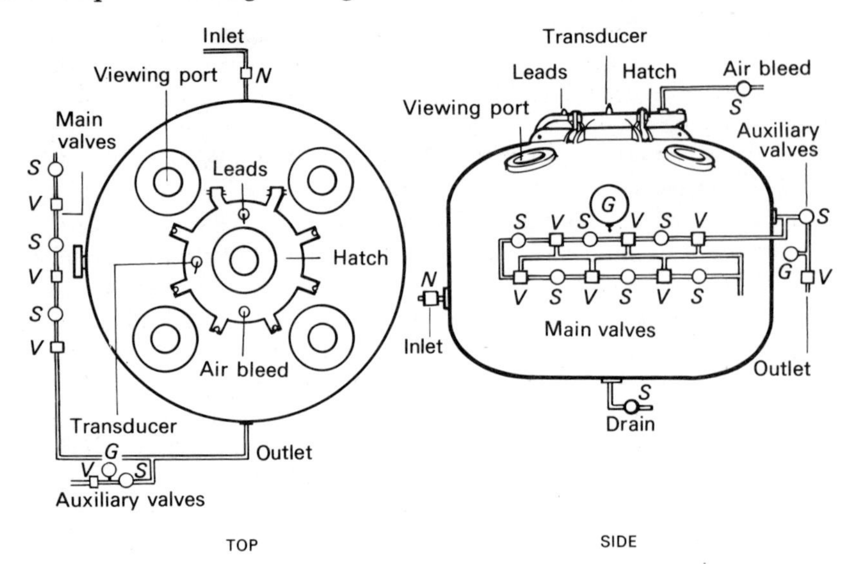

Fig. 7.56. Large volume, 20 atm pressure aquarium described by Blaxter and Tytler (1972). The vessel has an internal diameter of 90 cm. *G*, pressure gauge; *S*, shut-off valve; *V*, relief valve; *N*, non-return valve.

be established. On present evidence, humans might eventually be able to undertake work at depths of 1000 m or more, but it may not be practical or economic for them to do so.

One conclusion which emerges from this survey is that a wide range of deep sea techniques exist but we have yet to see their full development and exploitation. A major limitation in this field seems to be the world-wide lack of sea time for experimental biology on conventional research vessels. This situation should improve when it is more widely recognised that the deep sea poses so many fundamental problems in the life sciences.

REFERENCES

1

Agassiz, A. (1888). *Three cruises of the 'Blake'*, II. London: Low, Marston, Searle and Rivington.

Anon. (1970). Five miles of fish. *Nature, Lond.* **226,** 501–2.

Bertelsen, E. (1951). The ceratioid fishes. Ontogeny, taxonomy, distribution, and biology. *Dana Rept.* **7,** Paper 39, 5–276.

Bruun, A. F. (1957). In *Treatise on Marine Ecology and Paleoecology.* I *Ecology. Geol. Soc. Am. Memoir* **67,** ed. J. W. Hedgpeth.

Cannon, H. G. (1940). On the Anatomy of *Gigantocypris mülleri. Discovery Report* **19,** 185–244.

Certes, A. (1884). Note relative à l'action des hautes pressions sur la vitalité des micro-organisms d'eau douce et d'eau de mer. *C. r. Séanc. Soc. Biol.* **36,** 220–2.

Clarke, M. R. (1966). A review of the systematics and ecology of oceanic squids. *Adv. in mar. Biol.* **4,** 91–300.

Coutière, H. (1938). Notes sur la synonymie et la développment de quelques Hoplophoridae (Campagnes de la *Princess Alice*, 1904–5) *Rés. Camp. Scient. Monaco* **97,** 212–25.

Degens, E. T. & Ross, D. A. (1969). *Hot Brines and Recent Heavy Metal Deposits in the Red Sea.* New York: Springer-Verlag.

Degens, E. T. & Ross, D. A. (1970). The Red Sea hot brines. *Scient. Am.* **222,** April, 32–42.

Einarsson, H. (1945). Euphausiacea. 1. Northern Atlantic species. *Dana Rept.* **5,** Paper 217, 1–185.

Ekman, S. (1953). *Zoogeography of the Sea.* London: Sidgwick & Jackson.

Gaskell, T. F., Swallow, J. C. & Ritchie, R. M. (1953). Further notes on the greatest oceanic sounding and the topography of the Marianas Trench. *Deep-Sea Res.* **1,** 60–63.

Gunther, A. (1887). Report on the deep sea fishes collected by HMS *Challenger* during the years 1873–1876. *Report of the scientific results of the voyage of HMS Challenger* **22,** 1–329.

Gupta, B. L. & Little, C. (1970). Studies on Pogonophora. 4. Fine structure of the cuticle and epidermis. *Tissue & Cell* **2,** 637–96.

Hamann, S. D. (1957). *Physico-Chemical Effects of Pressure.* London: Butterworth.

Hardy, A. C. (1956). *The Open Sea.* London: Collins.

Hedgpeth, J. W. (1957). In *Treatise on Marine Ecology and Paleoecology.* I *Ecology. Geol. Soc. Am. Memoir* **67,** ed. J. W. Hedgpeth.

References

Heezen, B. C. & Hollister, C. D. (1971). *The Face of the Deep*. New York & London: Oxford Univ. Press.

Horne, R. A. (1969). *Marine Chemistry*. New York & London: Wiley–Interscience. p. 568.

Ivanov, A. V. (1963). *Pogonophora*. London: Academic Press. p. 478.

Little, C. & Gupta, B. L. (1968). Pogonophora: uptake of dissolved nutrients. *Nature, Lond.* **218**, 873–4.

McLellan, H. J. (1965). *Elements of Physical Oceanography*. Oxford & New York: Pergamon Press.

Marshall, N. B. (1954). *Aspects of Deep Sea Biology*. London: Hutchinson.

Munk, O. (1966). The ocular anatomy of some deep sea teleosts. *Dana Rept.* **13**, Paper 70.

Muromtsev, A. M. (1963). *The Principal Hydrological Features of the Pacific Ocean*. Jerusalem: Israel Program for Scientific Translations. p. 417.

Murray, J. & Hjort, J. (1912). *The Depths of the Ocean*. London: Macmillan.

Nybelin, O. (1957). Deep-sea bottom fishes. *Rept. Swedish Deep Sea Exped.* II. *Zoology*, No. 20, 250–345.

Poulsen, E. M. (1962). Ostracoda–Myodocopoda, Part 1, Cypridiniformes–Cypridinidae. *Dana Rept.* **10**, Paper 57, 1–414.

Pytkowicz, R. M. (1969). Chemical solution of calcium carbonate in sea water. *Am. Zool.* **9**, 673–9.

Pytkowicz, R. M. (1970). On the carbonate compensation depth in the Pacific Ocean. *Geochim. Cosmochim.* **34**, 836–9.

Regnard, P. (1891). *Recherches Expérimentales sur les Conditions Physiques de la Vie dans les Eaux*. Paris: Masson. pp. 1–500.

Richards, F. A. (1957). In *Treatise on Marine Ecology and Paleoecology*. 1 *Ecology*. *Geol. Soc. Am. Memoir* **67**, ed. J. W. Hedgpeth.

Rose, M. (1933). Copepodes Pélagiques. *Faune de France* **26**, Paris: Lechevalier.

Scheltema, R. S. & Scheltema, A. H. (1972). Deep sea biological studies in America (1846–1872), their contribution to the *Challenger* Expedition. *Proc. Roy. Soc. Edinb. B* **72**, 133–44.

Southward, E. C. & Southward, A. J. (1967). The distribution of Pogonophora in the Atlantic Ocean. *Symp. zool. Soc. Lond.* **19**, 145–58.

Sverdrup, H. U., Johnson, M. W. & Fleming, R. H. (1942). *The Oceans. Their physics, chemistry and general biology*. New York: Prentice Hall Inc.

2

Brown, D. E., Johnson, F. H. & Marsland, D. A. (1942). The pressure, temperature relations of bacterial luminescence. *J. cell. comp. Physiol.* **20**, 151–68.

Drost-Hansen, W. (1972). Effects of pressure on the structure of water in various aqueous systems. *Symp. Soc. exp. Biol.* **26**, 61–102.

Gerber, B. R. & Noguchi, H. (1967). Volume change associated with the G–F transformation of flagellin. *J. molec. Biol.* **26**, 196–210.

Horne, R. A. (1969). *Marine Chemistry. The structure of water and the chemistry of the hydrosphere*. New York: Wiley–Interscience.

Horne, R. A. & Courant, R. A. (1964). Application of Walden's Rule to the electrical conduction of seawater. *J. Geophys. Res.* **69**, 1971–7.

Horne, R. A. & Johnson, D. S. (1966). The viscosity of compressed seawater. *J. Geophys. Res.* **71**, 5275–7.

Ikkai, T. & Ooi, T. (1966). Actin: volume change on transformation of G-form to F-form. *Science, N.Y.* **152**, 1756–7.

Jaenicke, R. & Lauffer, M. A. (1969). Polymerisation–depolymerisation of Tobacco mosaic virus protein. Further studies on the role of water. *Biochemistry, N.Y.* **8**, 3083–92.

Johnson, F. H., Eyring, H. & Pollisar, M. J. (1954). *The Kinetic Basis of Molecular Biology*. New York: John Wiley & Sons.

Johnson, F. H. & Schlegel, F. M. (1948). Haemoglobin oxygenation in relation to hydrostatic pressure. *J. cell. comp. Physiol.* **31**, 421–5.

Josephs, R. & Harrington, W. F. (1968). On the stability of myosin filaments. *Biochemistry, N.Y.* **7**, 2834–47.

Kauzman, W., Bodensky, A. & Rasper, J. (1962). Volume changes in Protein Reactions. II. Comparison of ionisation reactions in proteins and small molecules. *J. Am. Chem. Soc.* **84**, (10), 1777–88.

Kettman, M. S., Nishikawa, A. H., Morita, R. Y. & Becker, R. R. (1966). Effect of hydrostatic pressure on the aggregation reaction of poly-L-valyl-ribonuclease. *Biochem. biophys. Res. Comm.* **22**, 262–7.

Laidler, K. J. (1951). Influence of pressure on the rates of biological reactions. *Archs. Biochem. Biophys.* **30**, 226–36.

Lauffer, M. (1964). Chapt. 5. Protein: Protein interaction: Endothermic polymerisation and biological processes. In *Symposium on Foods: Proteins and their reactions* ed. H. W. Schultz & A. F. Anglemier. Wesport, Connecticut: Avi Publishing Co.

McMeekin, T. L., Groves, M. L. & Hipp, N. J. (1954). Partial specific volume of the protein and water in β-lactoglobulin crystals. *J. Polymer. Sci.* **12**, 309–15.

Makino, S. & Noguchi, H. (1971). Volume and sound velocity changes associated with coil-B transition of poly(*S*-carboxymethyl L-cysteine) in aqueous solution. *Biopolymers.* **10**, 1253–60.

Masterton, W. L. (1954). Partial molal volumes of hydrocarbon in water. *J. chem. Physics* **22**, (11), 1830–3.

Murayama, M. (1966). Molecular mechanism of red cell 'sickling'. *Science, N.Y.* **153**, 145–9.

Murayama, M. & Hasegawa, F. (1970). Effect of hydrostatic pressure on the aggregation of reaction of human fibrin at 23 °C. *Fedn Proc. Fedn Am. Socs exp. Biol.* **29**, (2), 401 (Abstract).

Nishikawa, A. H., Morita, R. Y. & Becker, R. R. (1968). Effects of solvent medium on polyvalyl ribonuclease aggregation. *Biochemistry, N.Y.* **7**, 1506–13.

Payens, T. A. J. & Hermans, K. (1969). Effect of pressure on the temperature-dependent association of β-casein. *Biopolymers* **8**, 335–45.

Penniston, J. T. (1971). High hydrostatic pressure and enzymic activity: Inhibition of multimeric enzymes by dissociation. *Archs. Biochem. Biophys.* **142**, 322–32.

References

Rainford, P., Noguchi, H. & Morales, M. (1965). Hydrostatic pressure and the contractile system. *Biochemistry, N.Y.* **4,** 1958–65.

Rasper, J. & Kauzman, W. (1962). Volume changes in protein reactions. 1. Ionisation reactions of proteins. *J. Am. chem. Soc.* **84,** 1771–7.

Susuki, K. & Taniguchi, Y. (1972). Effect of pressure on biopolymers and model systems. *Symp. Soc. exp. Biol.* **26,** 103–24.

Suzuki, K., Taniguchi, Y. & Enomoto, T. (1972). The effects of pressure on the sol–gel transformations of macromolecules. *Bull. Chem. Soc. Japan* **45,** 336–8.

Suzuki, K. & Tsuchiya, M. (1969). Effects of hydrostatic pressure on the cloud point of surface active agents. *Bull. Inst. Chem. Res. Kyoto Univ.* **47** (4), 270–3.

Tuddenham, R. F. & Alexander, A. E. (1962). The effect of pressure on micelle formation in soap solutions. *J. Phys. Chem.* **66,** 1839–41.

Van Holden, K. E. (1966). The molecular architecture of multichain proteins. In *Molecular Architecture in Cell Physiology. Symposium of the Society of General Physiologists* ed. T. Hayashi & A. C. Szent-Gyorgyi. Englewood Cliffs, N.J.: Prentice Hall Inc.

Wells, J. M. (1972–73). Hydrostatic pressure and haemoglobin oxygenation. In *Proceedings of the Fifth Underwater Physiology Symposium*, ed. C. J. Lambertsen. In press.

Yayanos, A. A. (1972). Apparent molal volume of glycine, glycolamide, alanine, lactamide, and glycineglycine in aqueous solutions at 25 ° and high pressure. *J. Phys. Chem.* **76,** 1783–92.

Zamyatnin, A. A. (1972). Protein volume in solution. *Prog. Biophys. mol. Biol.* **24,** 107–23.

3

Albright, L. J. (1969). Alternate pressurization–depressurization effects on growth and net protein, RNA and DNA synthesis by *Escherichia coli* and *Vibrio marinus. Can. J. Microbiol.* **15,** 1237–40.

Albright, L. J. & Morita, R. Y. (1968). Effect of hydrostatic pressure on synthesis of protein, ribonucleic acid and deoxyribonucleic acid by the psychrophilic marine bacterium *Vibrio marinus. Limnol. Oceanogr.* **13,** 637–43.

Arnold, R. M. & Albright, L. J. (1971). Hydrostatic pressure effects on the translation stages of protein synthesis in a cell free system from *E. coli. Biochim. biophys. Acta* **238,** 347–54.

Certes, A. & Cochin, D. (1884). Action des hautes pressions sur la vitalité de la levure et sur les phénomènes des fermentations. *C. r. Séanc. Soc. Biol.* **36,** 639–40.

Chumak, M. D. (1959). Effect of high pressure on intensity of glucose consumption by pressure-tolerant bacteria. Trans. from *Dokladȳ, Akademii nauk USSR* **126,** 524–66.

Chumak, M. D. & Blokhina, T. P. (1964). Influence of high pressure on the accumulation of organic acids during glucose fermentation of barotolerant bacteria. *Mikrobiologiya* **33,** 230–5.

References

Distèche, A. (1972). Effects of pressure on the dissociation of weak acids. *Symp. Soc. exp. Biol.* **26**, 27–60.

Gross, J. A. (1965). Pressure-induced colour mutation of *Euglena gracilis*. *Science, N.Y.* **147**, 741–2.

Hammes, G. G. & Wu, C. W. (1971). Regulation of enzyme activity. *Science, N.Y.* **172**, 1205–11.

Hedén, C.-G. (1964). Effects of hydrostatic pressure on microbial systems. *Bact. Rev.* **28**, 14–29.

Hermolin, J. & Zimmerman, A. M. (1969). The effect of pressure on synchronous cultures of *Tetrahymena*: a ribosomal study. *Cytobios* **3**, 247–56.

Hill, E. P. & Morita, R. Y. (1964). Dehydrogenase activity under hydrostatic pressure by isolated mitochondria obtained from *Allomyces macrogynus*. *Limnol. Oceanogr.* **9**, 243–8.

Hochachka, P. W., Behrisch, H. W. & Marcus, F. (1971). Pressure effects on catalysis and control of catalysis by liver fructose diphosphatase from an off-shore benthic fish. *Am. Zool.* **11**, 439–49.

Hochachka, P. W., Schneider, D. E. & Moon, T. W. (1971). The adaptation of enzymes to pressure. I. A comparison of trout liver fructose diphosphatase with the homologous enzyme from an off-shore benthic fish. *Am. Zool.* **11**, 479–90.

Infante, A. A. & Krauss, M. (1971). Dissociation of ribosomes induced by centrifugation: evidence for doubting conformational changes in ribosomes. *Biochim. biophys. Acta* **246**, 81–99.

Landau, J. V. (1967). Induction, transcription and translation in *Escherichia coli*: a hydrostatic pressure study. *Biochim. biophys. Acta* **149**, 506–12.

Landau, J. V. (1970). Hydrostatic pressure on biosynthesis of macromolecules. In *High Pressure Effects on Cellular Processes*, ed. A. M. Zimmerman. New York & London: Academic Press. pp. 45–70.

Linderstrøm-Lang, K. & Jacobsen, C. F. (1941). The contraction accompanying enzymatic breakdown of proteins. *C.r. Trav. Lab. Carlsberg, Sér. Chim.* **24**, 1–47.

Lowe-Jinde, L. & Zimmerman, A. M. (1971). The incorporation of phenylalamine and uridine in *Tetrahymena*: a pressure study. *J. Protozool.* **18**, 20–3,

McElroy, W. D. (1952). Evidence for the occurrence of intermediates during mutation. *Science, N.Y.* **115**, 623–6.

Marquis, R. E., Brown, W. P. & Fenn, W. O. (1971). Pressure sensitivity of streptococcal growth in relation to catabolism. *J. Bact.* **105**, 504–11.

Marquis, R. E. & Fenn, W. O. (1969). Dilatometric study of streptococcal growth and metabolism. *Can. J. Microbiol.* **15**, 933–40.

Marquis, R. E. & ZoBell, C. E. (1971). Magnesium and calcium ions enhance barotolerance of streptococci. *Arch. Mikrobiol.* **79**, 80–92.

Moon, T. W., Mustafa, T. & Hochachka, P. W. (1971*a*). The adaptation of enzymes to pressure. II. A comparison of muscle pyruvate kinases from surface and midwater fishes with the homologous enzyme from an off-shore benthic species. *Am. Zool.* **11**, 491–502.

Moon, T. W., Mustafa, R. & Hochachka, P. W. (1971*b*). Effects of hydrostatic pressure on catalysis by different lactate dehydrogenase isozymes from tissues of an abyssal fish. *Am. Zool.* **11**, 473–8.

References

Morita, R. Y. (1957). Effect of hydrostatic pressure on succinic, formic, and malic dehydrogenases in *E. coli*. *J. Bact.* **74**, 251–5.

Morita, R. Y. (1965). In *The Fungi* 1, ed. G. C. Ainsworth & A. S. Sussman. New York & London: Academic Press.

Morita, R. Y. & Howe, A. R. (1957). Phosphatase activity by marine bacteria under hydrostatic pressure. *Deep-Sea Res.* **4**, 254–8.

Morita, R. Y. & ZoBell, C. E. (1955). Occurrence of bacteria in pelagic sediments collected during the Mid Pacific Expedition. *Deep-Sea Res.* **3**, 66–73.

Morita, R. Y. & ZoBell, C. E. (1956). Effect of hydrostatic pressure on the succinic dehydrogenase system of *E. coli*. *J. Bact.* **71**, 668–72.

Mustafa, T., Moon, T. W. & Hochachka, P. W. (1971). Effects of pressure and temperature on the catalytic and regulatory properties of muscle pyruvate kinase from an off-shore benthic fish. *Am. Zool.* **11**, 451–66.

Oppenheimer, C. H. & ZoBell, C. E. (1952). The growth and viability of sixty-three species of marine bacteria as influenced by hydrostatic pressure. *J. Mar. Res.* **11**, 10–18.

Paul, K. L. & Morita, R. Y. (1971). The effects of hydrostatic pressure and temperature on the uptake and respiration of amino acids by a facultatively psychrophilic marine bacterium. *J. Bact.* **108**, 835–43.

Pollard, E. C. & Weller, P. K. (1966). The effect of hydrostatic pressure on the synthetic processes in bacteria. *Biochim. biophys. Acta* **112**, 573–80.

Quigley, M. M. & Colwell, R. R. (1968). Properties of bacteria isolated from deep sea sediments. *J. Bact.* **95**, 211–20.

Simpson, G. G. (1964). Organisms and molecules in evolution. *Science, N.Y.* **146**, 1535–8.

Suzuki, K. & Taniguchi, Y. (1972). Effects of pressure on biopolymers and model systems. *Symp. Soc. exp. Biol.* **26**, 103–24.

Vacquier, V. (1962). Hydrostatic pressure has a selective effect on the copepod *Tigriopus*. *Science, N.Y.* **135**, 724–5.

Yayanos, A. A. (1972). Apparent molal volume of glycine, glycolamide, alanine, lactamide and glycylglycine in aqueous solution at 25 ° and high pressures. *J. Phys. Chem.* **76**, 1783–92.

Yayanos, A. A. & Pollard, E. C. (1969). A study of the effects of hydrostatic pressure on macromolecular synthesis in *Escherichia coli*. *Biophys. J.* **9**, 1464–82.

Zimmerman, A. M. & Silberman (1967). Studies on incorporation of $[H]^3$thymidine in *Arbacia* eggs under high hydrostatic pressure. *Expl Cell Res.* **46**, 469–76.

ZoBell, C. E. & Budge, K. M. (1965). Nitrate reduction by marine bacteria at increased hydrostatic pressures. *Limnol. Oceanogr.* **10**, 207–14.

ZoBell, C. E. & Cobet, A. B. (1962). Growth, reproduction and death rates of *Escherichia coli*. at increased hydrostatic pressures. *J. Bact.* **84**, 1228–36.

ZoBell, C. E. & Cobet, A. B. (1964). Filament formation by *E. coli* at increased hydrostatic pressure. *J. Bact.* **87**, 710–19.

ZoBell, C. E. & Hittle, L. L. (1969). Deep-sea pressure effects on starch hydrolysis by marine bacteria. *J. Oceanogr. Soc. Japan* **25**, 36–47.

ZoBell, C. E. & Johnson, F. H. (1949). The influence of hydrostatic pressure on the growth and viability of terrestial and marine bacteria. *J. Bact.* **57**, 179–89.

ZoBell, C. E. & Morita, R. Y. (1957). Barophilic bacteria in some deep sea sediments. *J. Bact.* **73**, 563–8.

ZoBell, C. E. & Oppenheimer, C. H. (1950). Some effects of hydrostatic pressure on the multiplication and morphology of marine bacteria. *J. Bact.* **60**, 771–81.

4

Abbott, B. L. & Baskin, R. J. (1962). Volume changes in frog muscle during contraction. *J. Physiol. Lond.* **161**, 379–91.

Barany, M. (1967). ATP-ase activity of myosin correlated with speed of muscle shortening. *J. gen. Physiol.* **50**, 197–218 (Supplement).

Bien, S. M. (1967). High hydrostatic pressure effects on *Spirostomum ambiguum*. *Calc. Tiss. Res.* **1**, 170–2.

Blaxter, J. H. S., Wardle, C. S. & Roberts, B. L. (1971). Aspects of the circulatory physiology and muscle systems of deep sea fish. *J. mar. biol. Ass. U.K.* **51**, 991–1006.

Brauer, R. W. (1972). Parameters controlling experimental studies of deep sea biology. In *Barobiology and the Experimental Biology of the Deep Sea*, ed. R. W. Brauer. Chapel Hill: University of North Carolina.

Brown, D. E. S. (1934*a*). The pressure coefficient of 'viscosity' in the eggs of *Arbacia punctulata*. *J. cell. comp. Physiol.* **5**, 335–46.

Brown, D. E. S. (1934*b*). The pressure–tension–temperature relations in cardiac muscle. *Am. J. Physiol.* **109**, 16. Abstract.

Brown, D. E. S. (1957). In *Influence of Temperature on Biological Systems*, ed. F. H. Johnson. American Physiological Society. pp. 83–110.

Brown, D. E. S. & Edwards, D. J. (1932). A contracture phenomenon in cross-striated muscle. *Am. J. Physiol.* **101**, 15. Abstract.

Brown, D. E. S., Guthe, K. F., Lawler, H. C. & Carpenter, H. P. (1958). The pressure, temperature and ion relations of muscle ATP-ase. *J. cell. comp. Physiol.* **52**, 59–77.

Brouha, A., Pequeux, A., Schoffeniels, E. & Distèche, A. (1970). The effects of high hydrostatic pressure on the permeability characteristics of the isolated frog skin. *Biochim. biophys. Acta* **219**, 455–62.

Cattell, M. (1935). Changes in the efficiency of muscular contraction under pressure. *J. cell. comp. Physiol.* **6**, 277–90.

Cattell, M. & Edwards, D. J. (1928). The energy changes of skeletal muscle accompanying contraction under high pressure. *Am. J. Physiol.* **86**, 371–82.

Cattell, M. & Edwards, D. J. (1930). The influence of hydrostatic pressure on the contraction of cardiac muscle in relation to temperature. *Am. J. Physiol.* **93**, 96–104.

Cattell, M. & Edwards, D. J. (1932). Conditions modifying the influence of hydrostatic pressure on striated muscle with special reference to the role of viscosity changes. *J. cell. comp. Physiol.* **1**, 11–36.

Digby, P. S. (1972). Detection of small changes in hydrostatic pressure by Crustacea and its relation to electrode action in the cuticle. *Symp. Soc. exp. Biol.* **26**, 445–72.

References

Distèche, A. (1972). Effects of pressure on the dissociation of weak acids. *Symp. Soc. exp. Biol.* **26,** 27–60.

Dreizen, P. & Kim, H. D. (1971). Contractile proteins of a benthic fish. I. The effects of temperature and pressure on myosin ATP-ase. *Am. Zool.* **11,** 513–21.

Drost-Hansen, W. (1972). Effects of pressure on the structure of water in various aqueous systems. *Symp. Soc. exp. Biol.* **26,** 61–101.

Ebbecke, U. (1935*a*). Über die Wirkungen hoher Drucke auf marine Lebewesen. *Pflügers Arch. ges. Physiol.* **236,** 648–57.

Ebbecke, U. (1935*b*). Muskelzung und Tetanus unter dem Einfluss der Kompression durch hohe Drucke. *Pflügers Arch. ges. Physiol.* **236,** 669–77.

Ebbecke, U. (1935*c*). Das Verhalten von Paramaecien unter der Einwirkung hohen Druckes. *Pflügers Arch. ges. Physiol.* **236,** 658–61.

Ebbecke, U. & Schaefer, H. (1935). Uber den Einfluss hoher Drucke auf den Aktionsstrom von Muskeln und Nerven. *Pflügers Arch. ges. Physiol.* **236,** 679–92.

Edwards, D. J. (1935). The action of pressure on the tension response of smooth muscle. *Am. J. Physiol.* **113,** 37. Abstract.

Edwards, D. J. & Brown, D. E. S. (1934). The action of pressure on the form of the electromyogram of auricle muscle. *J. cell. comp. Physiol.* **5,** 1–19.

Edwards, D. J. & Cattell, M. (1928). The stimulating action of hydrostatic pressure on cardiac function. *Am. J. Physiol.* **84,** 472–84.

Edwards, D. J. & Cattell, M. (1930). The action of compression on the contraction of heart muscle. *Am. J. Physiol.* **93,** 90–6.

Enns, T., Scholander, P. F. & Bradstreet, E. D. (1965). Effect of hydrostatic pressure on gases dissolved in water. *J. Phys. Chem.* **69,** 389–91.

Farquhar, M. G. V. & Palade, G. E. (1964). Functional organization of Amphibia skin. *Proc. Nat. Acad. Sci., US* **42,** 569–77.

Fenn, W. O. & Boschen, V. (1969). Oxygen consumption of frog tissues under high hydrostatic pressure. *Respiration Physiol.* **7,** 335–40.

Fontaine, M. (1928*a*). Sur les analogues existants entre les effects d'une tétanisation et ceux d'une compression. *C. r. hebd. Séanc. Acad. Sci., Paris* **186,** 99–101.

Fontaine, M. (1928*b*). Les fortes pressions et la consommation d'oxygène de quelques animaux marins. Influence de la taille de l'animal. *C. r. Séanc. Soc. Biol.* **99,** 1789–90.

Fontaine, M. (1929*a*). De l'action des fortes pressions sur la respiration des algues. *C. r. hebd. Séanc. Acad. Sci., Paris* **189,** 647–9.

Fontaine, M. (1929*b*). De l'augmentation de la consommation d'O_2 des animaux marins sous l'influence des fortes pressions. Ses variations en fonction de l'intensité de la compression. *C. r. hebd. Séanc. Acad. Sci., Paris* **188,** 460–61.

Fontaine, M. (1930). Recherches expérimentales sur les réactions des êtres vivants aux fortes pressions. *Annls Inst. Océanogr. Monaco* **8,** 5–99.

Gershfield, N. L. & Shanes, A. M. (1958). The influence of high hydrostatic pressure on cocaine and veratrine action in a vertebrate nerve. *J. gen. Physiol.* **42,** 647–53.

Grundfest, H. (1936). Effects of hydrostatic pressure upon the excitability, the recovery and the potential sequence of frog nerve. In *The Cold Spring Harbour Symposium on Quantative Biology.* IV. New York: The Biological Laboratory, Cold Spring Harbour. pp. 179–87.

Guthe, K. F. (1969). Hydrostatic pressure effects on rabbit and echinoderm myosin ATP-ase. *Archs. Biochem. Biophys.* **132,** 294–8.

Hardy, A. C. & Bainbridge, R. (1951). Effect of pressure on the behaviour of decapod larvae (Crustacea). *Nature, Lond.* **167,** 354–5.

Hêdén, C. G. (1964). Effects of hydrostatic pressure on microbial systems. *Bact. Rev.* **28,** 14–29.

Henderson, L. J. & Brink, F. N. (1908). The compressibilities of gelatine solutions and of muscle. *Am. J. Physiol.* **21,** 248–54.

Herman, L. & Dreizen, P. (1971). Electron microscope studies of skeletal and cardiac muscle of a benthic fish. I. Myofibrillar structure in resting and contracted muscle. *Am. Zool.* **11,** 543–57.

Ikkai, T. & Ooi, T. (1969). The effects of pressure on actomyosin systems. *Biochemistry, N.Y.* **8,** 2615–22.

Ikkai, T., Ooi, T. & Noguchi, H. (1966). Actin: volume change on transformation of G-form to F-form. *Science, N.Y.* **152,** 1756–7.

Johnson, F. H., Eyring, H. & Pollisar, M. J. (1954). *The Kinetic Basis of Molecular Biology.* New York: John Wiley.

Josephs, R. & Harrington, W. F. (1968). On the stability of myosin filaments. *Biochemistry, N.Y.* **7,** 2834–47.

Kennedy, J. R. & Zimmerman, A. M. (1970). The effects of high hydrostatic pressure on the microtubules of *Tetrahymena pyriformis. J. Cell. Biol.* **47,** 568–76.

Kim, H. D. & Dreizen, P. (1971). Contractile proteins of a benthic fish. II. Composition and ATP-ase properties of actomyosin. *Am. Zool.* **11,** 523–9.

Kitching. J. A. (1957*a*). Effects of high hydrostatic pressures on *Actinophyrs sol* (Heliozoa). *J. exp. Biol.* **34,** 511–17.

Kitching, J. A. (1957*b*). Effects of high hydrostatic pressures on the activity of flagellates and ciliates. *J. exp. Biol.* **34,** 494–510.

Kitching, J. A. (1964). The axopods of the sun animalcule *Actinophyrs sol* (Heliozoa). In *Primitive Motile Systems,* ed. R. D. Allen & N. Kamiya. New York & London: Academic Press. pp. 445–55.

Kitching, J. A. (1969). Effects of high hydrostatic pressure on the activity and behaviour of the ciliate *Spirostomum. J. exp. Biol.* **51,** 319–24.

Kitching, J. A. (1970). Some effects of high pressure on protozoa. In *High Pressure Effects on Cellular Processes,* ed. A. M. Zimmerman. New York & London: Academic Press. pp. 155–78.

Koefoed-Johnsen, V. & Ussing, H. H. (1958). The nature of the frog skin potential. *Acta Physiol. Scand.* **42,** 298–308.

Kono, I. (1958). Influence of high hydrostatic pressure upon the oxygen consumption of tissues. *Okayama Igakkai Zasshi* **70,** 4544. Abstract in English.

Laidler, K. J. & Beardell, A. J. (1955). Molecular kinetics of muscle ATP-ase III. Influence of hydrostatic pressure. *Arch. Biochem. Biophys.* **55,** 138–51.

Landau, J. V. (1961). The effects of high hydrostatic pressure on human cells in primary and continuous culture. *Exptl Cell Res.* **23,** 538–48.

Landau, J. V. & Marsland, D. (1952). Temperature pressure studies on the cardiac rate in tissue culture explants from the heart of the tadpole (*R. pipiens*). *J. cell. comp. Physiol.* **40,** 367–81.

References

Landau, J. V. & Peabody, R. A. (1963). Endogenous adenosine triphosphate levels in human amnion cells during application of high hydrostatic pressure. *Exptl Cell Res.* **29**, 54–60.

Landau, J. V. & Thibodeau, I. (1962). The micromorphology of *Amoeba proteus* during pressure induced changes in the sol–gel cycle. *Exptl Cell Res.* **27**, 591–4.

Landau, J. V., Zimmerman, A. M. & Marsland, D. A. (1954). Temperature–pressure experiments on *Amoeba proteus*; plasmagel structure in relation to form and movement. *J. cell. comp. Physiol.* **44**, 211–32.

Lincoln, R. J. (1971). Observations of the effects of changes in hydrostatic pressure and illumination on the behaviour of some planktonic crustaceans. *J. exp. Biol.* **54**, 677–88.

Macdonald, A. G. (1967). The effect of high hydrostatic pressure on the cell division and growth of *Tetrahymena pyriformis*. *Exptl Cell Res.* **47**, 569–80.

Macdonald, A. G. (1972). The role of high hydrostatic pressure in the physiology of marine animals. *Symp. Soc. exp. Biol.* **26**, 209–32.

Macdonald, A. G. (1973). Unpublished.

Macdonald, A. G. (1975). Locomotor activity and oxygen consumption in shallow water and deep sea invertebrates exposed to high hydrostatic pressures and low temperatures. In *Proceedings of the Fifth Underwater Physiology Symposium, 1972*. In press.

Macdonald, A. G. & Gilchrist, I. (1969). Life in the ocean depths. The physiological problems and equipment for the recovery and study of deep sea animals. *Proc. Oceanol. Internat., 1969*, Day 1. Brighton, UK.

Macdonald, A. G., Gilchrist, I. & Teal, J. M. (1972). Some observations on the tolerance of oceanic plankton to high hydrostatic pressure. *J. mar. biol. Ass. UK.* **52**, 213–23.

Macdonald, A. G. & Miller, K. W. (in press). Biological membranes at high hydrostatic pressure. In *Biophysical and Biochemical Perspectives in Marine Biology* III, ed. J. Sargent & D. Mallins. New York & London: Academic Press.

Marsland, D. A. (1938). The effects of high hydrostatic pressure upon cell division in *Arbacia* eggs. *J. cell. comp. Physiol.* **12**, 57–70.

Marsland, D. A. (1942). Protoplasmic streaming in plasmagel structure in relation to form and movement. In *The Structure of Protoplasm, Monograph of the Society of Plant Physiologists*. Iowa: Iowa Coll. Press.

Marsland, D. A. (1970). In *High Pressure Effects on Cellular Processes*, ed. A. M. Zimmerman. New York & London: Academic Press pp. 259–312.

Marsland, D. A. & Zimmerman, A. M. (1965). Structural stabilisation of the mitotic apparatus by heavy water in the cleaving eggs of *Arbacia punctulata*. *Exptl Cell Res.* **38**, 306–13.

Miki, H. (1960). Effect of high hydrostatic pressure on smooth muscle. *Okayama Igakkai Zasshi* **72**, 1613.

Murakami, T. H. & Zimmerman, A. M. (1970). A pressure study of galvanotaxis in Tetrahymena. In *High Pressure Effects on Cellular Processes*, ed. A. M. Zimmerman. New York & London: Academic Press. pp. 139–54.

Naroska, Von V. (1968). Vergleichende Untersuchungen über den Einfluss des hydrostatischen Druckes auf Überlebensfähigkeit und Stoffwechselintensität mariner Evertebraten und Teleosteer. *Kieler Meeresforsch.* **24**, 95–123.

Nishiyama, T. (1965). A preliminary note on the effect of hydrostatic pressure on the behaviour of some fish. *Bull. Fac. Fish. Hokkaido Univ.* **15**, 213–14.

Okada, K. (1954). Effects of high hydrostatic pressure on the permeability of plasma membranes. *Okayama Igakkai Zasshi* **66**, 2094.

Pease, D. C. (1946). Hydrostatic pressure effects upon the spindle figure and chromosome movement. II. Experiments on the meiotic divisions of *Tradescantia* pollen mother cells. *Biol. Bull. mar. biol. Lab., Woods Hole* **91**, 145–69.

Pease, D. C. & Kitching, J. A. (1939). The influence of hydrostatic pressure upon ciliary frequency. *J. cell. comp. Physiol.* **14**, 135–42.

Pease, D. C. & Marsland, D. A. (1939). The cleavage of *Ascaris* eggs under exceptionally high pressure. *J. cell. comp. Physiol.* **14**, 1–2.

Pequeux, A. (1972). Hydrostatic pressure and membrane permeability. *Symp. Soc. exp. Biol.* **26**, 483–4.

Podolsky, R. J. (1956). A mechanism for the effect of hydrostatic pressure on biological systems. *J. Physiol., Lond.* **132**, 38P.

Ponat, Von Anita (1967). Untersuchungen zur zellulären Druckresistenz verschiedener Evertebraten der Nord- und Ostsee. *Kieler Meeresforsch.* **23**, 21–47.

Regnard, P. (1887*a*). Les phénomènes de la vie sous les hautes pressions – La contraction musculaire. *C. r. Séanc. Soc. Biol.* **39**, 265–9.

Regnard, P. (1887*b*). Influence des hautes pressions sur la rapidité du courant nerveux. *C. r. Séanc. Soc. Biol.* **39**, 406–9.

Schlieper, C., Flügel, H. & Theede, H. (1967). Experimental investigations of the cellular resistance ranges of marine temperate and tropical bivalves: results of the Indian Ocean expedition of the German research association. *Physiol. Zoöl.* **40**, 345–60.

Sleigh, M. A. (1962). *The Biology of Cilia and Flagella*. Oxford: Pergamon Press. p. 242.

Spyropoulos, C. S. (1957*a*). The effects of hydrostatic pressure upon the normal and narcotised nerve fibre. *J. gen. Physiol.* **40**, 849–57.

Spyropoulos, C. S. (1957*b*). Response of single nerve fibres at different hydrostatic pressures. *Am. J. Physiol.* **189**, 214–18.

Teal, J. M. (1971). Pressure effects on the respiration of vertically migrating decapod Crustacea. *Am. Zool.* **11**, 571–6.

Teal, J. M. & Carey, F. G. (1967). Effects of pressure and temperature on the respiration of euphausiids. *Deep-Sea Res.* **14**, 725–33.

Tilney, L. G., Hiramoto, Y. & Marsland, D. (1966). Studies on the microtubules in Heliozoa. III. A pressure analysis of the role of these structures in the formation and maintenance of the axopodia of *Actinosphaerium nucleofilum* (Barrett). *J. Cell. Biol.* **29**, 77–95.

Vidaver, W. (1972). Effects of pressure on the metabolic processes of plants. *Symp. Soc. exp. Biol.* **26**, 159–74.

Whitt, C. S. & Prosser, C. L. (1971). Lactate dehydrogenase isozymes, cytochrome oxidase activity and muscle ions of the rat tail (*Coryphaenoides* sp.) *Am. Zool.* **11**, 503–11.

Yasuda, H. (1959). Effects of high hydrostatic pressure on the cardiac muscle. Part I. On the isolated frog heart. *Okayama Igakkai Zasshi* **71**, 5855.

Zimmerman, A. M. (1971). High pressure studies in cell biology. *Int. Rev. Cytol.* **30,** 1–47.

Zimmerman, A. M., Landau, J. V. & Marsland, D. A. (1957). Cell division: a pressure–temperature analysis of the effects of sulphydryl reagents on the cortical plasmagel structure and furrowing strength of dividing eggs (*Arbacia* and *Chaetopterus*). *J. cell comp. Physiol.* **49,** 395–435.

Zimmerman, A. M. & Rustad, R. C. (1965). Effects on high pressure on pinocytosis in *Amoeba proteus*. *J. Cell. Biol.* **25,** 397–400.

5

Alexander, R. McN. (1966). Physical aspects of swimbladder function. *Biol. Rev.* **41,** 141–76.

Alexander, R. McN. (1972). The energetics of vertical migration by fishes. *Symp. Soc. exp. Biol.* **26,** 273–94.

Barham, E. G. (1971). Deep sea fishes: lethargy and vertical orientation. In *Proc. Int. Symp. Biological Sound Scattering in the Ocean*, ed. G. B. Farquhar. Washington DC: Maury Centre Ocean Sci. Dept. Navy. pp. 100–18.

Bauer, B. B. (1967). In *Lateral Line Detectors*, ed. P. H. Cahn. Bloomington: Indiana Univ. Press.

Bayliss, L. E., Lythgoe, R. J. & Tansley, K. (1936). Some new forms of visual purple found in sea fishes, with a note on the visual cells of origin. *Proc. Roy. Soc. B.* **816,** 95–113.

Beebe, W. (1935). *Half Mile Down.* London: John Lane, The Bodley Head, p. 343.

Benson, A. A. & Lee, R. F. (1971). Wax esters. Major marine metabolic energy sources. *Proc. Biochem. Soc. & Soc. Biol. Chem.* (India) L 19 Bangalore.

Blaxter, J. H. S. (1970). Sensory deprivation and sensory input in rearing experiments. *Helg. wiss. Meeresuntershun.* **20,** 642–54.

Blaxter, J. H. S. & Tytler, P. (1972). Pressure discrimination in teleost fish. *Symp. Soc. exp. Biol.* **26,** 417–43.

Blaxter, J. H. S., Wardle, C. S. & Roberts, B. L. (1971). Aspects of the circulatory physiology and muscle systems of deep sea fish. *J. mar. biol. Ass. UK* **51,** 991–1006.

Blumer, M., Mullin, M. M. & Thomas, D. W. (1963). Pristane in the marine environment. *Helg. wiss. Meeresuntershun.* **10,** 187–201.

Boden, B. P., Kampa, E. M. & Abott, B. C. (1961). Photoreception of a planktonic crustacean in relation to light penetration in the sea. In *Progress in Photobiology*, ed. B. Christensen & B. Buchman. Amsterdam: Elsevier.

Breslau, L. R. & Edgerton, H. E. (1959). The luminescence camera. *J. Biol. Photog. Assoc.* **26,** 49–58.

Bronk, J. R., Harvey, E. N. & Johnson, F. H. (1952). The effects of hydrostatic pressure on luminescent extracts of the ostracod Crustacean, *Cypridina. J. cell. comp. Physiol.* **40,** 347–65.

Capen, R. L. (1967). Swimbladder morphology of some mesopelagic fishes in relation to sound scattering. *Research Report* 1447. Naval Electronics Laboratory, San Diego, California.

Chang, J. J. (1954). Analysis of the luminescent response of the ctenophore, *Mnemiopsis leidyi*, to stimulation. *J. cell. comp. Physiol.* **44**, 365–94.

Chang, J. J. & Johnson, F. H. (1959). The influence of pressure, temperature and urethane on the luminescent flash of *Mnemiopsis leidyi*. *Biol. Bull. mar. biol. Lab., Woods Hole* **116**, 1–14.

Chick, H. & Martin, C. J. (1913). The density and solution volume of some proteins. *Biochem. Jl* **7**, 92–6.

Clarke, G. L. (1936). On the depth at which fishes can see. *Ecology* **17**, 452–6.

Clarke, G. L. & Denton, E. J. (1962). Light and animal life. In *The Sea*, I, ed. M. N. Hill. New York & London: Wiley–Interscience.

Clarke, G. L. & Hubbard, C. J. (1959). Quantative records of the luminescent flashing of oceanic animals at great depths. *Limnol. Oceanogr.* **4**, 163–80.

Clarke, G. L. & Wertheim, G. K. (1956). Measurements of illumination at great depths and at night in the Atlantic Ocean, by means of a new bathyphotometer. *Deep-Sea Res.* **3**, 189–205.

Clarke, W. D. (1963). Function of bioluminescence in mesopelagic organisms. *Nature, Lond.* **198**, 1244–6.

Corner, E. D. S., Denton, E. J. & Forster, G. R. (1969). On the buoyancy of some deep sea sharks. *Proc. Roy. Soc. B* **171**, 415–29.

Cox, R. A. (1965). The physical properties of sea water. In *Chemical Oceanography*, ed. J. P. Riley & G. Skirrow. London: Academic Press.

D'Aoust, B. G. (1969). Hyperbaric oxygen: toxicity to fish at pressures present in their swimbladders. *Science, N.Y.* **163**, 576–8.

Dennell, R. (1955). Observations on the luminescence of bathypelagic Crustacea Decapoda of the Bermuda area. *J. Linn. Soc. (Zool.)* **42**, 393–406.

Denton, E. J. (1959). The contributions of the orientated photosensitive and other molecules to the absorption of whole retina. *Proc. Roy. Soc. B* **150**, 78–94.

Denton, E. J. (1961). The buoyancy of fish and cephalopods. *Progress in Biophysics* **11**, 178–234.

Denton, E. J. (1971). Examples of the use of active transport of salts and water to give buoyancy in the sea. *Phil. Trans. Roy. Soc. B* **262**, 277–87.

Denton, E. J. & Gilpin-Brown, J. B. (1971). Further observations on the buoyancy of *Spirula*. *J. mar. biol. Assoc. UK* **51**, 363–73.

Denton, E. J., Gilpin-Brown, J. B. & Shaw, T. I. (1969). A buoyancy mechanism found in cranchid squid. *Proc. Roy. Soc. B* **174**, 271–9.

Denton, E. J., Gilpin-Brown, J. B. & Wright, P. G. (1970). On the 'filters' in the photophores of mesopelagic fish and on a fish emitting red light and especially sensitive to red light. *J. Physiol. Lond.* **208**, 72–3P.

Denton, E. J., Liddicoat, J. D. & Taylor, D. W. (1970). Impermeable silvery layers in fish. *J. Physiol.* **207**, 64–5P.

Denton, E. J. & Marshall, N. B. (1958). The buoyancy of bathypelagic fishes without a gas filled swimbladder. *J. mar. biol. Assoc. UK* **37**, 753–67.

Denton, E. J. & Nicol, J. A. C. (1964). The choriodal tapeta of some cartilaginous fishes (Chondrichthytes) *J. mar. biol. Assoc. UK* **44**, 219–58.

Denton, E. J. & Shaw, T. I. (1962). The buoyancy of gelatinous marine animals. *J. Physiol, Lond.* **161**, 14P.

References

Denton, E. J. & Shaw, T. I. (1963). The visual pigments of some deep-sea elasmobranchs. *J. mar. biol. Assoc. UK* **43,** 65–70.

Denton, E. J. & Warren, F. J. (1956). Visual pigments of deep-sea fish. *Nature, Lond.* **178,** 1059.

Denton, E. J. & Warren, F. J. (1957). The photosensitive pigments in the retinae of deep sea fish. *J. mar. biol. Assoc. UK* **36,** 651–62.

Duedall, I. W. & Weyl, P. K. (1967). The partial equivalent volume of salts in sea water. *Limnol. Oceanogr.* **12,** 52–9.

Enns, T., Douglas, E. & Scholander, P. F. (1967). Role of the swimbladder rete of fish in secretion of inert gas and oxygen. *Adv. in Biol. and Med. Physics* **11,** 231–44.

Glasstone, S. (1964). *Textbook of Physical Chemistry.* London: Macmillan.

Goldsmith, T. H. (1972). The natural history of invertebrate visual pigments. In *Handbook of Sensory Physiology* VII, Part 1, ed. H. J. A. Dartnall. Berlin & New York: Springer-Verlag. p. 810.

Griffin, D. R. (1950). Underwater sounds and the orientation of marine animals, a preliminary survey. *Project N.R.* 162–429. Contract N6 O.N.R. 264, Tech. Rept. No. 3.

Griffin, D. R. (1958). *Listening in the Dark.* New Haven: Yale Univ. Press.

Gross, F. & Zeuthen, E. (1948). The buoyancy of plankton diatoms – a problem of cell physiology. *Proc. Roy. Soc. B* **135,** 382–9.

Haneda, Y. & Johnson, F. H. (1962). The photogenic organs of *Parapria canthus beryciformes* Franz and other fish with the indirect type of luminescent system. *J. Morphol.* **110,** 187–98.

Hara, T. J. (1971). Chemoreception. In *Fish Physiology* V, ed. W. S. Hoar & D. J. Randall. New York & London: Academic Press.

Heller, J. H., Heller, M. S., Springer, S. & Clarke, E. (1957). Squalene content of various shark livers. *Nature, Lond.* **179,** 919–20.

Horne, R. A. (1969). *Marine Chemistry.* New York: Wiley–Interscience.

Jerlov, N. G. (1968). *Optical Oceanography.* New York & London: Elsevier.

Kampa, E. M. (1955). Euphausiopsin, a new photosensitive pigment from the eye of euphausid crustaceans. *Nature, Lond.* **175,** 996–8.

Kanwisher, J. & Ebeling, A. (1957). Composition of the swimbladder gas in bathypelagic fishes. *Deep-Sea Res.* **4,** 211–7.

Kennedy, D. (1964). Chapt. 14, The photoreceptor process in lower animals. In *Photophysiology* II, ed. A. C. Giese. New York & London: Academic Press.

Kuhn, W., Ramel, A., Kuhn, H. J. & Marti, E. (1963). The filling mechanism of the swimbladder. *Experientia* **19,** 497–552.

Kunkle, J. S., Wilson, S. D. & Cota, R. A. (1969). *Compressed Gas Handbook, NASA* SP-3045. Washington D.C.

Kutchai, H. & Steen, J. B. (1971). The permeability of the swimbladder. *Comp. Biochem. Physiol.* **39***A*, 119–23.

Laverack, M. S. (1968). On the receptors of marine invertebrates. *Oceanogr. Mar. Biol. Ann. Rev.* **6,** 249–324.

Lewis, R. W. (1970). The densities of three classes of marine lipids in relation to their possible role as hydrostatic agents. *Lipids* **5,** 151–3.

Lomask, M. & Frassetto, R. (1960). Accoustic measurements in deep water using a bathyscaph. *J. Acoust. Soc. Amer.* **32,** 1028–33.

Lythgoe, J. A. (1972). The adaptations of visual pigments to the photic environment. In *Handbook of Sensory Physiology* VII Part 1, ed. H. J. A. Dartnall. Berlin & New York: Springer-Verlag.

Malins, D. C. & Barone, A. (1970). Glycerol ether metabolism: regulation of buoyancy in dogfish *Squalus acanthias*. *Science, N.Y.* **167**, 79–80.

Marshall, N. B. (1954). *Aspects of Deep Sea Biology*. London: Hutchinson's Scientific and Technical Publications.

Marshall, N. B. (1960). Swimbladder structure of deep-sea fishes in relation to their systematics and biology. *Discovery Report* **31**, 3–121.

Marshall, N. B. (1967). The olfactory organs of bathypelagic fishes. *Symp. Zool. Soc. Lond.* **19**, 57–70.

Marshall, N. B. (1971). *Explorations in the Life of fishes*. Cambridge, Mass.: Harvard Univ. Press.

Marshall, N. B. (1972). Swimbladder organisation and depth ranges of deep sea telecosts. *Symp. soc. exp. Biol.* **26**, 261–72.

McAllister, D. E. (1967). The significance of ventral bioluminescence in fishes. *J. Fish. Res. Bd Can.* **24**, 537–54.

McLellan, H. J. (1965). *Elements of Physical Oceanography*. New York & Oxford: Pergamon Press.

Menzies, R. J. & George, R. R. (1967). A re-evaluation of the concept of hadal or ultra-abyssal fauna. *Deep-Sea Res.* **14**, 703–23.

Munk, O. (1966). The ocular anatomy of some deep sea telecosts. *Dana Rept.* **13**, Paper 70.

Munz, F. W. (1958). Photosensitive pigments from the retinae of certain deep-sea fishes. *J. Physiol., Lond.* **140**, 220–35.

Munz, F. W. (1965). In *Colour Vision*, Ciba Foundation Symposium, ed. A. V. S. de Reuck & J. Knight. London: Churchill.

Neurath, N. & Bull, H. B. (1936). The denaturation and hydration of proteins. *J. biol. Chem.* **115**, 519–28.

Nevenzel, J. C. (1970). Occurrence, function and biosynthesis of wax esters in marine organisms. *Lipids* **5**, 308–19.

Nevenzel, J. C., Rodegker, W., Robinson, J. S. & Kayama, M. (1969). The lipids of some lantern fishes (family Myctophidae). *Comp. Biochem. Physiol.* **31**, 25–36.

Nicol, J. A. C. (1952*a*). Studies on *Chaetopterus variopedatus* (Renier). 1. The light producing glands. *J. mar. biol. Assoc. UK* **30**, 417–31.

Nicol, J. A. C. (1952*b*). Studies on *Chaetopterus variopedatus* (Renier). 2. Nervous control of light production. *J. mar. biol. Assoc. UK* **30**, 433–52.

Nicol, J. A. C. (1952*c*). Studies on *Chaetopterus variopedatus* (Renier). 3. Factors affecting the light response. *J. mar. biol. Assoc. UK* **31**, 113–44.

Nicol, J. A. C. (1958). Observations on luminescence in pelagic animals *J. mar. biol. Assoc. UK* **37**, 25–36.

Nicol, J. A. C. (1962). Animal luminescence. *Adv. Comp. Physiol. Biochem.* **1**, 217–73.

Nicol, J. A. C. (1969). Bioluminescence. In *Fish Physiology* III, ed. W. S. Hoar & D. J. Randall. New York & London: Academic Press.

Nielsen, J. G. & Munk, O. (1964). A hadal fish, *Bassogigas profundissimus*, with a functional swimbladder. *Nature, Lond.* **204**, 594–5.

Patton, S. & Thomas, A. J. (1971). Composition of lipid foams from swimbladders of two deep ocean fish species. *J. Lipid Res.* **12,** 331–5.

Phleger, C. F. (1971). Pressure effects on cholesterol and lipid synthesis by the swimbladder of an abyssal *Coryphaenoides* species. *Am. Zool.* **11,** 559–70.

Phleger, C. F. & Benson, A. A. (1971). Cholesterol and hyperbaric oxygen in swimbladders of deep sea fishes. *Nature, Lond.* **230,** 122.

Piccard, J. & Dietz, R. S. (1957). Oceanographic observations by the bathyscaph *Trieste* (1953–1956). *Deep-Sea Res.* **4,** 221–9.

Sand, O. & Hawkins, A. D. (1973). Acoustic properties of the cod swim bladder. *J. exp. Biol.* **58**, 797–820.

Sato, M. (1962). Studies on the pit organs of fishes. v. The structure and polysaccharide histochemistry of the cupula of the pit organ. *Annot. Zool. Jap.* **35,** 80–8.

Schevill, W. E., Backus, R. H. & Hersey, R. H. (1962). Sound production by marine animals. In *The Sea* I, ed. M. N. Hill. New York & London: Wiley–Interscience.

Scholander, P. F. (1954). Secretion of gases against high pressure in the swimbladder of deep sea fishes. II. The rete mirabile. *Biol. Bull. mar. biol. Lab., Woods Hole* **107,** 260–77.

Scholander, P. F. (1957). The wonderful net. *Scient. Am.* **196,** 96–107.

Scholander, P. F. & Van Dam, L. (1953). Composition of the swimbladder gas in deep sea fishes. *Biol. Bull. mar. biol. Lab., Woods Hole* **104,** 75–86.

Scholander, P. F. & Van Dam, L. (1954). Secretion of gases against high pressure in the swimbladder of deep sea fishes. I. Oxygen dissociation in the blood. *Biol. Bull. mar. biol. Lab., Woods Hole* **107,** 247–59.

Steen, J. B. (1970). The swimbladder as a hydrostatic organ. In *Fish Physiology* IV, ed. W. S. Hoar & D. J. Randall. New York & London: Academic Press.

Sie, H. C., Chang, J. J. & Johnson, F. H. (1958). Pressure–temperature-inhibitor relations in the luminescence of *Chaetopterus variopedatus* and its luminescent secretion. *J. cell. comp. Physiol.* **52,** 195–225.

Vlymen, W. J. (1970). Energy expenditure of swimming copepods. *Limnol. Oceanogr.* **15,** 348–56.

Wald, G., Brown, P. K. & Brown, P. S. (1957). Visual pigments and depth of habitat of marine fishes. *Nature, Lond.* **180,** 969–71.

Weale, R. A. (1955). Binocular vision and deep-sea fish. *Nature, Lond.* **175,** 996.

6

Anon. (1970). Five miles of fish. *Nature, Lond.* **226,** 501–2.

Banoub, M. W. & Williams, P. J. Le B. (1972). Measurements of microbial activity and organic material in the Mediterranean Sea. *Deep-Sea Res.* **19,** 433–44.

Barber, R. T. (1968). Dissolved organic carbon from deep water resists microbial oxidation. *Nature, Lond.* **220,** 274–5.

Belyaev, G. M. (1972). *Hadal Bottom Fauna of the World Ocean.* Israel program for scientific translations, p. 199.

Belyaev, G. M. & Vinogradova, N. G. (1961). Quantitative distribution of the bottom fauna in the northern half of the Indian Ocean. *Scripta Technica, Am. Geophys. Union* 136–41.

Berger, W. H. & Piper, D. J. W. (1972). Planktonic Foraminifera: differential settling, dissolution, and redeposition. *Limnol. Oceanogr.* **17**, 275–87.

Bernard, F. (1958). Plankton et benthos observés durant trois plongées en bathyscaphe au large de Toulon. *Ann Inst. Oceanogr. Monaco* **35**, 287–326.

Bernard, F. (1963). Density of flagellates and myxophyceae in the heterotrophic layers related to environment. In *Symposium on Marine Microbiology*, ed. C. H. Oppenheimer. Springfield, Illinois: Thomas. pp. 215–28.

Blaxter, J. H. S., Wardle, C. S. & Roberts, B. L. (1971). Aspects of the circulatory physiology and muscle systems of deep sea fish. *J. mar. biol. Assoc. UK* **51**, 991–1006.

Bogoyavlenskii, A. N. (1964). On the distribution of heterotrophic microorganisms in the Indian Ocean and in Antarctic waters. *Deep-Sea Res.* **11**, 105–8.

Carlucci, A. F. & Silbernagel, S. B. (1966). Bioassay of seawater. III. Distribution of vitamin B_{12} in the northern Pacific Ocean. *Limnol. Oceanogr.* **11**, 642–6.

Certes, A. (1884). Sur la culture, à l'abi des germes atmorphéniques, des eaux et des sédiments rapportes par les expéditions du *Travailleur* et du *Talisman*: 1882–1883. *C. r. Acad. Sci.* **98**, 690–3.

Childress, J. J. (1971). Respiratory rate and depth of occurrence of midwater animals. *Limnol. Oceanogr.* **16**, 104–6.

Clarke, C. L. & Breslau, L. R. (1959). Measurements of bioluminescence off Monaco and northern Corsica. *Bull. Inst. Océanogr. Monaco*, No. 1147, 1–31.

Clarke, M. R. (1966). A review of the systematics and ecology of oceanic squids. *Adv. in mar. Biol.* **4**, 91–300.

Conover, R. J. (1968). Zooplankton – Life in a nutritionally dilute environment. *Am. Zool.* **8**, 107–18.

Cooper, L. H. N. (1967). Stratification in the deep ocean. *Science Prog. (Oxford)* **55**, 73–90.

Dayton, P. K. & Hessler, R. R. (1972). Role of biological disturbance in maintaining diversity in the deep sea. *Deep-Sea Res.* **19**, 199–208.

Deevey, G. B. & Brooks, A. L. (1971). The annual cycle in quantity and composition of the zooplankton of the Sargasso Sea off Bermuda. II. The surface to 2000 m. *Limnol. Oceanogr.* **16**, 927–43.

Duursma, E. K. (1965). The dissolved organic constituents of seawater. In *Chemical Oceanography* I, ed. J. P. Riley & G. Skirrow. London & New York: Academic Press.

Ewing, M. & Thorndike, E. M. (1965). Suspended matter in deep ocean water. *Science, N.Y.* **147**, 1291–4.

Fournier, R. O. (1968). Observations of particulate organic carbon in the Mediterranean sea and their relevance to the deep living Coccolithophorid *Cylococcolithus fragilis*. *Limnol. Oceanogr.* **13**, 693–7.

Fournier, R. O. (1970). Studies on pigmented microorganisms from aphotic marine environments. *Limnol. Oceanogr.* **15**, 675–82.

Fournier, R. O. (1971). Studies on pigmented microorganisms from aphotic marine environments. II. North Atlantic distribution. *Limnol. Oceanogr.* **16**, 952–61.

Fournier, R. O. (1972). The transport of organic carbon to organisms living in the deep oceans. *Proc. Roy. Soc. Edinb. B* **73**, 203–12.

References

Fowler, S. W. & Small, L. F. (1972). Sinking rates of euphausid faecal pellets. *Limnol. Oceanogr.* **17,** 293–6.

Gordon, D. C. (1970*a*). A microscopic study of organic particles in the North Atlantic Ocean. *Deep-Sea Res.* **17,** 175–85.

Gordon, D. C. (1970*b*). Some studies on the distribution and composition of particulate organic carbon in the North Atlantic Ocean. *Deep-Sea Res.* **17,** 233–43.

Grey, M. (1956). The distribution of fishes found below a depth of 2000 m. *Fieldiana, Zool.* **36,** 77–93.

Grice, D. G. & Hulseman, K. (1965). Abundance, vertical distribution and taxonomy of calanoid copepods at selected stations in the north-east Atlantic. *J. Zool.* **146,** 213–62.

Hamilton, R. D., Holm-Hansen, O. & Strickland, J. D. H. (1968). Notes on the occurrence of living microscopic organisms in deep water. *Deep-Sea Res.* **15,** 651–6.

Hartman, O. & Emery, K. O. (1956). Bathypelagic coelenterates. *Limnol. Oceanogr.* **1,** 304–12.

Hobson, L. A. & Menzel, D. W. (1969). The distribution and chemical composition of organic particulate matter in the sea and sediments off the East coast of South America. *Limnol. Oceanogr.* **14,** 159–63.

Jahn, W. (1971). Deepest photographic evidence of an abyssal cephalopod. *Nature, Lond.* **232,** 487–8.

Jannasch, H. W. & Jones, E. G. (1959). Bacterial populations in sea water as determined by different methods of enumeration. *Limnol. Oceanogr.* **4,** 128–39.

Johannes, R. E. & Satomi, M. (1966). Composition and nutritive value of faecal pellets of a marine crustacean. *Limnol Oceanogr.* **11,** 191–7.

Johannes, R. E., Coward, S. J. & Webb, K. L. (1969). Are dissolved amino acids an energy source for marine invertebrates? *Comp. Biochem. Physiol.* **29,** 283–8.

Johnson, R. M., Schwent, R. M. & Press, W. (1968). The characteristics and distribution of marine bacteria isolated from the Indian Ocean. *Limnol. Oceanogr.* **13,** 656–64.

Johnston, R. (1972). The theories of August Pütter. *Trans. Roy. Soc. Edinb. B* **72,** 401–10.

Jørgensen, C. B. (1952). On the relation between water transport and food requirements in some marine filter feeding invertebrates. *Biol. Bull. mar. biol. Lab., Woods Hole* **103,** 356–63.

Jørgensen, C. B. (1966). *Biology of Suspension Feeding.* New York & Oxford: Pergamon Press.

Kriss, A. E. (1960). Micro-organisms as indicators of hydrological phenomena in seas and oceans. I. *Deep-Sea Res.* **6,** 88–94.

Kriss, A. E., Abyzov, S. S. & Mitzkevich, I. N. (1960). Micro-organisms as indicators of hydrological phenomena in seas and oceans. III. *Deep-Sea Res.* **6,** 335–45.

Kriss, A. E., Labedeva, M. N. & Mitzkevich, I. N. (1960). Micro-organisms as indicators of hydrological phenomena in seas and oceans. II. *Deep-Sea Res.* **6,** 173–83.

Lasker, R. (1966). Feeding, growth, respiration and carbon utilization of a euphausiid crustacean. *J. Fish Res. Bd Can.* **23,** 1291–317.

Leavitt, B. B. (1938). The quantitative vertical distribution of macrozooplankton in the Atlantic Ocean Basin. *Biol. Bull. mar. biol. Lab., Woods Hole* **74,** 376–94.

Little, C. & Gupta, B. L. (1968). Pogonophora: Uptake of dissolved nutrients. *Nature, Lond.* **218,** 873–4.

Little, C. & Gupta, B. L. (1969). Studies on Pogonophora. 3. Uptake of nutrients. *J. exp. Biol.* **51,** 759–73.

Little, C. & Gupta, B. L. (1970). Studies on Pogonophora. 4. Fine structure of the cuticle and epidermis. *Tissue & Cell* **2,** 637–96.

Lohmann, H. von. (1920). Die evölkerung des Ozeans mit Plankton nach den Engebnissen der Zentrifugenfänge während der Ausreise der *Deutschland* 1911. Zugleich ein Beitrag zur Biologie des Atlantischen Ozeans. *Arch. für Biontologie* **4,** 1–470 & 471–617.

Macdonald, A. G. (1975). Locomotor activity and oxygen consumption in shallow water and deep sea invertebrates exposed to high hydrostatic pressures and low temperature. In *Proceedings of the Fifth Underwater Physiology Symposium, 1972.* In press.

Macdonald, A. G., Gilchrist, I. & Teal, J. M. (1972). Some observations on the tolerance of oceanic plankton to high hydrostatic pressure. *J. mar. biol. Assoc. UK* **52,** 213–23.

Macleod, R. A. (1965). The question of the existence of specific marine bacteria. *Bact. Res.* **29,** 9–23.

Manwell, C., Southward, E. C. & Southward, A. J. (1966). Preliminary studies on haemoglobin and other proteins of the Pogonophora. *J. mar. biol. Assoc. UK* **46,** 115–24.

Menzel, D. W. (1967). Particulate organic carbon in the deep sea. *Deep-Sea Res.* **14,** 229–38.

Menzel, D. W. & Ryther, J. H. (1961). Zooplankton in the Sargasso Sea off Bermuda and its relation to organic production. *J. Cons., Cons. perm. int. Explor. Mer* **26,** 250–8.

Menzies, R. J. (1962). On the food and feeding habits of abyssal organisms as exemplified by the Isopoda. *Int. Revue ges. Hydrobiol.* **47,** 339–58.

Menzies, R. J., George, R. Y. & Rowe, G. T. (1973). *Abyssal Environment and Ecology of the World's Oceans.* New York: John Wiley & Sons.

Morita, R. Y. & ZoBell, C. E. (1955). Occurrence of bacteria in pelagic sediments during the Mid-Pacific Expedition. *Deep-Sea Res.* **3,** 66–73.

Neihof, R. A. & Loeb, G. I. (1972). The surface charge of particulate matter in seawater. *Limnol. Oceanogr.* **17,** 7–16.

Osterberg, C., Carey, A. G. & Curl, H. (1963). Acceleration of sinking rates of radionuclides in the ocean. *Nature, Lond.* **200,** 1276–7.

Østvedt, O. (1955). Zooplankton investigations from weather ship *M* in the Norwegian Sea, 1948–49. *Hvalråd. Skr.* **40,** 1–87.

Packard, T. T., Healey, M. L. & Richards, F. A. (1971). Vertical distribution of the activity of the respiratory electron transport system in marine plankton. *Limnol. Oceanogr.* **16,** 60–70.

Pamatmat, M. M. & Banse, K. (1968). Oxygen consumption by the sea bed. II. *In situ* measurements to a depth of 180 m. *Limnol. Oceanogr.* **14,** 250–9.

References

Parsons, T. R. & Strickland, J. D. H. (1962). On the production of particulate organic carbon by heterotrophic processes in sea water. *Deep-Sea Res.* **8,** 211–22.

Pearcy, W. G. & Laurs, R. M. (1966). Vertical migration and distribution of mesopelagic fishes off Oregon. *Deep-Sea Res.* **13,** 153–65.

Pomeroy, L. R. & Johannes, R. E. (1966). Total plankton respiration. *Deep-Sea Res.* **13,** 971–3.

Riley, G. A. (1970). Particulate and organic matter in sea water. *Adv. in mar. Biol.* **8,** 1–118.

Sanders, H. L. & Hessler, R. R. (1969). Ecology of the deep sea benthos. *Science, N.Y.* **163,** 1419–24.

Sanders, H. L., Hessler, R. R. & Hampson, G. R. (1965). An introduction to the study of deep-sea benthic fauna assemblages along the Gay-Head–Bermuda transect. *Deep-Sea Res.* **12,** 845–67.

Sieburth, J. M. (1971). Distribution and activity of oceanic bacteria. *Deep-Sea Res.* **18,** 1111–21.

Sokolova, M. N. (1959). On the distribution of deep-water bottom animals in relation to their feeding habits and the character of sedimentation. *Deep-Sea Res.* **6,** 1–4.

Southward, A. J. & Southward, E. C. (1968). Uptake and incorporation of labelled glycerine by pogonophores. *Nature, Lond.* **218,** 875–6.

Southward, A. J. & Southward, E. C. (1970). Observations on the role of dissolved organic compounds in the nutrition of benthic invertebrates. *Sarsia* **45,** 69–96.

Strickland, J. D. H. (1965). In *Chemical Oceanography* 1, ed. J. P. Riley & G. Skirrow. London & New York: Academic Press.

Sutcliffe, W. H., Baylor, E. R. & Menzel, D. W. (1963). Sea surface chemistry and Langmuir circulation. *Deep-Sea Res.* **10,** 233–43.

Teal, J. M. (1971). Pressure effects on the respiration of vertically migrating decapod Crustacea. *Am. Zool.* **11,** 571–6.

Teal, J. M. & Carey, F. G. (1967). Effects of pressure and temperature on the respiration of euphausids. *Deep-Sea Res.* **14,** 725–33.

Vaccaro, R. F. & Jannasch, H. W. (1966). Studies on heterotrophic activity in seawater based on glucose assimilation. *Limnol. Oceanogr.* **11,** 596–607.

Vinogradov, M. E. (1961). Food sources of the deep water fauna. Speed of decomposition of dead Pteropoda. *Scripta Technica, Am. Geophys. Union.* 136–41.

Vinogradov, M. E. (1962*a*). Feeding of the deep sea zooplankton. *Rapp. Proc. verb. Réun.* **153,** 114–20.

Vinogradov, M. E. (1962*b*). Quantitative distribution of deep-sea plankton in the Western Pacific and its relation to deep water circulation. *Deep-Sea Res.* **8,** 251–8.

Vlymen, W. J. (1970). Energy expenditure of swimming copepods. *Limnol. Oceanogr.* **15,** 348–56.

Wada, E. & Hattori, A. (1972). Nitrite distribution and nitrate reduction in deep sea waters. *Deep-Sea Res.* **19,** 123–32.

Wangersky, P. J. (1968). Distribution of suspended carbonate with depth in the ocean. *Limnol. Oceanogr.* **14,** 929–33.

Waterman, T. H., Nunnemacher, R. F., Chace, F. A. & Clarke, G. L. (1939). Diurnal vertical migrations of deep water plankton. *Biol. Bull. mar. biol. Lab., Woods Hole* **76,** 256–79.

Wheeler, E. H. (1967). Copepod detritus in the sea. *Limnol. Oceanogr.* **12,** 697–701.

Williams, P. J. Le B. (1970). Heterotrophic utilisation of dissolved organic compounds in the sea. 1. *J. mar. biol. Assoc. UK* **50,** 859–70.

Williams, P. M. (1971). In *Organic Compounds in Aquatic Environments*, ed. S. D. Faust & J. V. Hunter. New York: Dekker.

Wolff, T. (1960). The hadal community, an introduction. *Deep-Sea Res.* **6,** 95–124.

Wolff, T. (1970). The concept of the hadal or ultral abyssal fauna. *Deep-Sea Res.* **17,** 981–1003.

Wood, E. J. F. (1956). Diatoms in the ocean deeps. *Pacific Sci.* **10,** 377–9.

Yashnov, V. A. (1961). Vertical distribution of the mass of zooplankton in the tropical region of the Atlantic Ocean. *Scripta Technica, Am. Geophys. Union* 136–41.

Zenkevitch, L. A. (1969). In *The Pacific Ocean. Biology of the Pacific Ocean* II, Moscow: Nauka.

Zenkevitch, L. A., Barsanova, N. G. & Belyayev, G. M. (1960). Quantitative distribution of bottom fauna in the world ocean abyssal areas. Trans. in *US Joint Publications Res. Serv.* No. 2675. Washington: Office of Technical Services, Dept. of Commerce.

ZoBell, C. E. (1938). Studies on the bacterial flora of marine bottom sediments. *J. Sedim. Petrol.* **8,** 10–18.

ZoBell, C. E. (1942). Changes produced by micro-organisms in sediments after decomposition. *J. Sedim. Petrol.* **12,** 127–36.

ZoBell, C. E. (1946). *Marine Microbiology*. Waltham, Massachusetts: Chronica Botanica Co.

ZoBell, C. E. & Morita, R. Y. (1957). Barophilic bacteria in some deep sea sediments. *J. Bact.* **73,** 563–8.

ZoBell, C. E. & Morita, R. Y. (1959). Deep-sea bacteria. *Galathea Report Copenhagen* **1,** 139–54.

7

Anon, (1968). *Undersea Technology Handbook Directory.* Arlington, Va: Compass.

Aron, W. (1962). Some aspects of sampling the macroplankton. *Rapp. Proc. verb. Réun.* **153,** 29–38.

Aron, W., Ahlstrom, E. H., Bary, B. M., Bé, A. W. H. & Clarke, W. D. (1965). Towing characteristics of plankton sampling gear. *Limnol. Oceanogr.* **10,** 533–40.

Aron, W., Baxter, N., Noel, R. & Andrews, W. (1964). A description of a discrete depth plankton sampler with some notes on the towing behaviour of a 6-foot Isaacs–Kidd mid-water trawl and a one meter ring net. *Limnol. Oceanogr.* **9,** 324–33.

References

Backus, R. H. & Barnes, H. (1956–57). Television-echo sounder observations of mid-water sound scatterers. *Deep-Sea Res.* **4**, 116–23.

Baker, A de C. (1957). Underwater photographs in the study of oceanic squid. *Deep-Sea Res.* **4**, 126–9.

Barham, E. G. (1966). Deep scattering layer migration and composition: observations from a diving saucer. *Science, N.Y.* **151**, 1399–402.

Barnes, H. (1963). Underwater television. *Oceanogr. mar. Biol. An. Rev.* **1**, 115–28.

Barton, R. (1973). Armoured suit has 1000 ft capability. *Offshore Services* May, 18–21. Kingston-upon-Thames, UK: Spearhead Publications Ltd.

Beebe, W. (1935). *Half Mile Down.* London: John Lane, The Bodley Head, p. 343.

Bennett, P. B. (1969). Inert gas narcosis. In *The Physiology and Medicine of Diving and Compressed Air Work.* London: Baillière, Tindall & Cassell.

Bennett, P. B. (1970). Simulated oxygen–helium saturation diving to 1500 ft and the helium barrier. *J. R.N. Scient. Serv.* **26**, 91–106.

Bennett, P. B. & Elliott, D. H. (1969). (Eds.) *The Physiology and Medicine of Diving and Compressed Air Work.* London: Baillière, Tindall & Cassell.

Bernard, F. (1958). Plancton et benthos observés durant trois plongées en bathyscaphe au large de Toulon. *Annls Inst. Océanogr.* **35** (4).

Blaxter, J. H. S. & Currie, R. I. (1967). The effect of artificial lights on acoustic scattering layers in the ocean. *Symp. zool. Soc. Lond.* **19**, 1–14.

Blaxter, J. H. S. & Tytler, P. (1972). Pressure discrimination in teleost fish. *Symp. Soc. exp. Biol.* **26**, 417–43.

Brauer, R. W. (1968). Seeking man's depth level. *Ocean Industry* **3**, 28–33.

Brauer, R. W. (1972). Parameters controlling experimental studies of deep sea biology. In *Barobiology and the Experimental Biology of the Deep Sea*, ed. R. W. Brauer. Chapel Hill: Univ. of North Carolina.

Brauer, R. W., Dimov, S., Fructus, P., Gosset, A. & Nagnet, R. (1969). *Syndrome Neurologique et Eléctrographique des Hautes Pressions.* Paris. pp. 264–5.

Brauer, R. W., Jordan, M. R. & Way, R. O. (1972). The high pressure neurological syndrome in the squirrel monkey (*Saimiri sciureus*). In *Underwater Medicine*, ed. J. Corriol & P. Fructus. In press.

Brauer, R. W. & Way, R. O. (1970). Relative narcotic potencies of hydrogen, helium, nitrogen and their mixtures. *J. appl. Physiol.* **29**, 23–31.

Brauer, R. W., Way, R. O., Jordan, M. R. & Parrish, D. E. (1971). Experimental studies on the high pressure hyperexcitability syndrome in various mammalian species. *Proceedings of the Fourth Underwater Physiology Symposium*, ed. C. J. Lambertsen. New York & London: Academic Press. pp. 487–500.

Breslau, L. R., Clarke, G. L. & Edgerton, H. E. (1967). Optically triggered underwater cameras for marine biology. In *Deep Sea Photography*, ed. J. B. Hersey. Baltimore: The Johns Hopkins Press.

Burrows, D. W. (1969). Cableless underwater television link design and test results. *Proc. Oceanol. Internat., 1969*, Day 5. Brighton, UK.

Capen, R. L. (1967). Swimbladder morphology of some mesopelagic fishes in relation to sound scattering. *Navy Electronics Lab., San Diego, Rept.* 1447, 1–32.

Cattell, M. & Edwards, D. J. (1928). The energy changes of skeletal muscle accompanying contraction under high pressure. *Am. J. Physiol.* **86**, 371–82.

References

Chandler, M. T. & Vidaver, W. (1971). Stationery platinum electrode for measurement of O_2 exchange by biological systems under hydrostatic pressure. *Rev. Scient. Instr.* **42**, 143–6.

Chouteau, J. (1969). Saturation Diving: The Conshelf Experiments. In *The Physiology and Medicine of Diving and Compressed Air Work*, ed. P. B. Bennett & D. H. Elliott. London: Baillière, Tindall & Cassell.

Chouteau, J. (1971). Respiratory gas exchange in animals during exposure to extreme ambient pressures. *Proceedings of the Fourth Underwater Physiology Symposium*, ed. C. J. Lambertsen. New York & London: Academic Press.

Clarke, M. R. (1969). A new mid-water trawl for sampling discrete depth horizons. *J. mar. biol. Assoc. UK* **49**, 945–60.

Clutter, R. I. & Anraker, M. (1968). *Avoidance of sampler.* Monograph on oceanic methodology. UNESCO: Paris. pp. 57–86.

Culberson, C., Lester, D. R. & Pytkowicz, R. M. (1967). High pressure dissociation of carbonic and boric acids. *Science, N.Y.* **157**, 59–60.

Currie, R. I. & Foxton, P. (1957). A new quantitative plankton net. *J. mar. biol. Assoc. UK* **36**, 17–32.

D'Aoust, B. G. (1968). Apparatus for incubating and sampling tissue and/or cell free systems under elevated gas pressure. *Analyt. Biochem.* **26**, 85–91.

Devereux, R. F. & Winsett, R. C. (1953). Report on the Isaacs–Kidd mid-water trawl. *Scripps Inst.* Ref. 53–3, 1–21.

Distèche, A. (1962). Electrochemical measurements at high pressures. *J. Electrochem. Soc.* **109**, 1084–92.

Distèche, A. & Dubuisson, M. (1960). Mesures directes le pH aux grandes profondeurs sous marines. *Bull. Inst. Océanogr. Monaco*, No. 1174, 1–8.

Dossett, A. N. & Hempleman, H. V. (1972). Importance for mammals of rate of compression. *Symp. Soc. exp. Biol.* **26**, 355–61.

Feldman, S. (1969). The US navy deep submergence vehicles. *Proc. Oceanol. Internat., 1969*, Day 4. Brighton, UK.

Forster, G. R. (1964). Line-fishing on the continental slope. *J. mar. biol. Assoc. UK* **44**, 277–84.

Foxton, P. (1963). An automatic opening–closing device for large plankton nets and mid-water trawls. *J. mar. biol. Assoc. UK* **43**, 295–308.

Fulton, J. F. (1925). The latent period of skeletal muscle. *Q. J. exp. Physiol.* **15**, 349–66.

Gilchrist, I. (1972). Equipment for the recovery and study of deep sea animals. Thesis, Queen's University of Belfast, Dept. of Mech. Eng.

Grice, D. G. & Hulseman, K. (1968). Contamination in Nansen-type vertical plankton nets and a method to prevent it. *Deep-Sea Res.* **15**, 229–33.

Halsey, M. J. & Eger, E. I. (1971). The effect of pressure on the anaesthetic potency of nitrous oxide. *Fedn Proc. Fedn Am. Socs. exp. Biol.* **30**, 1375.

Hamilton, R. W., MacInnis, J. B., Noble, A. D. & Schreiner, H. R. (1966). *Saturation diving at 650 ft.* Technical memorandum B1 411. Ocean Systems Inc., Tonawanda, New York.

Harris, M. J. (1969). Acoustic command system. *Proc. Oceanol. Internat., 1969*, Day 3. Brighton, UK.

Harrisson, C. M. H. (1967). On methods of sampling mesopelagic fishes. *Symp. Zool. Soc. Lond.* **19**, 71–126.

References

Hersey, J. B. (1967). *Deep Sea Photography*. Baltimore: The Johns Hopkins Press.

Hersey, J. B., Backus, R. H. (1962). Sound scattering by marine organisms. In *The Sea*, 1, ed. M. N. Hill. New York & London: Wiley–Interscience.

Hersey, J. B., Backus, R. H. & Hellwig, J. (1962). Sound scattering spectra of deep scattering layers in the western North Atlantic. *Deep-Sea Res.* **8**, 196–210.

Holme, N. A. (1964). Methods of sampling the benthos. *Adv. in mar. Biol.* **2**, 171–260.

Johnson, F. H., Eyring, H., Steblay, R., Chaplin, J., Hubler, C. & Cherardi, G. (1945). The nature and control of reactions in bioluminescence. With special references to mechanisms of reversible and irreversible inhibitions by hydrogen and hydroxyl ions, temperature, pressure, alcohol, urethane and sulfanilimide in bacteria. *J. gen. Physiol.* **28**, 462–537.

Johnson, F. H. & Flagler, E. A. (1950). Hydrostatic pressure reversal of narcosis in tadpoles. *Science, N.Y.* **112**, 91–2.

Johnson, S. M. & Miller, K. W. (1970). Antagonism of pressure and anaesthesia. *Nature, Lond.* **228**, 75–6.

Jones, H. W. & Miles, H. T. (1969). The development of ultrasonic image convertors for underwater viewing and other applications. *Ocean Eng.* **1**, 479–96.

Kanwisher, J. K. (1962). Oxygen and carbon dioxide instrumentation. In *Marine Sciences Instrumentation* 1, ed. R. D. Gaul, D. D. Ketchum, J. T. Shaw & J. M. Snodgrass. New York: Plenum Press. pp. 334–9.

Kirkness, C. M. & Macdonald, A. G. (1972). Interaction between anaesthetics and high hydrostatic pressure in the cell division of *Tetrahymena pyrifomis*. *Exptl Cell Res.* **75**, 329–36.

Kono, I. (1958). Oxigraph. *Okayama Igakkai Sasshi.* **70**, 4521–33.

Kooyman, G. L. (1972). Deep diving behaviour and effects of pressure on reptiles, birds and mammals. *Symp. Soc. exp. Biol.* **26**, 295–311.

Kullenberg, B. (1957). On the shape and length of the cable during a deep sea trawling. *Rept. Swedish Deep Sea Exped.* II. *Zoology*, No. 2, 31–44.

Kylstra, J. A. (1967). Advantages and limitations of liquid breathing. *Proceedings of the Third Underwater Physiology Symposium*, ed. C. J. Lambertsen. Baltimore: Williams & Wilkins. pp. 341–50.

Kylstra, J. A., Nantz, R., Crowe, J., Wagner, W. & Saltzman, H. A. (1967). Hydraulic compression of mice to 166 atmospheres. *Science, N.Y.* **158**, 793–4.

Lambertsen, C. J. (1975). Collaborative investigation of limits of human tolerance to pressurisation with helium, neon and nitrogen. Simulation of density equivalent to helium–oxygen respiration at depths to 2000, 3000, 4000 and 5000 feet of seawater. *Proceedings of the Fifth Underwater Physiology Symposium*, C. J. Lambertsen. In press.

Landau, J. V. & Thibodeau, L. (1962). The micro-morphology of *Amoeba proteus* during pressure-induced changes in the sol gel cycle. *Exptl Cell Res.* **27**, 591–4.

Lanphier, E. H. (1969). Pulmonary function. In *The Physiology and Medicine of Diving and Compressed Air Work*, ed. P. B. Bennett & D. H. Elliott. London: Baillière, Tindall & Cassell.

Lanphier, E. H. (1972). Human respiration under increased pressures. *Symp. Soc. exp. Biol.* **26**, 379–94.
Lever, M. J., Miller, K. W., Paton, W. D. M. & Smith, E. B. (1971). Pressure reversal of anaesthesia. *Nature, Lond.* **231**, 368–71.
Macdonald, A. G. (1965). The effect of high hydrostatic pressures on the oxygen consumption of *Tetrahymena pyriformis* W. *Exptl. Cell Res.* **40**, 78–84.
Macdonald, A. G. (1967). The effect of high hydrostatic pressure on the cell division and growth of *Tetrahymena pyriformis*. *Exptl Cell Res.* **47**, 569–80.
Macdonald, A. G. (1970). Life at high pressure. *Proc. Challenger Soc.* **4**, (2) 86–7.
Macdonald, A. G. (1975). Locomotor activity and oxygen consumption in shallow and deep sea invertebrates exposed to high hydrostatic pressures and low temperature. In *Proceedings of the Fifth Underwater Physiology Symposium*, ed. C. J. Lambertsen. In press.
Macdonald, A. G. & Gilchrist, I. (1969). Recovery of deep seawater at constant pressure. *Nature, Lond.* **222**, 71–2.
Macdonald, A. G. & Gilchrist, I. (1972). An apparatus for the recovery and study of deep sea plankton at constant temperature and pressure. In *Barobiology and the Experimental, Biology of the Deep Sea*, ed. R. W. Brauer. Chapel Hill: Univ. of North Carolina.
MacInnis, J., Dickson, J. G. & Lambertsen, C. J. (1967). Exposure of mice to a helium–oxygen atmosphere at pressures to 122 atmospheres. *J. appl. Physiol.* **22**, 694–8.
Macklen, P. T. & Mead, J. (1968). Factors determining maximum expiratory flow in dogs. *J. appl. Physiol.* **25**, 159–69.
Marsland, D. (1950). The mechanism of cell division: temperature–pressure experiments on the cleaving eggs of *Arbacia punctulata*. *J. cell. comp. Physiol.* **36**, 205–27.
Miller, K. W. (1972). Inert gas narcosis and animals under high pressure. *Symp. Soc. exp. Biol.* **26**, 363–78.
Milliman, J. D. & Manheim, F. T. (1968). Observations in deep scattering layers off Cape Hatteras, USA. *Deep-Sea Res.* **15**, 503–7.
Morrison, J. M. (1975). Physiological studies during a deep simulated oxygen–helium dive to 1500 ft. In *Proceedings of the Fifth Underwater Physiology Symposium*, ed. C. J. Lambertsen. In press.
Nansen, F. (1915). Closing nets for vertical hauls and for horizontal towing. *Pub. de circonstance Cons. perm int. Explor. Mer.* **67**, 3–8.
Naroska, Von V. (1968). Vergleichende Untersuchungen über den Einfluss des hydrostatischen Druckes auf Überlabensfähigkeit und Stoffwechselintensität mariner Evertebraten und Teleosteer. *Kieler Meeresforsch.* **24**, 95–123.
Pamatmat, M. M. & Banse, K. (1968). Oxygen consumption by the sea bed. II. *In situ* measurements to a depth of 180 m. *Limnol. Oceanogr.* **14**, 250–9.
Peres, J. M. (1965). Aperçu sur les resultats de deux plongées effectuées dans le ravin de Puerto Rico par le bathyscaphe *Archimède*. *Deep-Sea Res.* **12**, 883–91.
Phleger, C. F. & Soutar, A. (1971). Free vehicles and deep-sea biology. *Am. Zool.* **11**, 409–18.
Piccard, A. (1956). In *Balloon and Bathyscaphe*. London: Cassell. p. 188.

References

Piccard, J. & Dietz, R. S. (1961). *Seven Miles Down.* London: Longmans.

Pode, L. (1951). *Rept. No. 687.* Navy Dept. David Taylor Model Basin.

Rather, R. L., Goerland, V., Hersey, J. B., Vine, A. C. & Dakin, F. (1965). Improved towline design for oceanography. *Under Sea Technology.* (Separate reprint.)

Rowe, G. T. & Menzies, R. J. (1967). Use of sonic techniques and tension recordings as improvements in abyssal trawling. *Deep-Sea Res.* **14,** 271–4.

Skutt, R. H., Fell, R. B. & Hagstrom, E. C. (1972). A multichannel ultrasonic underwater telemetry system. In *Biotelemetry*, ed. H. P. Kimmich & J. A. Vos. Leiden: M. V. Meander.

Smith, E. B. (1969). The role of exotic gases in the study of narcosis. In *The Physiology and Medicine of Diving and Compressed Air work*, ed. P. B. Bennett & D. H. Elliott. London: Baillière, Tindall & Cassell.

Spyropoulos, C. S. (1957*a*). The effects of hydrostatic pressure upon the normal and narcotised nerve fibre. *J. gen. Physiol.* **40,** 849–57.

Spyropoulos, C. S. (1957*b*). Response of single nerve fibres at different hydrostatic pressures. *Am. J. Physiol.* **189,** 214–8.

Stachiw, J. D. (1970). Materials for the sea, Part 2. Acrylic plastic design. *Ocean Industry*, (August) 55–63.

Sundnes, G. (1962). A pressure aquarium for experimental use. *Fiskeridir. Skr. Ser. Havunders.* (Reports on Norwegian Fishery and Marine Investigations) **13,** 1–7.

Talkington, H. R. (1969). Useful undersea work by remote controlled systems. *Proc. Oceanol. Internat., 1969*, Day 5. Brighton, UK.

Tavernier, P. & Prache, P. (1952). Influence de la pression sur la resistivité d'une thermistance. *J. Phys. Rad.* **13,** 423–6.

Teal, J. M. & Carey, F. G. (1967). Effects of pressure and temperature on the respiration of euphausiids. *Deep-Sea Res.* **14,** 725–35.

Tranter, D. J. & Smith, P. E. (1968). *Filtration Performance.* Monograph on oceanic methodology. UNESCO: Paris. pp. 27–56.

Tucker, D. G. (1967). *Sonar in Fisheries.* London: Fishing News (Books) Ltd.

Wenzel, J. G. (1969). *Deep Quest*, a research submarine system. *Proc Oceanol. Internat., 1969*, Day 4. Brighton, UK.

Weston, D. E. (1967). Sound propagation in the presence of bladder fish. In *Underwater Acoustics* II, ed. V. M. Albers. New York: Plenum Press.

Whitfield, M. (1969). Multicell assemblies for studying ion-selective electrodes at high pressures. *J. Electrochem. Soc.* **116,** 1042–6.

Wolff, T. (1961). The deepest recorded fishes. *Nature, Lond.* **190,** 283.

Wood, J. D. (1969). Oxygen toxicity. In *The Physiology and Medicine of Diving and Compressed Air Work*, ed. P. B. Bennett & D. H. Elliott. London: Baillière, Tindall & Cassell.

Wood, J. D. (1971). Oxygen toxicity in neuronal elements. In *Proceedings of the Fourth Underwater Physiology Symposium*, ed. C. J. Lambertsen. New York & London: Academic Press. p. 9–17.

ZoBell, C. E. (1941). Apparatus for collecting water samples from different depths for bacterial analysis. *J. Mar. Res.* **4,** 173–88.

ZoBell, C. E. (1959). Thermal changes accompanying the compression of aqueous solutions to deep sea conditions. *Limnol. Oceanogr.* **4,** 463–71.

INDEX

Index

Index

Index

Index

Index

Index